LOW ENERGY LOW CARBON ARCHITECTURE:
RECENT ADVANCES & FUTURE DIRECTIONS

Sustainable Energy Developments

Series Editor

Jochen Bundschuh
University of Southern Queensland (USQ), Toowoomba, Australia
Royal Institute of Technology (KTH), Stockholm, Sweden

Volume 12

Low Energy Low Carbon Architecture
Recent Advances & Future Directions

Editor

Khaled A. Al-Sallal

Department of Architectural Engineering, UAE University, Al-Ain, UAE

CRC Press
Taylor & Francis Group
Boca Raton London New York

CRC Press is an imprint of the
Taylor & Francis Group, an **informa** business

A BALKEMA BOOK

Cover photo

Al Bahr Towers in Abu Dhabi (The book cover photo is a courtesy of professional photographer Ali Jamal, 2016). The most distinguishing feature of the iconic Al Bahr towers in Abu Dhabi is its world's largest sun-responsive façade shades. Each tower is protected by a shading skin of umbrella-like elements that automatically open and close depending on the intensity of sunlight. The design concept was inspired by the "mashrabiya", geometrically-designed wooden lattice screens that have been used to fill windows of traditional Arabic architecture. This intelligent façade is dynamically controlled by a building management system. The adjustable shades help achieve 50% less solar gain than a comparable tower, with zero compromises in natural lighting. Al Bahr Towers is designed by AHR (formerly Aedas UK), in accordance with the US Green Building Council LEED rating system, targeted for Silver rating. The development was placed second at the Emporis Skyscraper Award in 2012 and was featured on the CTBUH's "Innovative 20" list of buildings that "challenge the typology of tall buildings in the 21st century".

Published by: CRC Press/Balkema
P.O. Box 11320, 2301 EH Leiden, The Netherlands
e-mail: Pub.NL@taylorandfrancis.com
www.crcpress.com – www.taylorandfrancis.com

First issued in paperback 2020

ISBN 13: 978-0-367-57484-0 (pbk)
ISBN 13: 978-1-138-02748-0 (hbk)

Visit the Taylor & Francis Web site at
http://www.taylorandfrancis.com

and the CRC Press Web site at
http://www.crcpress.com

Typeset by MPS Limited, Chennai, India

Library of Congress Cataloging-in-Publication Data

Names: Al-Sallal, Khaled A., editor.
Title: Low energy low carbon architecture : recent advances & future directions editor, Khaled A. Al-Sallal.
Description: Boca Raton : Taylor & Francis, 2016. | Series: Sustainable energy developments ; volume 12 | Includes bibliographical references and index. | Description based on print version record and CIP data provided by publisher; resource not viewed.
Identifiers: LCCN 2015045082 (print) | LCCN 2015041810 (ebook) | ISBN 9781315624020 (ebook) | ISBN 9781138027480 (hardcover : alk. paper)
Subjects: LCSH: Sustainable architecture. | Carbon dioxide mitigation. | Buildings–Energy conservation. | Sustainable buildings.
Classification: LCC NA2542.36 (print) | LCC NA2542.36 .L69 2016 (ebook) | DDC 720/.47–dc23

About the book series

Renewable energy sources and sustainable policies, including the promotion of energy efficiency and energy conservation, offer substantial long-term benefits to industrialized, developing, and transitional countries. They provide access to clean and domestically available energy and lead to a decreased dependence on fossil fuel imports and a reduction in greenhouse gas emissions.

Replacing fossil fuels with renewable resources affords a solution to the increased scarcity and price of fossil fuels. Additionally, it helps to reduce anthropogenic emission of greenhouse gases and their impacts on climate change. In the energy sector, fossil fuels can be replaced by renewable energy sources. In the chemistry sector, petroleum chemistry can be replaced by sustainable or green chemistry. In agriculture, sustainable methods can be used that enable soils to act as carbon dioxide sinks. In the construction sector, sustainable building practice and green construction can be used, replacing, for example, steel-enforced concrete by textile-reinforced concrete. Research and development and capital investments in all these sectors will not only contribute to climate protection but will also stimulate economic growth and create millions of new jobs.

This book series will serve as a multidisciplinary resource. It links the use of renewable energy and renewable raw materials, such as sustainably grown plants, with the needs of human society. The series addresses the rapidly growing worldwide interest in sustainable solutions. These solutions foster development and economic growth while providing a secure supply of energy. They make society less dependent on petroleum by substituting alternative compounds for fossil-fuelbased goods. All these contribute to minimize our impacts on climate change. The series covers all fields of renewable energy sources and materials. It addresses possible applications not only from a technical point of view, but also from economic, financial, social, and political viewpoints. Legislative and regulatory aspects, key issues for implementing sustainable measures, are of particular interest.

This book series aims to become a state-of-the-art resource for a broad group of readers including a diversity of stakeholders and professionals. Readers will include members of governmental and non-governmental organizations, international funding agencies, universities, public energy institutions, the renewable industry sector, the green chemistry sector, organic farmers and farming industry, public health and other relevant institutions, and the broader public. It is designed to increase awareness and understanding of renewable energy sources and the use of sustainable materials. It also aims to accelerate their development and deployment worldwide, bringing their use into the mainstream over the next few decades while systematically replacing fossil and nuclear fuels.

The objective of this book series is to focus on practical solutions in the implementation of sustainable energy and climate protection projects. Not moving forward with these efforts could have serious social and economic impacts. This book series will help to consolidate international findings on sustainable solutions. It includes books authored and edited by world-renowned scientists and engineers and by leading authorities in economics and politics. It will provide a valuable reference work to help surmount our existing global challenges.

Jochen Bundschuh
(Series Editor)

Editorial board

Table of contents

List of contributors

Khaled A. Al-Sallal — Department of Architectural Engineering, UAE University, Al-Ain, UAE, E-mail: k.sallal@uaeu.ac.ae

Raymond J. Cole — School of Architecture and Landscape Architecture, University of British Columbia, Vancouver, Canada, E-mail: ray.cole@ubc.ca

Alfredo Fernández-González — School of Architecture, University of Nevada, Las Vegas, USA, E-mail: alfredo.fernandez@unlv.edu

Walter Grondzik — Department of Architecture, Ball State University, Muncie, Indiana, USA, E-mail: gzik@polaris.net

Kenneth Ip — School of Environment and Technology, University of Brighton, Brighton, UK, E-mail: k.ip@brighton.ac.uk

Robert J. Koester — AIA, Department of Architecture & Center for Energy Research/Education/Service (CERES), Ball State University, Muncie, Indiana, USA, E-mail: rkoester@bsu.edu

Timothy Lentz — Architecture Department, Iowa State University, Ames, Iowa, USA, E-mail: trlentz@gmail.com

Ardeshir Mahdavi — Department of Building Physics and Building Ecology, Vienna University of Technology, Vienna, Austria, E-mail: amahdavi@tuwien.ac.at

Andrew Miller — School of Environment and Technology, University of Brighton, Brighton, UK, E-mail: bsb.andrewmiller@gmail.com

Amy Oliver — Faculté de l'Aménagement, Université de Montréal, Montréal, Canada, E-mail: Amy.oliver@umontreal.ca

Ulrike Passe — Architecture Department, Iowa State University, Ames, Iowa, USA, E-mail: upasse@iastate.edu

Foreword by Larry Degelman

Before October 1973, minimizing a building's construction cost was regarded as a noble goal and satisfactory in serving "due diligence" to the owner. In that year, however, the world was awakened to what decades later is still a festering problem for every living creature on the planet – the oil embargo set consumers into shock and researchers and scientists scurrying to figure out how to cope with limited energy resources. One result of that oil embargo was sharply rising fuel prices, so owning and operating costs suddenly became an equally important consideration and many times could outweigh the initial costs of construction. Thus, the concept of life-cycle costs had been embedded into the thinking of building planners and designers – as is pointed out clearly by one of the contributing authors of this book. Engineers started to produce sophisticated software tools to help designers and owners to predict accurately annual fuel consumption and thereby enabling one to choose better building products that could reduce a building's annual operating costs. Government agencies and municipalities got into the act by producing codes and standards aimed at reducing every building's energy consumption. But, was it to be all about costs? Important as that is, it could no longer be the sole building design objective. We've come to recognize that serious damage is being done to our entire atmosphere by greenhouse gas production. We see that our whole environment is in serious jeopardy. We've come to know what atmospheric pollution and greenhouse gases really mean and that much of the problem is created by human activity. So, what was regarded in 1973 as a mere "inconvenience" is now recognized as a broader issue of resource sustainability, atmospheric preservation and human sustenance.

In the building profession, energy conservation and sustainability have become familiar buzz words, and modeling and simulation of building operations are commonplace. Today, too often the topic of sustainability ends there – i.e., as an exercise in minimizing the life-cycle costs for the owner. Many international energy conservation codes are based on this principle, suggesting that the premise of minimizing the life-cycle cost of a building project is a sufficient goal. A major paradigm shift is urgently needed. Governments and society in general are awaking to the realization that our environment is no longer an infinite supply of healthy air provided by nature. Now, we recognize that the environment we live in is being adversely affected by human activities. Not only is this manifested in our choices of our methods of travel but also in the way we approach building designs.

Whatever solutions can be contributed by the building professions, they cannot be targeted at single-minded remedies. There needs to be a multi-faceted attack using innovative solutions and techniques. Energy is not the only issue; energy codes are not the only solutions (though they are a start); water consumption needs to be reduced; carbon-based fuel use needs to be reduced;

renewable energy sources need to be bolstered (wind, solar, recycling, etc.) Important as these are, what else is needed in the new paradigm is an attitude change, which is probably the only way to a new and sustainable future. This book makes a valuable contribution in that it covers all these approaches through unique and meaningful contributions of prominent scholars from across the globe. Each contributor delves deeply into the essence of what it will take to bring about true sustainability in the world, at least from the viewpoint of the building design professions.

My first experience with the primary author (Khaled Al-Sallal) was in 1991–1995 at Texas A&M where he was my advisee while he conducted his Ph.D. studies in Architecture. His work specialized in energy conservation approaches for buildings and software development to accompany his efforts to put computational relevance to his tests on energy conservation measures. In his current role as professor at the UAE, he has published widely in venues such as conferences of the International Building Performance Simulation Association (IBPSA) and other environmental issues conferences. I commend him for his skill and foresight in assembling an excellent team of contributors for this book on carbon emissions of buildings. He opens the discussions by talking about rapid urbanism that has occurred over the decades and how this has led to an unbalanced environment. His opening chapter deals with the topics of carbon emissions and how these have been inflated by unsustainable practices of humans over the decades. He also reveals how this has caused degradation of the environment in which we live. His second chapter deals with one facet of building design that can help alleviate the carbon production through reduced use of energy to cool our buildings. He reveals the potential in using passive (enhanced building design approaches, with good examples) and low energy cooling techniques (including the increased efficiency of HVAC equipment.) The vicious cycle of increased cooling use, resulting energy production by the power plants, increased atmospheric greenhouse gas production, depletion of ozone layer, and climate change/global warming result in increased cooling loads, and so the cycle repeats.

I'm pleased to see that Dr. Al-Sallal had the foresight to include an important chapter on daylight/daylighting design. This area is sometimes overlooked in favour of emphasizing only the building envelope and HVAC systems. Daylight is so important due to its dual role of providing a pleasing naturally lit environment and replacing some of the need for electric lights, resulting in reduction of the mechanical cooling energy with concomitant reduced greenhouse gas production. His chapter focuses on new advances in daylight sustainable design strategies and performance metrics and evaluation. A special section is devoted to the discussion of daylighting challenges, which has identified some gaps in research that need to be scrutinized in order to make real achievements in sustainable design. Another section explains daylight performance metrics and highlights the importance of using climate-based metrics. The chapter also discusses briefly the established general models for simulating the daylight sky and highlights the importance of dynamic daylighting simulation. Lastly, the closing sections of the chapter focus on important information for designers such as daylighting strategies, rules of thumb, and advanced daylighting systems. This is a very insightful chapter throughout its entirety.

The other contributors in this book are internationally known for their work as leaders in pushing the envelope in a variety of ground-breaking approaches for reducing building energy usage and saving the planet. Chapter 4 focuses on reducing energy use through passive heating of building spaces. The author points out that passive heating has been around for a couple thousand years, but during our industrial revolution we've found an excuse for ignoring the passive techniques in favor of fossil fuel combustion and mechanical engines that produce heat. She reminds us that the time has come to return to these passive concepts that are as valid now as they were millennia ago, and are becoming a necessity once again. Also included are very good examples and references for the reader to catch up on passive design principles.

In Chapter 5, the author addresses the human factors role in sustainable architecture. Here is included the important influences of ecology, motivation, and life style. Human beings actually intervene in the environment, resulting in resource depletion, environmental emissions, and waste generation, all creating an oversaturation of the planet's capability to sustain future generations. This chapter concludes by reminding us of the role of people in bringing about sustainable architecture. Pointed out are the especially important roles that designers, educators, researchers, developers, managers, etc. can play in this effort.

The authors of Chapter 6 discuss the building construction materials as yet another dimension of the building sustainability issue. This opens the topic of embodied energy and the manufacturing processes (with their concomitant emissions of carbon dioxide and those implications for climate change) and the requirement for systems that incorporate less waste. Building stakeholders can make significant contributions by promoting the use of natural building materials and the recycling and reuse of building components. The conclusions illustrate two examples of use of natural building materials.

Chapter 7 addresses the issues surrounding water use. And, what about landscaping? Environmental concerns do not stop at the building line. This chapter alerts readers to the severe impacts that the increase in human population and urbanization have had on the depletion of global water and the ramifications this might have on all sorts of sustainability – not only buildings. Buildings and parking lots are guilty of creating impermeable surfaces, which diverts water from underground storage.

Chapter 8 is about energy-efficient HVAC systems and systems integration. Though the most energy efficient climate control options would be those based upon passive heating and cooling principles, some climates make this very difficult, if not impossible, to achieve. Today, it is generally expected that an HVAC system will provide an environment for human comfort. The author points out that we should recognize that HVAC equipment is not about to be abandoned. This chapter discusses the important of designing the proper equipment from human factors, ecological, and economic viewpoints. Comfort is important for health as well as productivity. In addition to continually improved equipment efficiencies, another emerging trend is lifetime commissioning to ensure that equipment is performing the way the designer and owner had intended for human comfort needs and for energy savings.

Chapter 9 discusses approaches that take advantage of on-site renewable energy. This chapter has lots of good examples. This topic is of lesser importance to single family structures but presents enormous opportunities for commercial and institutional projects like colleges, universities, manufacturing plants and the like. These building types can easily employ on-site renewable energy because they have under singular control all the elements needed for whole-systems design. Also, these systems present a golden opportunity to make major impacts on reducing greenhouse gas emissions through their built-in waste heat recovery and recycling systems.

The concluding chapter, Chapter 10, jumps into the new realm of paradigm shifts. The first chapters discuss specific strategies and opportunities that directly affect the building's efficiency and impacts on the environment. What else can help? This last chapter explores shifts in capabilities that can indirectly shape designs. Given, there is a multiplicity of trades and professions involved in building design, so this makes the design process complex and challenging. Team integration is the key to good project execution. Life cycle costing is helpful; green building design is helpful. Lately, we're hearing more and more about net-zero buildings – ones that give back annually as much energy as they have consumed. Shifting from green to regenerative (e.g., PV) may prove even more helpful. Building automation (through information and communication technologies) is another useful tool increasing in popularity.

In summary, I feel it necessary to mention the progressive steps taken by some of the recent energy standards that are more in tune with the approaches addressed in this book. For example, while the first eight or so versions of ASHRAE Standard 90.1 were target at lowering energy cost, a more recent standard (ASHRAE 189.1) is targeted at the design of high-performance green buildings. In fact, this standard only has one chapter devoted to energy efficiency. Five chapters address site sustainability, water use efficiency, indoor environmental quality (IEQ), impact on the atmosphere/materials/ resources, and building construction and plans. This standard seems like a step in the right direction and in perfect concert with the approaches followed in this book. Possibly all these efforts will all lead to a net-zero energy, sustainable future.

Larry O. Degelman, P.E.
Professor Emeritus of Architecture
Texas A&M University
November 2015

Editor's foreword

Buildings have been reported at attributing 48% of all energy consumption in the USA and 20–40% of the total energy consumption in Europe and other developed countries. Around three-quarters of the electricity in the U.S. is used just to operate buildings. Also, about half of the greenhouse gas emissions in the U.S. and 30–40% of the total carbon emissions in the UK are attributable to the built environment. Production of construction materials is primarily dependent on conventional energy sources in many parts of the world. In some countries like India, around 80% of the emissions from the construction sector are generated as a result of the products/industrial processes associated with energy intensive building materials (i.e., cement, lime, steel, brick and aluminum). At present, the buildings energy is predominantly provided from fossil fuel sources. The increasing scarcity and correspondingly increasing costs of fossil fuels and freshwater, and demand of reducing greenhouse gas emissions, demands a modernisation of the building sector towards more energy and water efficiency and replacement of fossil fuel use by energy from renewable energy sources. The need to slash CO_2 emissions is continuing to mount with growing levels of legislation and incentives to preserve our environment. To meet these targets, considerable research into energy efficiency and sustainability is underway. The book identifies the role of buildings in global carbon emission and provides comprehensive review of the developments in sustainable design strategies and building systems and materials.

The opening chapter identifies the important role of city planning and buildings in the global carbon emission and provides a review of the global efforts and solutions to tackle the carbon emission problem. Chapter 2 provides a broad literature review of passive and low energy cooling systems, addressing systems that have strong and tangible effect on architectural design and detailing systems' components and operation. Chapter 3 presents a review of daylight/dayighting design research with focus on recent advances in sustainable daylight performance evaluation and design methods and technologies. Chapter 4 provides a case study discussion for the rules of thumb applied in designing the passive heating strategies of the 2009 U.S. DOE Solar Decathlon homes and a student contribution to the Living Aleutian Home Design Competition through analysis of architectural design and actual performance data and by comparing how well the houses conformed to the rules and succeeded in keeping these prototypes comfortable. Chapter 5 provides a broad and critical framework to address human factor issues in sustainable architecture by approaching it in the context of the complex and consequential relationships involving people, buildings, and environment. A critical remark was given at the end of the chapter concerning a specific group of people, namely those involved in research, education, development, production, and management activities pertaining to building delivery and operation processes. Chapter 6 develops an understanding of the key issues affecting the sustainable provision of construction materials to meet increasing demands and concludes with two examples of natural building materials illustrating their potential for use in building construction. Chapter 7 provides a discussion of a new decentralized approach, called the Living Oasis, that makes the case for an integrated site/building water management approach to directly reuse reclaimed water onsite, thereby reducing the consumption of potable water in buildings. The case studies presented in Chapter 7 suggest that direct water reuse in the U.S. is both feasible and could be highly beneficial particularly in areas susceptible to drought and/or water shortages. Chapter 8 provides a discussion of the key issues affecting Energy-efficient HVAC systems and systems integration. Chapter 9 provides a case study story for the strategies implemented by Ball State University to employ on-site renewable energy, which depended on a reasonable assessment of the potential magnitude of energy production by differing technologies and an understanding of the "order of demand" resulting from the patterns of the day-to-day presence of students, staff, faculty, and administrators. Finally, the closing chapter examines the shift from green building design to regenerative approaches that, while deploying many current green strategies and technologies, accept a much broader, holistic framing of design responsibilities. It also investigates the ways in which human and automated intelligence, and the role of Information and Communication Technologies (ICT) may reinforce or challenge the regenerative agenda.

I believe that this book will provide the readers with a thorough understanding of sustainable design methods and green technological options to improve energy and water efficiency, to use sustainable building materials, and to replace fossil fuels through energy from renewable energy resources, hence reducing energy consumption and greenhouse gas emissions and at the same time creating healthier and more comfortable and enjoyable environment.

I hope that this book will help all readers, in the professional, academics and non-specialists, as well as key institutions that are working in architecture, urban planning, and green building. It will be useful for leading decision and policy makers, municipalities and building and energy sector representatives and administrators, business leaders, power producers, and architects and building engineers from industrialized and developing countries as well. It is expected that this book will become a standard, used by educational institutions, and Research and Development establishments involved in the respective issues.

Khaled A. Al-Sallal
(editor)
December 2015

About the editor

Khaled A. Al-Sallal is currently a professor of architectural engineering and the director of the Daylighting Simulation Laboratory at UAE University. His area of expertise is sustainable design with emphasis on building energy. He has a Ph.D. from Texas A&M University and a Master's degree from Arizona State University. His teaching and research has focused on building performance and simulation, carbon-neutral design, daylighting, and climatic responsive architecture. In 2011, he established the state-of-the-art Daylighting Simulation Laboratory at UAE University that runs under his technical management. He has taken major roles in many research and consulting projects that introduced new sustainable building design ideas and technologies to the Middle East, with a total value of U.S. $952,000 granted in 16 different projects in the last decade. Among his innovations, he developed a novel methodology to measure and simulate trees interception of natural lighting in architectural spaces useful for sustainable design applications; and he created a new concept of high performance cool skylights for hot climates; which is currently being investigated. He has supervised several Ph.D. and M.Sc. students and helped them to complete their dissertations/theses and publish their work in reputable journals. He has produced numerous publications in international refereed journals and scientific conference proceedings, and wrote several chapters in edited books by reputable scientific publishers (Taylor & Francis, Elsevier, Hogrefe & Huber). He has given numerous lectures and presentations in various conferences, meetings, and scientific forums. He is currently the Editor-in-Chief of the EJER and served in the editorial board of four scientific journals. He has been active in professional and scientific societies. He is a member of the IBPSA Board of Directors and founder and president of IBPSA-UAE Affiliate (IBPSA-UAE) since 2005. He is the WSSET Vice President, Middle East since 2009. He is an active full member of ASHRAE, Vice President of ASHRAE Falcon Chapter: since 2011, and a Member of Board of Governors since 2009. He received several awards including: Best Research Project Award of Engineering, 2007 and Faculty Award of Excellence, 2000 in UAEU.

CHAPTER 1

Energy and carbon emissions of buildings

Khaled A. Al-Sallal

1.1 FAST URBANIZATION AND INCREASING ECOLOGICAL FOOTPRINTS

The growth of the industrial societies in many parts of the world has resulted in fast urbanization with a highly consuming, resources-depleting lifestyle. The backbone of sustaining this massive growth is the continuous streaming of energy that has been drastically increasing, as power demands have been increasing since the industrial revolution. Since most of the energy is basically produced by conventional resources which are non-renewable and finite, this has led to many environmental and health dangers. Hence, the increasing level of anthropogenic carbon emissions and the resulting degradation of the environment that we witness today is a result of many unsustainable practices that were originated to fulfill the needs of fast urbanization and excessively enhanced lifestyles.

In some parts of the world, such as the Gulf Cooperation Council (GCC) countries, the fast growth of societies supported by rich energy resources, mainly oil and gas, has also resulted in expansive urbanization and reliance on increased levels of fossil fuel supply to support completely new forms of excessively enhanced lifestyles. This, in turn, results in high ecological footprints (World Wildlife Foundation, 2010). The highest ecological footprints (measured in global hectares per capita; gha/capita) by far are found in countries like the United Arab Emirates (10.68 gha/capita; this is the world's highest average ecological footprint), Qatar (10.51 gha/capita), and the United States of America (8.00 gha/capita) (Global Footprint Network, 2010). If everyone in the world lived their life like an average person of these countries, the biocapacity of more than 4.5 earths would be required to support the needs of the global population. Governments today must realize the critical need to control anthropogenic emissions; which can be achieved through adopting short-term and long-term plans that promote increasing awareness amongst all members of society, and encouraging the adoption of measures that combat this danger.

Buildings and the construction industry today consume massive amounts of energy mainly produced by burning fossil fuels and the production of electrical power to run operations at different stages of the building lifecycle. This includes manufacturing building materials, handling construction processes, and operating building systems (lighting, heating, cooling, ventilation, etc.) and appliances. Because buildings are the largest consumers of energy among all sectors, they are also the largest contributor to climate change. Building stock is projected to grow extensively over the next three decades. This can create a remarkable opportunity to achieve significant emissions reductions in the building sector, if we follow greener approaches in designing and constructing buildings. Pressure to slash CO_2 emissions continues to mount, with growing levels of legislation and incentives to preserve our environment. To meet these targets, considerable research into energy efficiency is underway and some of the recent findings can be found in this book. The review in this chapter identifies the role of buildings in global carbon emission and casts light on several global responses to tackle this important issue.

1.2   BACKGROUND

1.2.1   *Climate issues*

The reports of the Intergovernmental Panel on Climate Change (IPCC) show a high level of concern about anthropogenic (human caused) emissions of greenhouse gases and the consequent impact these emissions have on the environment and society (IPCC, 2007a). Although there is controversy regarding this issue, one thing that cannot be argued is the documented fact of the steadily rising CO_2 levels in the atmosphere. The recorded data of the monthly mean atmospheric carbon dioxide at Mauna Loa Observatory, Hawaii shows this trend. The data constitute the longest record of direct measurements of CO_2 in the atmosphere, reported in a dry mole fraction (i.e., the number of molecules of carbon dioxide divided by the number of molecules of dry air multiplied by one million, ppm) (Keeling *et al.*, 1976; Thoning *et al.*, 1989). One can see in the 2013 data how the CO_2 levels of concentrations exceeded 400 ppm for the first time. These measurements started originally by C. David Keeling of the Scripps Institution of Oceanography in March of 1958 at a facility of the National Oceanic and Atmospheric Administration (Keeling *et al.*, 1976; Scripps Institution of Oceanography, 2014). NOAA continued its own CO_2 measurements in May of 1974, and they have run in parallel with those made by Scripps since then (NOAA/ESRL, 2014; Thoning *et al.*, 1989).

1.2.2   *Global carbon emissions/emissions from fossil fuels*

Emissions from fossil fuels vary considerably by country and by lifestyle. Industrial countries such as the USA and China produce the highest levels of emissions while rich countries with small populations such as the Arab Gulf states have the highest levels of emissions per capita.

Global carbon emissions from fossil fuels have significantly increased since 1900 with estimations showing an increase by over 16 times between 1900 and 2008, and by about 1.5 times between 1990 and 2008 (Boden *et al.*, 2010; CDIAC, 2010). Data show that since 1751, approximately 337 billion metric tons of carbon have been released to the atmosphere from the consumption of fossil fuels and cement production. Half of these accumulated emissions have occurred since the mid 1970s. The global fossil-fuel carbon emission estimate in 2007 was 8365 million metric tons of carbon; which represented at that time an all-time high and a 1.7% increase from 2006. Other key findings in Boden *et al.* (2010) can be listed as follows:

- Liquid and solid fuels accounted for 76.3% of the global emissions from fossil-fuel burning and cement production in 2007.
- Combustion of gas fuels (e.g., natural gas) accounted for 18.5% (1551 million metric tons of carbon) of the total emissions from fossil fuels in 2007 and reflects a gradually increasing global utilization of natural gas.
- Emissions from cement production (377 million metric tons of carbon in 2007) have more than doubled since the mid 1970s and now represent 4.5% of global CO_2 releases from fossil-fuel burning and cement production.

More recent data for the global carbon emissions (see Table 1.1) has been posted on November 20, 2013 (CDIAC, 2013; Global CO_2 Budget 2013). The key findings are listed here:

- Atmospheric carbon dioxide levels increased in 2012 at a faster rate than the average over the past 10 years because of a combination of continuing growth in emissions and a decrease in land carbon sinks from very high levels in the previous two years.
- The 2012 carbon dioxide emissions breakdown is coal (43%), oil (33%), gas (18%), cement (5.3%) and gas flaring (0.6%).
- Global emissions due to fossil fuel alone are set to grow in 2013 at a slightly lower pace of 2.1% than the average 3.1% since 2000, reaching a level that is 61% above emissions in 1990.
- Growth rates for major emitter countries in 2012 were 5.9% (China), −3.7% (USA), −1.3% (EU28), and 7.7% (India).

Table 1.1. Data for global carbon emissions (fossil fuels, cement, land-use change). Source: Global CO_2 Budget (2013), CDIAC Global Carbon Project (2013).

Year	Carbon emissions[1]
2012	9.7 billion metric tons per year (+2.1%)
2011	9.47 billion metric tons per year
2010	9.19 billion metric tons per year
2009	8.74 billion metric tons per year
2008	8.77 billion metric tons per year
2007	8.57 billion metric tons per year
2006	8.37 billion metric tons per year

[1]To convert carbon to carbon dioxide (CO_2), multiply the numbers above by 3.67.

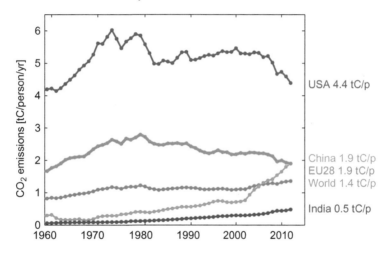

Figure 1.1. Average per capita emissions in 2012 for the top fossil fuel emitter countries. China is growing rapidly and US is declining fast. Courtesy of Oak Ridge National Laboratory, US Dept. of Energy. Source: CDIAC Data (2014), Le Quéré *et al.* (2013), Global Carbon Project: CDIAC (2013).

- Land and oceans are losing their ability to absorb man-made carbon dioxide. According to Dr. Mike Raupach of CSIRO: "*A continuation of the emissions growth trends observed since 2000 would place the world on a path to reach 2 degrees Celsius above pre-industrial times in 30 years*".

The average per capita emissions in 2012 for the top fossil fuel emitter countries are shown in Figure 1.1 (Global Carbon Project: CDIAC, 2013). China is growing rapidly and the US is declining fast. If all the world population led lifestyles similar to the North American patterns of consumption, then emissions would increase more than three-fold. Global emissions are standing at around 35.5 billion metric tons of CO_2 per annum. With a population estimated at around 7 billion, this amounts to an average of approximately 1.4 metric tons of carbon per person (5.1 tons of CO_2 per person). The maximum concentration of atmospheric CO_2 considered safe is 450 parts per million (ppm), as argued by Hillman and Fawcett (2004). This is equivalent to reaching global average emissions of around 2.1 tons of CO_2 per person per year before 2050. This requires reductions in the order of 70% in Europe, 70% in China, 87% in North America

Table 1.2.   Average per capita emissions in 2012 for the top fossil fuel emitter countries and the required reductions by 2050. Source: CDIAC Data (2014), Le Quéré *et al.* (2013), Global Carbon Project: CDIAC (2013).

Country/region	Metric tons of carbon/person	Metric tons of CO_2/person	Required reductions by 2050[1]
USA	4.4	16.12	87%
Europe	1.9	6.96	70%
China	1.9	6.96	70%
India	0.5	1.83	−12%
World	1.4	5.12	59%

[1]Based on global average emissions of around 2.1 metric tons of CO_2/person.

and 59% in the world (see Table 1.2). To preserve a planet similar to that on which civilization developed and to which life on earth is adapted, Hansen *et al.* (2008) suggested an initial objective of reducing atmospheric CO_2 to 350 ppm, with the target to be adjusted as scientific understanding and empirical evidence of climate effects accumulate.

1.2.3   World Energy Outlook

The International Energy Agency (IEA) warned in its flagship publication World Energy Outlook WEO-2011 edition that the world would lock itself into an insecure, inefficient and high-carbon energy system unless a bold change of policy direction was made (IEA, 2011a). *"As each year passes without clear signals to drive investment in clean energy, the "lock-in" of high-carbon infrastructure is making it harder and more expensive to meet our energy security and climate goals"* said Fatih Birol, IEA Chief Economist (IEA, 2011b). According to the IEA, there is still time to act but the window of opportunity is closing and hence governments need to introduce stronger measures to drive investment in efficient and low-carbon technologies. The primary energy demand increases by one-third between 2010 and 2035, with 90% of the growth in non-OECD (OECD: The Organisation for Economic Co-operation and Development) economies (IEA, 2011a). If recent government commitments are implemented in a careful manner as assumed by the New Policies Scenario, the cumulative CO_2 emissions can be controlled by the year 2035 to become three-quarters of the total from the past 110 years. The New Policies Scenario is a scenario in the World Energy Outlook that takes account of broad policy commitments and plans that have been announced by countries, including national pledges to reduce greenhouse-gas emissions and plans to phase out fossil-energy subsidies; even if the measures to implement these commitments have yet to be identified or announced. This broadly serves as the IEA baseline scenario. Applying it can lead to a long-term average temperature rise of 3.5°C (instead of 6°C, if the new policies are not implemented).

China, as the world's largest energy consumer, will consume nearly 70% more energy than the United States by 2035, even though by then, per capita demand in China is still less than half the level in the United States (IEA, 2011b). Renewables increase from 13% to 18% in 2035 and the share of fossil fuels in global primary energy consumption falls from around 81% today to 75% in 2035. The growth in renewables is supported by subsidies that rise from US$ 66 billion in 2010 to US$ 250 billion in 2035. By contrast, subsidies for fossil fuels amounted to US$ 409 billion in 2010.

The WEO presents a 450 Scenario, which sets out an energy pathway consistent with the goal of limiting the global increase in temperature to 2°C by limiting concentration of greenhouse gases in the atmosphere to around 450 parts per million of CO_2. Four-fifths of the total energy-related CO_2 emissions permitted up to 2035 in the 450 Scenario are already locked-in by existing capital stock, including power stations, buildings, and factories. Delaying action is a false economy; as for every US$ 1 of investment in cleaner technology that is avoided in the power sector before

2020, an additional US$ 4.30 would need to be spent after 2020 to compensate for the increased emissions.

According to the IEA 2013 edition of the World Energy Outlook (IEA, 2013), global energy demand rises by one-third by 2035 and energy-related carbon-dioxide emissions are projected to rise to 20% by 2035, leaving the world on track for a long-term average temperature increase of 3.6°C, far above the internationally-agreed 2°C climate target. Energy demand in OECD countries barely rises, and by 2035, is less than half that of non-OECD countries, with low-carbon energy sources meeting around 40% of the growth in global energy demand. Higher shifts in the global energy demand in Asia will be seen, but China moves towards a back seat in the 2020s as India and countries in Southeast Asia take the lead in driving consumption higher. The WEO-2013 highlights the importance of energy efficiency. Two-thirds of the economic potential for energy efficiency is set to remain untapped in 2035 unless market barriers can be overcome. The report also emphasizes the importance of carefully designed subsidies to renewables, which totaled US$ 101 billion in 2012 and expands to US$ 220 billion in 2035 to support the anticipated level of deployment.

The Efficient World Scenario (EWS) has been developed for the World Energy Outlook 2012 (WEO-2012). It enables the quantification of the implications for the economy, the environment, and energy security of a major step change in energy efficiency. The central assumption of the EWS is that policies are put in place to allow the market to realize the potential of energy efficiency measures that are economically viable. The method depended on two steps to calculate the economic potential, which varies by sector and region:

- First, the technical potentials were determined, identifying key technologies and measures to improve energy efficiency by sector, in the period through to 2035. This process involved analysis, over a number of sub-sectors and technologies, of a substantial amount of data and information from different sources. For the buildings sector (Fig. 1.2), the analysis depended on consultation with a large number of companies, experts, and research institutions at national and international levels. It also involved conduction of an extensive literature search to catalogue the technologies that are now in use in different parts of the world, as well as judgment of their probable evolution.
- In the second step, energy efficiency measures, which are economically viable, are identified. The criterion adopted was the amount of time an investor might be reasonably willing to wait to recover the cost of an energy efficiency investment (or the additional cost, where appropriate) through the value of undiscounted fuel savings. Acceptable payback periods were calculated as averages over the outlook period and took account of regional and sector-specific considerations.

1.3   BUILDINGS ENERGY AND EMISSIONS

1.3.1   *Building sector energy*

Buildings are the largest users of energy according to the US Climate Action Report (USDS, 2010) produced by the United States Department of State. That is because buildings are huge in number, size, and distribution, as well as the appliances and heating and cooling systems that operate within them. Hence their influence on energy consumption and greenhouse gas emissions (GHG emissions) is crucial.

Energy consumption by buildings accounts for around 20–40% of the total energy consumption in Europe and other developed countries. In the Unites States, the built environment is responsible for about 48% of all energy consumption and greenhouse gas emissions (Architecture 2030, 2011). The estimated value of building energy sector in the US was revealed by Architecture 2030, based on data from the US Energy Information Administration (EIA). Architecture 2030 combined relevant building data contained in each energy sector to make a single Building Sector; which revealed the extraordinary influence buildings have on energy use, greenhouse gas emissions, and

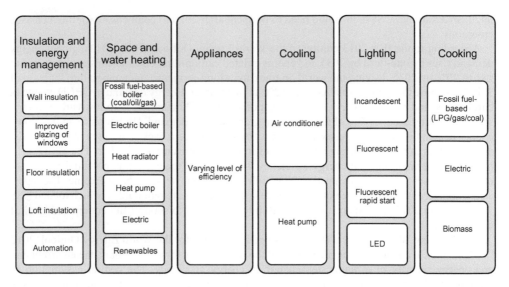

Figure 1.2.   Technical potentials identifying key technologies to improve energy efficiency in the building
sector considered in the efficient world scenario (adapted from WEO-2012: IEA, 2012).

the economy. Prior to 2006, traditional energy data gathering and reporting methods distributed
the impact of buildings across various sectors, including industrial, commercial, residential,
transportation, and electricity.

Around 75% (74.9%) of all the electricity produced in the US is used just to operate buildings.
Globally, these estimated figures are even greater, which makes buildings the largest source of both
energy consumption and greenhouse gas emissions in the world. Moreover, recent reports by EIA
indicate that the energy consumption of the building sector is expected to grow faster than that of
industry or transportation. Between the year 2012 and 2030, building sector energy consumption
is expected to increase by 4.74 quadrillion British thermal units (qBtu) ($\approx 5.0 \times 10^{18}$ J). Energy
consumption by industry is expected to grow by 3.33 qBtu ($\approx 3.5 \times 10^{18}$ J) and transportation
sector energy is expected to grow by 0.37 qBtu ($\approx 0.39 \times 10^{18}$ J). To put these projections into
perspective, 1 qBtu ($\approx 1.055 \times 10^{18}$ J) is equal to the delivered energy of 37 nuclear power plants
(at 1000 MW each), or 235 coal-fired power plants (at 200 MW each).

1.3.2   Emissions from construction industry

The construction sector in India emits about 22% of the Indian economy's total annual emission of
CO_2 (Dakwale and Ralegaonkar, 2012). Excluding the quantities caused by electricity use, direct
emissions from the building sector in 2004 were about three gigatons (Gt) of CO_2, 0.4 $GtCO_2e$ of
CH_4 (methane), 0.1 $GtCO_2e$ of N_2O (nitrous oxide) and 1.5 $GtCO_2e$ of halocarbons (including
chlorofluorocarbons [CFCs] and hydrochlorofluorocarbons [HCFCs]). When the emissions from
electricity use were included, energy-related CO_2 emissions were 8.6 Gt/yr, that is almost a quarter
of global total CO_2 emissions (IPCC, 2007b).

The production of construction materials is primarily dependent on conventional energy sources
in many parts of the world. In some countries like India, around 80% of the emissions from the
construction sector are generated as a result of the products/industrial processes associated with
energy intensive building materials (i.e., cement, lime, steel, brick and aluminum) (Reddy and
Jagadish, 2003).

The increased capital investment in infrastructure projects in China in recent years has fueled
a new carbonizing dragon of the energy-intensive construction industry. A study showed that the

explosive growth in China's construction sector started to drive the country's steep increase in carbon emissions, reversing a long-term trend in which consumption and exports were the dominant factors (Minx *et al.*, 2011). Until 2002, the most critical factor driving Chinese CO_2 emissions was the growth of consumption and factory production for exports. However, researchers found that improved energy efficiency from 2002 to 2007 actually offset the rise in emissions from increased consumption; but emissions continued to skyrocket during that period – with the average annual CO_2 emissions growth rivaling the UK's total CO_2 emissions – largely because of growth in the construction sector and related energy-intensive products such as steel and concrete.

In a recent study on CO_2 Emissions of Construction Industry in China (Cong *et al.*, 2014), it was presented that the macro-building CO_2 emissions were 3.7 billion metric tons, accounting for 44.7% of China's CO_2 emissions in 2010. Among the macro-building CO_2 emissions, the averaged annual percentage of materials production, new construction, and operation phase were 40, 3, and 57%, respectively.

1.3.3 *Building sector and climate change*

In the United States, commercial and residential buildings are responsible for approximately 30% of the greenhouse gas emissions. Recent reports show that nearly half of all energy consumption and greenhouse gas emissions in the US and 30–40% of total carbon emissions in the UK are attributable to the built environment. This should give a clear indication to decision makers and building professionals of the magnitude of building performance impact on degrading the environment (compared to other sectors) and why current building industry practices need to be changed or improved. The good news however is that buildings have also been identified as the economic sector with the best potential for cost-effective mitigation of greenhouse gas emissions (Climate Change, 2007). Therefore, the building industry can and should take responsibility for reducing greenhouse gas emissions, primarily through a reduction in energy consumption.

The United States relies on electricity to meet a significant portion of its energy demands, especially for lighting, electric motors, heating, and air conditioning. In 2007, the emissions attributable to electricity consumption for lighting, heating, cooling, and operating appliances in the US residential and commercial sectors were 72 and 79%, respectively. This made these two sectors accountable for 21 and 18%, respectively, of CO_2 emissions from fossil fuel combustion. The remaining emissions were caused by consumption of natural gas and petroleum for heating and cooking. Electricity generators consumed 36% of US energy generated from fossil fuels, and emitted 42% of total CO_2 from fossil fuel combustion in 2007.

Because buildings are the largest consumers of energy, they are also the largest contributor to climate change. This fact might be surprising as many people still think that transportation and industry have greater impact on climate change and greenhouse gas emissions, due to the high attention given to them. In 2012, the building sector was responsible for nearly half (44.7%) of US CO_2 emissions (Architecture 2030, 2011). By comparison, transportation accounted for 34.3% of CO_2 emissions and industry just 21.1%. According to the US Energy Information Administration, coal is responsible for 74.3% of the CO_2 emissions produced by electricity generation, natural gas is responsible for 24.4% and petroleum is responsible for just 0.8%. It is also remarkable to realize that around three-quarters (73.4%) of the electricity the US consumed in 2012, went just to operate the buildings people live and work in every day. By comparison, industry uses 26.6% and transportation, less than 0.02% (Electric Power Annual 2012: EIA, 2013). The statistics of the year 2011 shown in the same report were very similar. Most of the buildings' energy is produced from burning fossil fuels, making this sector the largest emitter of greenhouse gases on the planet and the single leading contributor to anthropogenic (human forcing) climate change.

Fossil fuels supply 84% of total US consumption, and 76% of building sector energy consumption (Architecture 2030, 2011). CO_2 emissions from the building sector are projected to increase between 2012 and 2030, remaining the largest source of US CO_2 emissions. Fossil fuels consumption in the US is expected to grow by a proportion of 9.8% between 2010 and 2030; around 34% of this growth will be attributable to the building sector. It is the burning of fossil fuels to generate energy that results in the production of carbon dioxide and other greenhouse gases that are now fuelling dangerous climate change. As of 2010, the total US building was equal to 25.5 billion m^2 (275 billion square feet). Every year around 0.6% of the total US building area is demolished down, 1.8% is renovated, and 1.8% is built as new. With these estimated figures, approximately 75% of the built environment will be either new or renovated in 25 years. This transformation over the years represents a historic opportunity for the architecture and building community to create low energy low carbon buildings that help to avoid dangerous climate change (Architecture 2030, 2011).

1.3.4 *Commercial buildings*

Commercial buildings represent just less than one-fifth of US energy consumption, with office space, retail space, and educational facilities representing about half of commercial sector energy consumption (Buildings Energy Data Book: DOE, 2012). A 10% drop in energy expenditures in the commercial building sector is evidenced due to the global economic recession of 2008. The value of new commercial construction also declined by 22% – the largest percentage drop in the last 30 years. This has had a positive effect on carbon dioxide emissions, which decreased by 6%. Between 1980 and 2009, commercial floor space and primary energy consumption grew by 58% and 69%, respectively. They will continue to grow at slower rates between 2009 and 2035, 28% and 22%, respectively, according to the EIA projections. Average energy prices, on the other hand, have been, and are expected to remain, relatively stable. Commercial buildings in total consumed 17.9 quads of primary energy in 2009. This represented 46% of building energy consumption and 18.9% of US energy consumption. In comparison, the residential sector consumed 21.0 quads of primary energy, equal to 22.3% of US energy consumption. The top three end-uses in the commercial sector are space lighting, space heating, and space cooling, which represent half of commercial primary energy consumption (Annual Energy Outlook 2012: EIA, 2012).

Space conditioning and lighting accounted for almost half of commercial energy consumption in the US in 2010. Space heating consumed 27% of site energy in the commercial sector in 2010 (the largest among end-uses) while lighting and cooling consumed 14 and 10%, respectively. Office and retail buildings represented the greatest proportions of commercial floor space – 17% and 16% – and 19% and 18%, respectively, of commercial sector energy consumption in 2003 (the most recent year for which such data is available). Some of these end-use splits vary considerably by building type. Lighting and space conditioning are the most energy intensive end uses in mercantile and office buildings. According to the EIA's 2003 Commercial Buildings Energy Consumption Survey (CBECS-2003: EIA, 2003), a breakdown of electrical energy consumption (in Btu or J) in office buildings in 2003 indicates that lighting accounts for the most use (39%), followed by space cooling (14%) and ventilation (9%). This gives a clear indication that the energy problem must be approached by first implementing energy efficient design strategies related to daylighting and cooling/ventilation measures, and then satisfying the remaining energy requirements with renewable sources.

1.3.5 *Operating energy and embodied energy in buildings*

There are two kinds of energy related to the construction and operation of any building during its life cycle. The first one is the operating energy that is needed to maintain a comfortable indoor environment through operating building systems such as heating, cooling, lighting, and ventilation. The second one is the embodied energy that is directly related to the embodied carbon content of basic building material. The embodied energy is consumed in several manners related to building material during production, on-site construction, and final demolition and disposal. The continuous advancement in building systems' technologies and energy efficient appliances has made the cut down of operating energy possible. Cutting down the embodied is more complicated and requires a different approach that depends on 10 parameters. These are: system boundaries, methods of EE analysis, geographic location of study area, primary and delivered energy, age of data sources, source of data, completeness of data, technology of manufacturing processes, feedstock energy consideration and temporal representativeness (Dixit *et al.*, 2010). Chapter 6 (Sustainable Construction Materials) in this book discusses in more depth the issues related to embodied energy and embodied carbon of building materials.

In a study for a small house in southeast England, the total embodied CO_2 quantities were calculated and predictions for operational CO_2 emissions were obtained from a 100-year dynamic thermal modeling simulation under a medium-high emissions climate change scenario. In the heavier weight cases, the calculated initial embodied CO_2 was found to be higher by up to 15% of the lightweight case value. However, these differences are offset early in the life cycle due to the savings in operational CO_2 emissions. The results of the study showed a total savings of up to 17% in life cycle CO_2 for the heaviest weight case (Hacker *et al.*, 2008). Thormark (2006) studied the potential of material recycling and combustion in reducing the embodied energy in conventional buildings. The results showed a reduction of $\sim$10–15% made through this approach. Another study by Roberts recommended the use of prefabricated building elements on modular units, joined together to create larger or smaller homes, as a sustainable construction technique (Roberts, 2008).

1.3.6 *Buildings indirect emissions*

Buildings are also responsible for contributing indirectly to greenhouse gas emissions. There are many forms of indirect emissions such as the energy required to source and transport materials for construction, the fuel used to move people to and from work, the energy required to transport and treat water for drinking and flushing toilets, and the fuel expended to recycle and dispose of occupant waste streams. Keeping track of all these indirect sources of emissions and quantifying them accurately is very difficult if not impossible. It is still important to address them though by keeping energy efficiency measures in mind when decisions are made about water use, waste generation, transportation patterns, land use, storm water flows, and a myriad of other issues. All these have indirect impacts on climate change and hence taking them into consideration is important.

1.4   GLOBAL ACTIONS TO REDUCE BUILDING EMISSIONS

In the building community, many international initiatives were developed to improve buildings' energy and environmental performance. Their common goal is to achieve a dramatic reduction in the GHG emissions of the building sector by changing the way buildings and developments are planned, designed and constructed. It is not within the scope of this chapter to mention all these global efforts and explain how they are related to each other; or how they started or developed. It is recommended to review them in other literature, such as Appleby (2011) that gave a comprehensive description of these ongoing efforts. These global efforts can be divided into several directions; some of which are explained below.

1.4.1 *Setting sustainable design targets*

The increased greenhouse gas emissions gave incentives for the birth of initiatives at the beginning of the 21st century such as the Architecture 2030 Challenge and the European Commission's, Concerted Action Energy Performance of Buildings Directive. These initiatives helped to issue challenges to the industry to design and implement buildings that had significantly lower energy consumption compared to current typical designs.

Architecture 2030 was initiated in the US by Edward Mazria in 2003, and has a chief goal of net zero energy and net zero carbon buildings by the year 2030. This goal is realized by improving building design and technologies to achieve substantial reductions of energy consumption on a sliding scale from 2010 through 2030. The Architecture 2010 Imperative adopted by the American

Institute of Architects (AIA) was the near-term focus of the challenge. It sets a goal of having buildings routinely built by the year 2010 that would show a 50% improvement in energy efficiency compared to the 1999 version of ASHRAE Energy Standard for Buildings Except Low-Rise Residential Buildings (ANSI/ASHRAE/IESNA Standard 90.1-1999). The AIA 2030 Challenge sets the following targets:

- That all new buildings and major renovations be designed to meet a fossil fuel, GHG emitting, energy consumption performance standard of 50% of the regional average for that building type.
- An equal amount of existing building area should be renovated annually to meet a fossil fuel, energy-consumption performance standard of 50% of the regional average.
- That the fossil fuel reduction standard for all new buildings be increased to carbon neutral by 2030 (using no fossil-fuel GHG-emitting energy to operate).

The AIA 2030 Challenge sets targets for building energy in terms of percentage improvement of the energy use intensity (EUI). EUI is expressed in energy use per area per year (or $GJ/m^2/year$). These targets are specified for different locations and different commercial space uses and building types. It starts by 50 or 60% improvement targets and sets more challenging targets for the future years such as 70% by the year 2015, 80% by the year 2020, 90% by the year 2025, and finally carbon neutral by the year 2030. Through design strategies, technologies and systems, and off-site renewable energy, buildings can be designed and constructed today to meet the 2030 Challenge targets.

Europe also has been making significant progress towards improving energy efficiency and other sustainability aspects in buildings since the Directive on the Energy Performance of Buildings (EPBD) was placed in effect throughout the European Union in January 4, 2006 (EC, 2002). The objective is to properly design new buildings and renovate existing buildings in a manner that will use the minimum non-renewable energy, produce minimum air pollution as a result of the building operating systems, and minimize construction waste. All EU member states are obligated by this directive to bring into force national laws, regulations, and administrative provisions for setting minimum requirements on the energy performance of new and existing buildings, and establishing procedures for energy performance certification. This includes buildings that are subject to major renovations. Assessment of the existing facilities is required and advice on possible improvements and alternative solutions is provided.

According to the EPBD, minimum energy performance requirements are set for new buildings and for the major renovation of large existing buildings in each EU member state. The energy performance certificate (EPC) was introduced as a result of the EPBD. The EPC is required when buildings are constructed, sold, or rented out. The purpose of the certificate is to document the energy performance of the building expressed as a numeric indicator that allows benchmarking. It also includes recommendations for cost-effective improvement of the energy performance, and will be valid for up to ten years.

Technical, functional, and economical feasibility must all be tackled in coordination to upgrade energy performance and meet the minimum requirements. Sometimes alternative energy supply systems (e.g., decentralized energy supply systems based on renewable energy, combined heat and power, district or block heating or cooling, heat pumps, etc.) should be considered especially in the case of large new buildings. Inspection of building systems and installations such as boilers and air-conditioning units is compulsory on a regular basis. The EPC helps to promote the efforts especially in public buildings in which it is required to display the EPC to the public.

1.4.2 *Energy efficiency standards*

To address the Architecture 2010 Imperative, significant effort was put into modifying ASHRAE Standard 90.1 (ANSI/ASHRAE/IESNA Standard 90.1-1999) to drastically improve energy efficiency (Lawrence *et al.*, 2013). The 2010 version of Standard 90.1 basically meets the AIA challenge for 2010 by introducing requirement changes. ASHRAE has also produced the

ANSI/ASHRAE/USGBC/IES Standard 189.1-2009, Standard for the design of high-performance green buildings (ANSI/ASHRAE/IESNA Standard 189.1-2009), which has energy efficiency levels comparable to the 90.1.

1.4.3   *Energy design guidelines*

To meet the Architecture 2030 Challenge (i.e., net zero energy and net zero carbon buildings by the year 2030), ASHRAE will need to continually update Standard 90.1 and Standard 189.1. To help this be accomplished, ASHRAE has produced the ASHRAE Advanced Energy Design Guide series, which covers prescriptive measures that result in significant energy efficiency improvements. The first series dealt with measures that could achieve a 30% savings over Standard 90.1. A continuation of that series that will lead to a 50% energy improvement is currently in process. Other actions are also planned to produce guides for achieving net zero energy performance.

1.4.4   *Green building standards and codes*

The development of green-building standards envisioned for adoption as part of building codes has a major role in promoting the implementation of green-building practices. ASHRAE produced the ANSI/ASHRAE/USGBC/IES Standard 189.1-2009, standard for the design of high-performance green buildings (ANSI/ASHRAE/IESNA Standard 189.1-2009), which has energy efficiency levels comparable to the 90.1. The International Green Construction Code (IgCC) is a code intended to be enforced primarily by building officials (on a mandatory basis) to drive green building into everyday practice. It was developed by the International Code Council (ICC) in association with cooperating sponsors: American Society for Testing and Materials (ASTM) and the American Institute of Architects (AIA). The transition of green design from voluntary implementation to being part of the mainstream design was greatly helped by the release of Standard 189.1 and the IgCC, and the adoption of Standard 189.1 as the reference for design by organizations such as branches of the US military (Lawrence *et al.*, 2013).

1.4.5   *Assessment methods and building-rating systems*

The development of green building rating systems such as LEED in the US and Canada, BREEAM in U.K., GreenStar in Australia, and Estidama in Abu Dhabi have played a vital role by offering motivating incentives for the introduction of green building practices. Prior to the development of these rating systems, green-building concepts were considered only when a client believing in sustainability was luckily found. There are many green building-rating systems at the moment that were created to suit the environmental, economical, and sociocultural conditions/needs of different countries. The ones mentioned above are just a few examples. Applyby (2011) gave a comprehensive review of the currently existing green building rating systems.

1.4.6   *Building low carbon strategies*

Designing a new green project would require estimating the CO_2 equivalent emissions footprint of the building through energy consumption and other sources. These calculations are usually simple and based on the use of emissions factors. For existing buildings, computing the reduction in emissions associated with energy conservation measures is being proposed (Lawrence *et al.*, 2013). The factors used for calculating the greenhouse gas emissions should be based on source energy and not on energy consumed on site alone. A good reference source for emissions factors is contained in a National Renewable Energy Laboratory (NREL) report released in 2007 (Deru and Torcellini, 2007).

When searching for solutions to improve building sustainability, it is important to give priority to tackle the problems that contribute the most to building carbon emissions. One of the major problems especially in commercial buildings is the common use of excessively glazed envelopes.

This kind of envelope results in substantial thermal loads. To make considerable reductions of building carbon emissions, it is necessary to minimize building energy by improving the thermal performance of the envelope. A saving of carbon emission of 31–36% can be achieved via retrofitting and the selection of appropriate U-factors for building envelope materials (Rijksen et al., 2010). Also, reuse, recycling and regeneration of building energy together can save up to 10% of total energy and subsequently mitigate further emissions. Another common problem leading to high carbon emissions, which is caused mainly by the high electrical power demands especially in office buildings, is related to the architectural design and how it is integrated with lighting design. Most of the designs we usually adopt for the architectural form, the floor-plate, and other design factors necessitate the use of electrical lighting, rather than natural lighting, of the indoor space. This results in increasing electrical lighting loads, which also contribute in increasing internal heat gains and cooling loads.

Building systems are usually categorized according to their functions and their relation and impact on natural resources such as energy and water. These categories include passive systems, active systems, integrated passive and active systems, water conservation strategies, wastewater and sewage recycling systems, and power generation systems. Creating a green or an environmentally sustainable building would require maximizing reliance on passive-mode systems while minimizing reliance on active systems. This could be achieved through appropriate design of the built-form configuration, site layout, facade design, solar-control devices, passive daylight devices, envelope materials, vertical landscaping, and passive cooling. Yeang (1999) provided detailed guidelines on how to design these components with special focus on tower buildings.

Kharecha et al. (2010) classified the available strategies that can substantially reduce building sector GHG emissions into three (Table 1.3):

- The planning and design strategies include: building shape, orientation and color, spatial layout, window shape and orientation, daylighting, natural ventilation, exterior shading, vegetation and microclimate control, and passive heating and cooling systems.
- The building envelope and material and equipment selection include: adequate insulation values, radiant barriers, low-emissivity (low-e) coatings and argon-filled glazing, thermal break windows and systems and movable insulation, sunlight and daylight fixtures and systems, cool and green roofs, occupancy and CO_2 sensors, and daylighting controls and photo sensors.
- The added technologies category include: solar hot water heating, photovoltaic systems, micro-wind electric generation, community scale solar thermal, wind and biomass electric generation, and combined heat and power systems.

Improvement of policies, regulations and fuel switching measures to consider emission mitigation can result in a carbon emission saving of up to 25% according to Dakwale and Ralegaonkar (2012). Shifting power generation technology from conventional sources to cogeneration or hybrid technology can result in a substantial reduction in carbon emission. For example, the combination of conventional and solar technology for power generation results in a carbon emission reduction of 40% when compared with conventional methods (Dakwale and Ralegaonkar, 2012).

1.5 CONCLUSIONS

Fast urbanization and higher need of habitat for more people in urban areas in addition to enhanced lifestyle of the people demands increasing levels of energy every year. Most of the world energy is still produced by non-renewable sources of power generation that are major contributors of greenhouse gas (GHG) emission out of which, carbon dioxide (CO_2) is crucial. Buildings consume more energy than any other sector (nearly half of all energy produced in the USA and 20–40% of the total energy consumption in Europe). Around 75% of all the electricity is used just to operate buildings in the USA. Reducing energy demand through building efficiency is significantly cheaper than producing the same amount of energy by coal or nuclear power. Available strategies that would substantially reduce building sector GHG emissions can be divided

Table 1.3. Available strategies that can substantially reduce building sector GHG emissions. (Source: Kharecha *et al.*, 2010).

Planning and design strategies	Building envelope and material and equipment selection	Added technology
Building shape, orientation and color	Adequate insulation values	Solar hot water heating
Spatial layout	Radiant barriers	Photovoltaic systems
Window shape and orientation	Low-e coatings and argon gas filled glazing	Micro-wind electric generation
Daylighting	Thermal break windows and systems and movable insulation	Community scale solar thermal, wind and biomass electric generation
Natural ventilation	Sunlight and daylight fixtures and systems	Combined heat and power systems
Exterior shading	Cool roofs	
Vegetation and microclimate control	Green roofs	
Passive solar heating systems	Occupancy and CO_2 sensors	
Night-vent and night-sky radiation cooling systems	Daylighting controls and photo sensors	
Double envelope systems	Energy management systems	
Common wall design strategies	High efficiency equipment, lighting and appliances	
Building and unit density	Geothermal heat pump	
Mixed-use development	Air-to-air heat exchangers and heat recovery systems	
Pedestrian and transit oriented development (reduced miles traveled)	Building commissioning	

into three general categories: (i) the planning and design strategies, (ii) the building envelope and material and equipment selection, and (iii) the added technologies category. The notion of creating "low energy low carbon architecture" can only be realized through maximizing reliance on passive and low energy systems. Technically, this is done through appropriate design of the built-form configuration, site layout, facade design, solar-control devices, daylight strategies, envelope materials, vertical landscaping, passive cooling and heating strategies, water efficient systems, and onsite renewable energy systems. The following chapters include rich discussions on the strategies and technologies that help to achieve "low energy low carbon architecture".

REFERENCES

ANSI/ASHRAE/IESNA Standard 90.1-1999: American Society of Heating, Refrigerating and Air-Conditioning Engineers, Inc., Atlanta, GA, 1999.

ANSI/ASHRAE/IESNA Standard 189.1-2009: American Society of Heating, Refrigerating and Air-Conditioning Engineers, Inc., Atlanta, GA, 2009.

Appleby, P.: Integrated sustainable design of buildings. Earthscan, London, UK, 2011.

Architecture 2030: 2011, http://architecture2030.org/ (accessed 1/11/2013).

Boden, T.A., Marland, G. & Andres, R.J.: Global, regional, and national fossil-fuel CO_2 emissions. Carbon Dioxide Information Analysis Center, Oak Ridge National Laboratory, US Department of Energy, Oak Ridge, TN, 2010.

CDIAC: Data, last updated 8 January 2014, http://cdiac.ornl.gov/trends/emis/meth_reg.html; http://cdiac.ornl.gov/trends/emis/top2010.tot (accessed 14/1/2014).

CDIAC: Global fossil-fuel CO_2 emissions. 2010, http://cdiac.ornl .gov/trends/emis/tre_glob.html (accessed 1/11/2013).

CDIAC: Global Carbon Project. 2013, http://cdiac.ornl.gov/GCP/carbonbudget/2013/ (accessed 29/11/2013).

Climate Change 2007: *Mitigation of Climate Change, Contribution of Working Group III to the Fourth Assessment Report of the Intergovernmental Panel on Climate Change.* Edited by: B. Metz, O.R. Davidson, P.R. Bosch, R. Dave & L.A. Meyer, Cambridge University Press, Cambridge, UK and New York, NY, 2007.

Cong, X., Mu, H., Li, H. & Liu, C.: Analysis on CO_2 emissions of construction industry in China based on life cycle assessment. Proceedings of the 8th International Symposium on Heating, Ventilation and Air Conditioning. *Lecture Notes in Electrical Engineering* Volume 263, Springer, 2014, pp. 499–506.

Dakwale, V.A. & Ralegaonkar, R.V.: Review of carbon emission through buildings: threats, causes and solution. *Int. J. Low-Carbon Technol.* 7 (2012), pp. 143–148.

Deru, M. & Torcellini, P.: Source energy and emissions factors for energy use in buidlings. Technical report NREL/TP_550-38617, National Renewable Energy Laboratory, Golden, CO, 2007.

Dixit, M.K., Fernandez, J.L., Lavy, S. & Culp, C.H.: Identification of parameter for embodied energy measurement: a literature review. *J. Energy Build.* 42 (2010), pp. 1238–1247.

DOE: 2011 Buildings energy data book. Chapter 3. US Department of Energy, prepared for the DOE Office of Energy Efficiency and Renewable Energy by D&R International, 2012, http://buildingsdatabook. eren.doe.gov/ChapterIntro3.aspx (accessed 1/11/2014).

EC (European Commission): Energy Performance of Buildings, Directive 2002/91/EC of the European Parliament and of the Council, Official Journal of the European Communities, Brussels, Belgium, 2002.

EIA: CBECS 2003. US Energy Information Administration (EIA)/Department of Energy (DoE), http://www.eia.gov/consumption/commercial/data/2003/ (accessed 1/11/2014).

EIA: Annual Energy Outlook 2012; Early Release, US Energy Information Administration (EIA)/Department of Energy (DoE), January 2012.

EIA: Electric Power Annual 2012 Report, December 2013, Table 2.2. Sales to ultimate customers in thousand mega-watt-hours, US Energy Information Administration/Department of Energy (DoE), 2013, http://www.eia.gov/electricity/annual/pdf/epa.pdf (accessed 1/11/2014).

Global CO_2 Budget 2013: http://co2now.org/Current-CO_2/CO_2-Now/global-carbon-emissions.html (accessed 24/11/2013).

Global Footprint Network: *The ecological footprint atlas 2010.* Oakland, CA, 2010.

Hacker, J.N., De Saulles, T.P., Minson, A.J. & & Holmes, M.J.: Embodied and operational carbon dioxide emissions from housing: a case study on the effects of thermal mass and climate change. *J. Energy Build.* 40 (2008), pp. 375–384.

Hansen, J., Sato, M., Kharecha, P., Beerling, D., Berner, R., Masson-Delmotte, V., Pagani, M., Raymo, M., Royer, D.L. & Zachos, J.C.: Target atmospheric CO_2: where should humanity aim? *Open Atmos. Sci. J.* 2 (2008), pp. 217–231.

Hillman, M. & Fawcett, T.: *How we can save the planet.* Penguin, London, UK, 2004.

IEA: World Energy Outlook 2011. International Energy Agency, Paris, France, 2011a, http://www.iea. org/publications/freepublications/publication/WEO2011_WEB.pdf (accessed 1/3/2014).

IEA: Press releases. International Energy Agency, France, Paris, 2011b, http://www.iea.org/ newsroomandevents/pressreleases/2011/november/name,20318,en.html (accessed 06/03/2014).

IEA: Efficient World Scenario: Policy Framework, World Energy Outlook 2012 (WEO-2012). International Energy Agency, Paris, France, http://www.worldenergyoutlook.org/media/weowebsite/energymodel/ documentation/Methodology_EfficientWorldScenario.pdf (accessed 1/3/2014).

IEA: Press releases. International Energy Agency, Paris, France, 2013, http://www.iea.org/ newsroomandevents/pressreleases/2013/november/name,44368,en.html (accessed 06/03/2014).

IPCC (Intergovernmental Panel on Climate Change): Mitigation of climate change: contribution of Working Group III to the Third Assessment report of the Intergovernmental Panel on Climate Change. Cambridge University Press, 2007a. (IPCC reports can be found at www.ipcc.ch).

IPCC (Intergovernmental Panel on Climate Change): The physical science basis; contribution of Working Group I to the 4th Assessment Report of the Intergovernmental Panel on Climate Change (IPCC). 2007b.

Keeling, C.D., Bacastow, R.B., Bainbridge, A.E., Ekdahl, C.A. Guenther, P.R. & Waterman, L.S.: Atmospheric carbon dioxide variations at Mauna Loa Observatory, Hawaii. *Tellus* 28 (1976), pp. 538–551.

Kharecha, P., Kutscher, C., Hansen, J. & Mazria, E.: Options for near-term phase-out of CO_2 emissions from coal use in the United States. *Environ. Sci. Technol.* 44:11 (2010), pp. 4050–4062.

Lawrence, T., Darwich, A.K., Boyle, S. & Means, J. (eds): *ASHRAE GreenGuide: design, construction, and operation of sustainable buildings*. Fourth edition. ASHRAE, Atlanta, GA, 2013.

Le Quéré, C., Peters, G.P., Andres, R.J., Andrew, R.M., Boden, T.A., Ciais, P., Friedlingstein, P., Houghton, R.A., Marland, G., Moriarty, R., Sitch, S., Tans, P., Arneth, A., Arvanitis, A., Bakker, D.C.E., Bopp, L., Canadell, J.G., Chini, L.P., Doney, S.C., Harper, A., Harris, I., House, J.I., Jain, A.K., Jones, S.D., Kato, E., Keeling, R.F., Klein, Goldewijk, K.K., Körtzinger, A., Koven, C., Lefèvre, N., Maignan, F., Omar, A., Ono, T., Park, G.-H., Pfeil, B., Poulter, B., Raupach, M.R., Regnier, P., Rödenbeck, C., Saito, S., Schwinger, J., Segschneider, J. , Stocker, B.D., Takahashi, T., Tilbrook, B., van Heuven, S., Viovy, N., Wanninkhof, R., Wiltshire, A. & Zaehle, S.: Global carbon budget 2013. *Earth System Science Data* 6 (2014), 2014, pp. 235–263, http://www.earth-syst-sci-data.net/6/235/2014/essd-6-235-2014.pdf (accessed 1/11/2014).

Minx, J.C., Baiocchi, G., Peters, G.P., Weber, C.L., Guan, D. & Hubacek, K.: A "carbonizing dragon": China's fast growing CO_2 emissions revisited. *Environ. Sci. Technol.* 45:21 (2011), pp. 9144–9153.

NOAA/ESRL: US Department of Commerce, 2014, www.esrl.noaa.gov/gmd/ccgg/trends/ (accessed 21/12/2013).

Reddy, BVV & Jagadish, K.S.: Embodied energy of common and alternative building materials and technologies. *J. Energy Build.* 35 (2003), pp. 129–137.

Rijksen, D.O., Wisse, C.J. & van Schijndel, A.W.M.: Reducing peak requirements for cooling by using thermally activated buildings. *J. Energy Build.* 42 (2010), pp. 298–304.

Roberts, S.: Effects of climate change on the built environment. *J. Energy Policy* 36:12 (2008), pp. 4552–4557.

Scripps Institution of Oceanography: Scripps CO_2 Program, last updated 2014, scrippsco2.ucsd.edu (accessed 21/12/2013).

Thoning, K.W., Tans, P.P. & Komhyr, W.D.: Atmospheric carbon dioxide at Mauna Loa Observatory 2: analysis of the NOAA GMCC data, 1974–1985. *J. Geophys. Res.* 94 (1989), pp. 8549–8565.

Thormark, C.: The effect of material choice on the total energy need and recycling potential of a building. *Build. Environ.* 41:8 (2006), pp. 1019–1026.

USDS: US Climate Action Report 2010. US Department of State, Global Publishing Services, Washington DC, June 2010, http://www.state.gov/g/oes/rls/rpts/car/index.htm (accessed 21/12/2014).

World Wildlife Foundation: Living planet report 2010: biodiversity, biocapacity and development. Switzerland, 2010.

Yeang, K.: *The green skyscraper: the basis for designing sustainable intensive buildings.* Prestel Verlag, Munich, Germany, 1999.

CHAPTER 2

Passive and low energy cooling

Khaled A. Al-Sallal

2.1   INTRODUCTION

The top four end-uses in the commercial sector are space lighting (20%), space heating (16%), space cooling (14.5%), and ventilation (9%). All together they represent 60% of commercial primary energy consumption (EIA, 2012). In hot climate countries, energy needs for cooling can amount to two or three times those for heating on an annual basis. In very hot climates such as in the UAE, 40% of the total cooling energy can be utilized to offset heat gains from walls and roofs, and it could reach to 75% when combined with the glazing effect (Aboul-Naga *et al.*, 2000). The extensive use of air conditioning is associated with several problems:

- *Environmental*: extensive use of air conditioning has caused a shift of the energy consumption to the summer season with increased peak electricity demands. This imposes an additional strain on electrical national grids and requires development of extra new power plants. The increased energy production leads to exploiting the finite fossil fuels causing atmospheric pollution and climatological changes. Ozone layer depletion is caused by the refrigerants of the air conditioning units such as CFCs and HFCs. Heat rejection during the air conditioning process increases the phenomenon of the "urban heat island". The heat island effect has a great impact in exacerbating cooling energy requirements in warm to hot climates in summer. For U.S. cities with populations larger than 100,000, the peak electricity load increases 2.5 to 3.5% for every °C increase in temperature (Akbari *et al.*, 1992). It is estimated that 3 to 8% of the urban electricity demand in the U.S. is used to compensate for the heat island effect alone. For the USA, the electricity costs for the summer heat island alone could be over US$ 1 million per hour, or over US$ 1 billion per year (Akbari *et al.*, 1992).
- *Indoor air quality*: people working in air-conditioned spaces suffer from increased levels of illness symptoms such as lethargy, headache, blocked or runny noses, dry or sore eyes, dry throat and dry skins, and asthma. This is also known as sick building syndrome. Also, occupants often are not satisfied with indoor comfort conditions.
- *Economic*: installation of air conditioning units adds considerable amounts to the initial cost of the building construction, followed by additional costs due to operation and maintenance of the air conditioning systems.

This chapter is meant to be a comprehensive reference. Yet, passive cooling in buildings is a huge area and it would be impossible to cover all its architectural and engineering aspects in one article. Hence, this chapter is concerned only with those passive and low-energy cooling systems that have strong and tangible effect on the architectural design (e.g., building form, spatial arrangement, structural components, etc.) in order to maintain thermal comfort for the building occupants, with no or limited expenditure of an auxiliary source of energy such as electricity. For other cooling techniques that require higher energy and operation of mechanical systems, the reader can refer to the chapter in this book entitled "*Energy-efficient HVAC systems and systems integration*".

The purpose of this chapter is to provide a comprehensive framework to address the design and operation of several passive and low energy cooling systems. The methodology included revisiting the fundamental literature of these systems and reviewing some of their recent advances. Toward this end, the chapter is structured into several sections. These are as follows:

- Section 2.1 (this section) outlines the problems associated with the extensive use of air conditioning, which justifies why passive cooling should be considered.
- Section 2.2 provides a theoretical background on fundamental topics related to passive cooling. This includes examining the issues that affect thermal comfort, identifying avenues of heat loss and the factors that affect heat transfer rates through these avenues, identifying the sources of coolness (or heat sinks) available in the natural environment, and scrutinizing sources of heat gain and discussing the methods that help to minimize their effect.
- Section 2.3 provides general classifications of passive cooling strategies and explains how to minimize heat gains through architectural design.
- Section 2.4 provides a review of passive cooling systems applied in vernacular architecture especially in the Middle East and North Africa.
- Section 2.5 illustrates the passive cooling techniques that are mainly effective in hot arid regions with regards to the system's components and operation methods. The systems in this chapter are divided into three classifications: systems with no thermal storage, systems with daily storage of coolness, and systems with seasonal storage of coolness.
- Section 2.6 explains the systems that can promote airflow for comfort cooling. These systems are applicable in both hot arid and warm humid climates.
- Section 2.7 is about passive cooling in warm humid climates. It theoretically explains why sources of coolness are limited in hot humid climates. It also illustrates the methods that can help to overcome the limitations of the humid climates. This includes the systems that promote airflow for comfort cooling and the methods that rely on desiccants to reduce air moisture.
- Section 2.8 draws the conclusions.

The reader of this chapter might also need to review other important topics related to passive cooling such as:

- Forced ventilation using ceiling fans or whole-house fans: this topic has minimal effect on architectural design (except for the space height when using ceiling fans) and hence is not covered here thoroughly.
- Systems employing seasonal storage of coolness other than earth-coupled buildings such as the use of earth-air heat exchangers, seasonal storage of coolness in water, and seasonal storage of coolness in the form of ice. These systems are basically operated by mechanical equipment and hence are not covered here.
- Dehumidification systems in hot humid climates: these systems are covered only partially in this chapter, as they are mechanical by nature.
- In addition to the references cited in the chapter, further useful resources about research centers or groups, other literature, and case studies are also provided in the appendices.

2.2   FUNDAMENTALS

2.2.1   *Thermal comfort*

Thermal comfort can be defined as the conditions at which there is no sensation of discomfort due to cold, heat, excessive skin wetness or dryness, air stuffiness, or air moving at high speeds. To be comfortable, the body has to lose the necessary amount of heat, while maintaining the following conditions:

- Keep skin temperature within specified limits. Different parts of the body might require different comfort ranges of temperature.
- Avoid skin wetness due to sweating, or dryness due to much moisture loss.
- Avoid air stuffiness or annoying high-speed air movement.

2.2.2 *Avenues of heat loss*

There are three major avenues of heat loss from the body. These are evaporation, convection, and radiation. The heat balance under steady state conditions can be expressed by the following equation:

$$H = E \pm C \pm R \tag{2.1}$$

where H is the total heat loss, E is the heat loss by evaporation, C is the heat loss by convection, and R is the heat loss by radiation.

The factors that affect the heat transfer rates through these avenues are as follows:

- The partial pressure of water vapor in the air (P_v), or its corresponding saturation temperature, the dew point temperature (T_{dp}). This affects heat loss by evaporation.
- The air dry-bulb temperature (T_{db}). This affects heat loss by convection.
- The air velocity (V_a). This affects heat loss by convection and evaporation.
- The mean radiant temperature of the surfaces surrounding the human body (T_{mr}). This affects heat loss by radiation.
- The insulating effect of clothing (I_{cl}).
- Metabolism (M).

Hence, thermal comfort is a function of all these six factors or it can be expressed as:

$$\text{Thermal comfort} = f(T_{dp}, T_{db}, V_a, T_{mr}, I_{cl}, M) \tag{2.2}$$

The first four factors in the above list (i.e., T_{dp}, T_{db}, V_a, T_{mr}) are the only ones that can be manipulated by the designer in order to maintain thermal comfort since the designer cannot control either clothing or metabolism of building occupants.

2.2.3 *Natural sources of coolness (heat sinks)*

Cooling a building by passive means requires identifying the sources of coolness (or heat sinks) available in the natural environment where the building is located. Natural sources of coolness can be listed as follows:

- The ambient water vapor, which controls the rate of evaporation into the air and evaporative cooling of moist surfaces.
- The ambient air, which may be at a temperature low enough to reject heat by convection.
- The upper atmosphere or the sky, which may be used by building hot surfaces to reject heat by thermal radiation.
- The wind, which may be in speeds high enough to maintain a desired airflow in the indoor space to promote convective and evaporative cooling.

2.2.4 *Reduction of heat gains in buildings*

Heat gains into a building can be categorized into the following (Fig. 2.1):

- Heat gain by solar radiation through window glazing, $Q_{\text{solar windows}}$.
- Heat gain by solar radiation through opaque surfaces such as walls and roof, $Q_{\text{solar walls/roof}}$.
- Heat gain by transmission through the building envelope due to difference in temperature between outside and inside air, $Q_{\text{transmission}}$.
- Internal heat gain generated by building equipment, $Q_{\text{equipment}}$.
- Internal heat gain generated by lighting fixtures, Q_{light}.
- Internal heat gain generated by building occupants, Q_{people}.

The total heat gain of a building is the sum of all these heat gains:

$$Q = Q_{\text{solar windows}} + Q_{\text{solar walls/roof}} + Q_{\text{equipment}} + Q_{\text{light}} + Q_{\text{transmission}} + Q_{\text{people}} \tag{2.3}$$

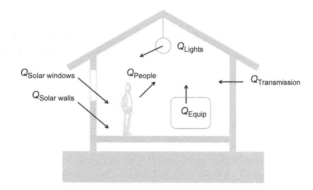

Figure 2.1.   External and internal heat gains of a building.

Designing a building with effective passive cooling requires scrutinizing all sources of heat gain and trying to minimize their effect using various and integrated strategies:

- Reduce $Q_{\text{solar windows}}$ by solar shading and minimizing the glazing area. Shading can be provided by natural elements like trees and other plants, by indoor fittings like curtains and venetian blinds, and by outdoor shading devices like overhangs or fins.
- Reduce $Q_{\text{solar walls/roof}}$ by using low absorbance materials. This can be done by several methods: by painting the external surfaces of the building (walls and roof) with light colors, by cladding the walls and covering the roof with light colored materials, and by using a green roof.
- Reduce $Q_{\text{transmission}}$ by minimizing heat transmission through the building envelope surfaces.
- Reduce $Q_{\text{equipment}}$ by more conscious use of equipment (use least heat generating equipment).
- Reduce Q_{light} by using more efficient lighting fixtures.

$Q_{\text{transmission}}$ can be expressed as follows:

$$Q_{\text{transmission}} = \Sigma Q_{\text{transmission}} \text{ (for all surfaces)} \tag{2.4}$$

$$Q_{\text{transmission}} = UA(T_{\text{out}} - T_{\text{in}}) \tag{2.5}$$

where U (or U-value) is the overall heat transfer coefficient of the building envelope component (wall, roof, slab), A is the surface area of the envelope component, T_{out} is the outdoor ambient air temperature, and T_{in} is the indoor air temperature. The U value can be calculated by the following equation:

$$U = \frac{1}{\frac{1}{h_o} + \frac{x_a}{k_a} + \frac{x_b}{k_b} + \frac{x_c}{k_c} + \ldots + \frac{1}{h_i}} \tag{2.6}$$

where h_o and h_i are the convective heat transfer coefficients for the outside and inside air, respectively, x_a, x_b, x_c, ... are the thicknesses of different envelope components, and k_a, k_b, k_c, ... are the thermal conductivities of these components.

To reduce heat transmission into the building one can:

- Reduce the U value, which can be done by increasing the thicknesses of different envelope components (x_a, x_b, x_c, ...) and/or reducing their thermal conductivities (k_a, k_b, k_c, ...). Practically, this is done by adding thermal insulation to the envelope components and choosing materials with higher R-values.
- Reduce surface area (A) of envelope components by proper design.
- Reduce the difference between T_{out} and T_{in}. T_{out} can be reduced by creating a better microclimate that helps to mitigate the outdoor conditions through building form design (e.g., courtyard) and/or landscape design.

2.3 PASSIVE COOLING AND ARCHITECTURAL DESIGN

2.3.1 *General classifications of passive cooling techniques*

Application of passive cooling techniques in buildings has been proved to be extremely effective and can contribute greatly to the decrease of the cooling load of buildings (Santamouris and Assimakopoulos, 1997). These techniques can be classified as follows:

- *Preventive techniques*: protection from solar and heat gains should involve the following measures: landscaping, the use of outdoor and semi outdoor spaces, building form, layout and external finishing, solar control and shading of building surfaces, thermal insulation, and control of internal gains.
- *Modulation of heat gain techniques*: this has to do with the capacity for heat storage in the building structure. This delay strategy depends on the decreasing of peaks in cooling load and modulation of internal temperature with heat discharge at a later time.
- *Heat dissipation techniques*: these techniques depend on the potential for disposal of excess heat by natural means. This requires two main conditions: (i) the availability of an appropriate environmental heat sink and (ii) the appropriate thermal coupling and sufficient temperature differences for the transfers of heat from indoor spaces to a heat sink. The main processes and heat sinks required can include radiative cooling using the sky, evaporative cooling using air and water, and convective cooling, using air and ground soil.

2.3.2 *Minimizing heat gains through architectural design*

The main sources of the external heat gains are solar radiation, which is transmitted indoors through the glazing or is absorbed by the opaque elements of the building envelope such as the walls and the roof, and then conducted to the indoor spaces. The internal heat gains include metabolic heat produced by occupants, artificial lighting, appliances, and occupants' activities such as cooking or bathing.

Considering any cooling system in the building design will add extra costs including the construction/installation costs and energy costs (when active or hybrid systems are used). Passive cooling systems can help to deduct or minimize the energy costs but they will still cost money for construction. The construction cost of the system can be reduced if the system is downsized; and this can be done only if the cooling load (or heat gain) is minimized. Hence, the logical approach is to minimize heat gain before considering applying any cooling system. This requires setting clear design objectives at the outset and following a sound architectural design process that eventually results in minimizing building heat gains, as explained below.

2.3.2.1 *Improving the microclimate around the building*
This requires mitigating the ambient hot weather conditions in order to eventually improve indoor comfort and reduce cooling load. The microclimate can be improved by the following:

- *Site selection, site planning, and building siting*: during the site selection stage, the architect should select a site that helps to reduce heat gains. Some sites are better than others with regards to environmental factors such as receiving less quantities of solar radiation and having better access to desirable wind orientation. The site planning, site zoning, and building siting should also contribute to protect the building from excessive solar radiation and provide shading and access to cool breezes. The topographical features and adjacent buildings and vegetation should also be exploited to improve the microclimate around the building.
- *Proper landscaping*: choosing appropriate types of trees and distributing them properly on the site can help to minimize solar radiation and provide shading. Trees and other plants can also help to reduce the ambient temperature through the process of evapotranspiration and create areas for outdoor activities. Using limited water bodies such as pools, fountains, and water channels shaded by plants can improve evaporative cooling and create beautiful outdoor environments. This encourages occupants to spend some hours outside the building and hence helps to reduce the cooling load inside.

2.3.2.2 Form configuration of the building

The designer should develop a building form that helps to shade itself and reduce exposure to the sun. Compact forms in hot dry climates help to reduce surface exposure to harsh outdoor conditions (high solar radiation and high ambient temperature) and prevent heat gains. However, the compact form is useful only during the daytime. At the nighttime, the larger area of surface exposure can help to lose heat by radiation to the upper atmosphere and by convection to cool breezes. Givoni (1994a) suggested using a form that includes porches with closing insulated shutters that help increase the surface area when the shutters are opened. The form should also help to promote cool breezes to channel through the building's openings and improve cooling of the indoor. The spread-out form can be very effective in hot humid climates. The introvert form configuration or the courtyard design can provide many advantages especially in hot dry climates (discussed later in Sections 2.4 and 2.5.1).

2.3.2.3 Arrangement of the building spaces

The spatial zoning of the building should respond to the environmental influences by utilizing desirable environmental factors (e.g., cool breezes) while blocking undesirable ones (e.g., hot and dusty wind). The courtyard design helps to employ an effective spatial arrangement for functional and sociocultural requirements to serve climatic and thermal comfort needs as well. Al-Hawsh (meaning courtyard in Arabic) is a good exemplar of a sustainable design configuration that holds a balance between climatic and sociocultural requirements (discussed later in Section 2.4).

2.3.2.4 Shading the building envelope and openings

Solar gain through the opaque and glazing surfaces is a major component of building heat gains. Controlling solar gain through the proper design and distribution of solid and glazed walls/roof is therefore of primary importance in all hot climates. One should avoid using large areas of openings in hot dry climates especially on the west and east orientations because of the low solar altitude. In hot humid climates, large openings can be very useful in promoting cool breezes into the building; yet the openings have to be shaded by extended porches or shading devices. South openings require external shading treatment using horizontal shading devices. West and east orientations require external vertical shading devices. In very hot climates, egg-crate shading devices are recommended in order to prevent all kinds of solar radiation including the direct, the diffuse, and the reflected components. Another important strategy is to manipulate the solar-optical properties of the glazed surfaces to cut down heat gains.

2.3.2.5 Color of the envelope

The color of the envelope surfaces has significant effect on the impact of solar radiation on heat gains and the resulted indoor temperatures. White roofs are highly recommended in hot climates. Givoni (1994b) conducted an experimental work in Haifa comparing between a roof painted with white and another one using the same material (i.e., Ytong or lightweight concrete $600 \, \text{kg/m}^3$) painted with gray, both of different thicknesses (7, 12, 20 cm). The results showed a big difference in the external surface temperature between the two cases ($27.5°C$ for the white roof versus $69°C$ for the gray roof). The indoor ceiling temperatures for the white roof were about $25.5°C$ for all tested thicknesses while it was 45, 39, and $33°C$ for the gray roof cases of thicknesses 7, 12, and 20 cm, respectively. He also conducted a similar experiment in Haifa to compare between a wall painted with white and another one painted with gray. The tests were conducted for two materials (solid concrete and Ytong, 12 and 22 cm thick). For the gray wall case, the indoor temperatures were constantly above the outdoor. The average indoor air temperatures were $4.5°C$ above the average outdoor temperature; and the indoor maxima were above the outdoor maximum ($8°C$ for the 12 cm concrete). For the white wall case, the air indoor temperatures were below the average outdoor temperature most of the daytime hours; and the indoor maxima were about $2°C$ below the outdoor maximum.

2.3.2.6 Thermal insulation of the envelope

Building envelopes should be designed to limit transmittance of heat to the interior by using high resistance thermal insulation to minimize heat transfer by conduction. Radiant heat transfer can also be minimized by using a low-emissivity material (radiant barrier) next to an air gap in a multi-layer roof assembly. It helps to reflect the radiation causing reduction of the inner layer temperature and the radiant temperature of the internal spaces, while at night, it blocks radiant exchange, thus reducing the required cooling load.

2.3.2.7   *Control of internal gains*

In public and large buildings, the internal heat gains have greater impact on the building cooling load than the ambient conditions even in cold regions. These kinds of buildings have large numbers of people and many sources of internal heat gains and hence are considered as internal load dominated buildings. There are many measures to control internal heat gains. The most important ones are as follows:

- Maximize reliance on daylighting and use electrical lighting only when needed. This will require from designers to investigate methods and use control technologies (such as automated sensors) that help to integrate natural and electrical lighting.
- Use high performing energy-efficient equipment and appliances and locate them in places that help to minimize their effect on the internal gains.
- Allocate the spaces according to the occupation densities and types of activities (Santamouris and Assimakopoulos, 1997).

2.4   PASSIVE COOLING IN VERNACULAR ARCHITECTURE

2.4.1   *Vernacular architectural approaches in hot climates*

Vernacular architecture in hot arid regions shows how people throughout history developed a number of effective approaches to provide relief from the severe hot climates. Some of these approaches are as follows:

- *Clustering the buildings*: when buildings are clustered and some of their outer surfaces are attached together, the area of their exposure to the hot weather is minimized. This helps to reduce transmission of heat to the indoor spaces. It also helps to reduce dust that is usually a common problem in hot arid regions.
- *Limiting openings*: limiting the number/area of windows and external doors to the outdoor environment helps to reduce heat gain caused by solar radiation transmitted through the glazing (i.e., $Q_{solar\ windows}$ and $Q_{transmission}$) and heat gain transmitted through the building envelope ($Q_{transmision\ windows}$). It also helps to reduce dust penetration to inside.
- *Building with thick adobe walls*: building the walls with a material such as adobe that has a high thermal mass capacity helps to store night coolness for several hours then the coolness is released to the interior space during the hot daytime hours when it is needed.
- *Creating a cool microclimate*: when building form is shaped to make its mass surround a courtyard (i.e., an introvert form) that has trees/shrubs and perhaps a water pool or fountain, this helps to create a microclimate that is much cooler than the normal climate conditions.
- *Adapting the climate with sensible traditions*: traditional people living in vernacular architecture follow a dynamic lifestyle that matches the climate changes/cycles. For instance, during hot summer afternoon hours people live in the cool basements, which are cooled by the low ground temperature, and during the cool summer nights they move to the roof to sleep there.
- *Passive cooling systems*: people throughout a long history learned how to develop and build several effective cooling systems that run naturally (without fuel or consumed energy). Examples of these passive cooling systems are wind towers, wind catchers, curved/domed roofs, water cisterns, and ice makers. Bahadori (1978; 1979; 1986) presented a comprehensive review about passive cooling systems used mainly in the vernacular Iranian architecture, in addition to other regions. Fathy (1986) presented a review about passive cooling systems used in the vernacular architecture of the Arab houses in Egypt.

2.4.2   Vernacular architecture of the Arab Gulf

2.4.2.1   Urban and architectural context

The desert architecture in the Arabian Gulf was basically the result of three main factors: the hot dry/humid climate, the people's socio-cultural life and the locally available construction methods and materials. Usually this type of architecture is characterized as a high-density plan where buildings are close to each other or attached, penetrated by arrow alleys (called Sikka in the local dialect) that were shaded for most of the day. These alleys are oriented either to promote sea breezes in the coastal settlements or to limit dusty winds in the inland settlements.

In the coastal settlements such as Al Fahidi Historical District in Dubai or Al Murijah District in Sharjah, these alleys mostly run from north to south and end at the sea, thus permitting the prevailing north winds to pass through. 'Introvert plan' houses comprising a series of rooms built around a central courtyard, open to the sky, is the form configuration of residential buildings in this area. This plan satisfies the social conditions and in particular privacy for the various elements of the extended family and flexibility of space, which can be adapted to the changing requirements of the extended family. The courtyard also acts as a wind generating tool in the house, the hot air ascends and cooler air replaces it from the surrounding rooms. This movement creates constant air circulation in the house and provides a pleasant living microclimate for the inhabitants. Another important feature in this architecture is the use of wind towers that helps to passively cool the occupants. The wind-towers of these vernacular houses give the only variation in height to the outline of the district and thus creates a more interesting and beautiful skyline.

2.4.2.2   Approaches for solar shading

The vernacular settlements in the Arab Gulf are compact organizations to maximize solar shading and keep the extreme solar radiation away from buildings' structures and occupants. This is achieved through the following:

- *Form-space proportion*: increasing the density of the built area.
- *Size of the alleys*: the alley's width is narrow and does not exceed 3 m.
- *Building height and alley width proportion*: increasing the building height and reducing the alley width ($h/w = 2{:}1{-}1{:}1$).
- *Public social squares*: the size and shape proportion promotes solar shading by the surrounding buildings and vegetation.
- *Covered alleys*: alleys in the public zones such as in the market are shaded by palm groves.

At the architectural scale, several methods are used to provide shading. These are as follows:

- The courtyard maximizes shading by its walls and vegetation for most of the day hours.
- The air-puller wall (Masqat) provides total shading by its solid walls.
- The traditional window provides total shading by its wooden shutters.

2.4.2.3   Approaches for airflow

The main alleys in the coastal settlements in the Arab Gulf run along the north-south direction. This helps the desirable north and north-west breezes to infiltrate into the city masses. The houses are oriented to catch the pleasant sea breezes. This is an effective traditional design solution at the urban scale, which promotes a passive cooling effect through the following:

- Convective cooling of the building structures by dissipating heat from the building mass.
- Cooling by ventilation for the building occupants by dissipating heat from their bodies.
- Reducing the effect of excessive humidity by stirring air currents.

At the architectural scale, several methods were used to provide passive cooling:

- The courtyard provides convective cooling by acting as a wind-generating tool in the house, the hot air ascends and cooler air replaces it from the surrounding rooms. It promotes evaporative cooling by the use of vegetation, lowering the air temperature.

- The wind tower helps to passively cool private spaces in the traditional house by its four triangular vertical tunnels that attract air from all directions and accelerate its speed.
- The air-puller wall allows airflow to circulate through the rooms and remove hot air through convective cooling, while keeping the privacy of the indoor space.

2.4.3   *Passive cooling devices in vernacular architecture*

2.4.3.1   *Wind tower (baud-geer)*

The baud-geer (meaning the wind catcher in Persian) is an architectural passive cooling and ventilation device that was used in the traditional architecture of Iran and some Arab Gulf countries. The wind tower operates by changing the temperature and thus the density of the air in and around the tower. The difference in density creates a draft pulling air either up or down through the tower. The operation of the wind tower depends on the wind conditions and the time of day (Fig. 2.2; Bahadori, 1987). There are four operation conditions:

- *When there is no wind at nighttime*: the wind tower operates like a chimney. During the daytime hours when the sun strikes the walls of the wind tower, these walls will absorb heat. Eventually the heat is transferred to inside the tower, and the air layers next to these walls are warmed up. A natural convection is created due to difference in air temperature; and the hot air inside the tower rises up to exit the tower, while it pulls the air in lower levels inside the building spaces. The hot air is replaced by fresh night cool air that enters the building from the windows.
- *When there is no wind during the day*: the wind tower operates like a reverse chimney. The top part of the wind tower is cooled down at night and stays relatively cold during the early hours of the daytime. The air layers in contact with the walls inside the tower are cooled down. A natural convection is created but in a reverse direction (compared to nighttime operation) due to difference in air temperature; and the cool air inside the tower falls down and channels into the rooms. The hot air in the rooms is replaced by cool air coming from the tower. When the temperature of the tower reaches that of the ambient air, the tower begins to operate again like a chimney.
- *When there is wind at night*: the wind tower operates as a wind catcher. It will catch the wind at the top opening of the tower. The higher the tower, the more effective it operates due to higher wind speeds. Lower wind towers are less effective because of the wind resistance caused by the ground and building masses that result in slowing down wind speeds.
- *When there is wind during the day*: the rate of air circulation is increased as a result of combined effects of wind catching (similar to the wind nighttime operation) and solar chimney (similar to the no-wind daytime operation).

2.4.3.2   *Evaporative cooling in wind towers*

The wind towers in the vernacular architecture of Iran coupled evaporative cooling with sensible cooling to improve performance and provide higher levels of thermal comfort. Several methods were invented to do this:

- By placing a small pool with a fountain at the bottom of the wind tower.
- By placing the tower at a large distance (approx. 50 m) from the building with an underground tunnel running from the bottom of the tower to the basement of the building (Fig. 2.3). The ground over the tunnel is planted. When the soil is watered, the tunnel walls around the tunnel are kept damp and the air coming through it is cooled evaporatively and sensibly.
- Wind tower in conjunction with underground stream (Fig. 2.4): when the air gets into the tower, its velocity increases. When it moves from the vertical tower to the door opening, which is smaller in cross section, it slows down and its pressure decreases. This point is very close to the outlet opening of the stream shaft located inside the building and hence the pressure here is also decreased. The depression in pressure at this point continues until enough pressure differential is generated between this point and the top outer opening of the stream shaft. This results in the sucking of ambient air from the outer opening of the stream shaft. The air is channeled through the stream shaft to pass over the water stream and is cooled evaporatively before entering the building.

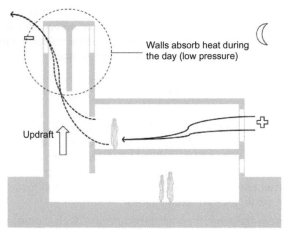

When there is no wind at night, the tower operates like a chimney.

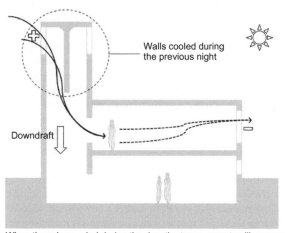

When there is no wind during the day, the tower operates like a reverse chimney. The rate of circulation is increased when there is wind.

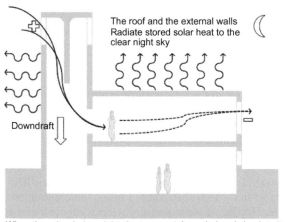

When there is wind at night, the tower catches wind and circulate it.

Figure 2.2.   How wind conditions affect wind tower (baud-geer) modes of operation.

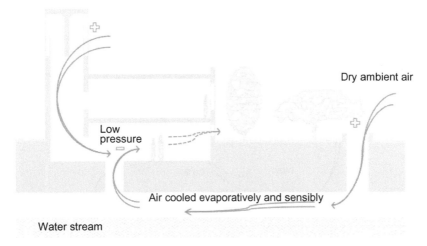

Figure 2.3. Coupling the wind tower (baud-geer) with an underground tunnel to promote evaporative and sensible cooling (adapted from Bahadori, 1978).

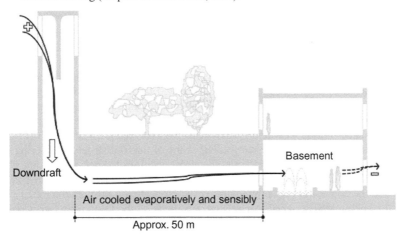

Figure 2.4. Coupling the wind tower (baud-geer) with a water stream to promote evaporative and sensible cooling (adapted from Bahadori, 1978).

2.4.3.3 *Egyptian wind catcher*

The malqaf (meaning the wind catcher in Arabic) is an architectural passive cooling and ventilation device that was used in traditional architecture in Egypt. Other forms of wind catchers under different names also existed in other countries in the Middle East. It helps to catch cool wind from levels above the building where the air is cooler and less dusty to cool the indoor spaces of a building (usually a large house). Sometimes, windows alone cannot help to provide sufficient airflow for cooling due to several reasons, such as lack of access to proper wind orientations, especially in high density urban areas where wind access can be blocked. Also, windows are designed to provide other environmental and human functions such as natural lighting and view; this might create conflict on the required window size (either smaller or larger than the optimum size for ventilation). Hence, wind catchers became very popular and useful in bringing cool and less dusty air to indoor spaces in these kinds of urban environments.

A good example that illustrates the operation of the malqaf is the qa'a of Mohib Addin Ashaafi Al-Muwaqqi in Cairo which dates back to the 14th century A.D. It depends primarily on air movement by pressure differential (Fig. 2.5). The design of the building also helps to generate a

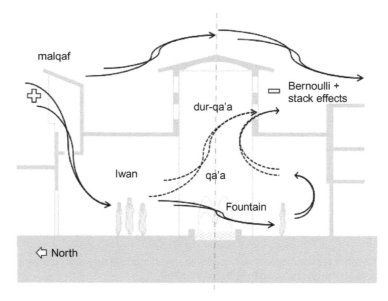

Figure 2.5.   Operation of the Egyptian malqaf in the qa'a of Mohib Addin Ashaafi Al-Muwaqqi (adapted from Fathy, 1986).

stack effect using a convective cooling effect, which improves performance especially during the daytime when wind velocities are relatively slow (Fathy, 1986). The malqaf channels cool breezes from the north down into the central space (qa'a) through an intermediate zone (Iwan) due to the increased pressure at the entrance of the malqaf. It flows through the Iwan to the upper part of the central space (dur-qa'a) where it finds windows (covered with screens) near the ceiling. The airflow then exits from these windows. The shape of the roof structure covering the central space helps to create the Bernoulli effect at the outlet openings to suck the air out. The top part of the roof also helps to create a stack effect due to an air temperature differential between this exit point and the cooler inlet opening (caused by solar heat gain through the roof.) This further improves the airflow performance. An attempt was made by the famous architect Hassan Fathy to develop the design of the traditional Egyptian malqaf and improve its performance (Fig. 2.6). In this concept, Fathy (1986) added wetted baffles and coupled the tower with a wind-escape so as to maximize evaporative cooling effectiveness and promote airflow.

2.4.3.4   *Wind scoops in the traditional house of Baghdad*

Wind scoops can be found in Egypt (Cairo), Iraq (Baghdad), and Pakistan (Hyderabad and Herat). Although wind scoops' function is similar to wind towers, they are smaller than wind towers and they look different. A wind scoop faces one direction and does not extend as high as the wind tower. The wind scoop of Baghdad (Fig. 2.7) has the following dimensions: 90–120 cm width, 60 cm depth, 90 cm inlet opening height, 90–120 cm opening height over roof level (Al-Azzawi, 1969; Fardeheb 1987). The top part above the roof level that has the inlet opening is made of brick, wood, or even metal material; and has a 45° inclined top surface to maximize capturing wind. During the nighttime hours, cool wind is channeled in and the coolness is stored in the thick walls (with high thermal mass). During the daytime hours, wind is also captured and is cooled by the cool walls; consequently pushing out the warm air inside the rooms to the courtyard. The courtyard collects all the warm air being removed from the rooms and get rids of the heat by radiation to the sky and by convection to the ambient air, especially during the night hours. Wind scoops or malqafs are still being used in traditional houses in Baghdad. Every room has its own wind scoop and large rooms can have more than one. The air temperature inside a room cooled by a malqaf can be 5°C cooler than the outside (Fardeheb, 1987; Lezine, 1971).

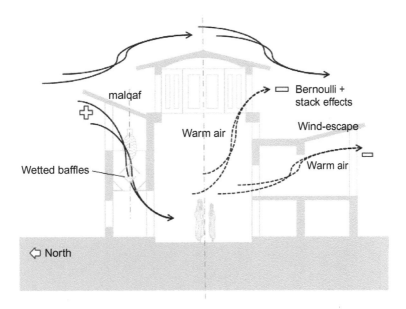

Figure 2.6.   Improving the performance of the Egyptian malqaf by adding wetted baffles and a wind-escape to maximize evaporative cooling and promote airflow (adapted from Fathy, 1986).

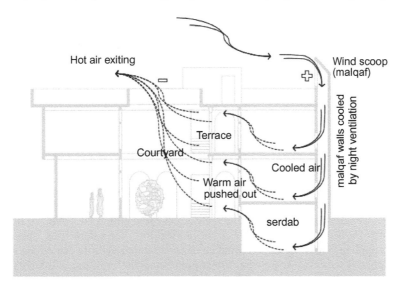

Figure 2.7.   The wind scoop of the traditional house of Baghdad (adapted from Al-Azzawi, 1969).

2.4.3.5   Ventilated basements (or serdabs) in the traditional house of Baghdad

Wind scoops were used in Baghdad to ventilate rooms and also the basements (Fig. 2.7). The basement (or the serdab) was the coolest space in the traditional house. Al Douri monitored five different rooms in a traditional Baghdadi house including the serdab. A peak reduction in temperature (21%) was found at 3:00 pm inside the serdab, which confirms that it is the coolest room in the house during the summer (Danby, 1984; Fardeheb, 1987).

Wind catchers can also play an important role in ventilation and dehumidification. In a measurement study about performance of serdabs with and without wind catchers, air temperatures

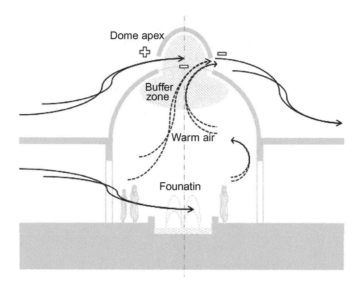

Figure 2.8.   Airflow in curved roofs (adapted from Bahadori, 1978).

and relative humidity (*RH*%) were monitored. In serdabs without a wind catcher, the air temperature ranged from 25–28°C and the RH was 40–55% during the day while 60–65% at night. In serdabs with a wind catcher, the temperature was surprisingly higher (27°C at midnight to 32°C in the afternoon) but the relative humidity range was lower (25% at 4:00 pm to 35% at 2:00 am).

2.4.3.6   *Curved/domed roofs*
The curved roof is usually applied when a larger space and extra height is needed. This kind of space is traditionally used as a central space for living and socializing. Usually building roofs lose heat by convection more than by radiation. Flat and curved roofs absorb about the same amount of solar radiation. Yet, the curved roof loses heat more effectively than the flat roof since it has a larger surface area. The roof loses most of its absorbed heat by convection. The hot air generated inside the house and by the surrounding environment transfers gradually by convective currents to the top level of the indoor space, which is under the curved roof of the living area. This trapped hot air creates a thermal buffer that helps to limit heat transmission from outside to inside (Fig. 2.8). When air flows over a curved roof the velocity of air at the apex increases and the pressure decreases. The opening at the apex of the curve allows the hot air to flow out. The curved roof is formed either as a vault (i.e., cylindrical form) or a dome (i.e., spherical form). The vault is applied in locations where wind comes predominantly from one direction; while the dome is applied in locations where wind comes predominantly from several directions.

2.4.3.7   *The cistern*
The cistern is another type of vernacular domed structure that was used in the past in Iran to store cold water cold for several months. In winter, the cistern tank is filled with cold water. In summer, the domed roof is warmed up by the sun and outside heat. Heat transmits gradually through the dome structure and accumulates in the air layers under the dome. This also warms up the top layer of the cistern water. Before the lower layers of the cistern water are affected by the heat gain, the top layer evaporates and the air admitted through the wind tower helps to pull the water vapor away. If the dome has a vent (i.e., opening at the dome apex) the air flows down through the wind tower, passes over the top layer of the water, carries the water vapor and blows it out of the vent. If there is no vent, the air carrying the water vapor blows out through the leeward side of the wind tower.

2.4.3.8 *The ice maker*

A shallow pond is used to make the ice depending on radiation losses to the clear sky. The pond has a strip shape, 10–20 m width. The long side of the pond runs on the east-west axis with several hundreds of meters in length. The pond is shaded from the sun using walls, a tall vertical wall on the south side to protect from the middle of the day high altitude sun plus two shorter ones on the east and the west sides to protect from the low altitude sun during the early morning and late afternoon hours. On cloudless winter nights, each pond is filled with water. Water loses heat to the sky by radiation. Some heat gains are also received by conduction with the lower soil and convection with the air. The heat loss by radiation is sufficient to freeze the water in the pond. On the following day, the ice is cut up, transported to the storage pits that are located 10–15 m deep under the ground. The high thermal resistance of the soil surrounding the pit helps to keep the ice frozen for several months; and hence can be used during the summer when it is needed the most.

2.5 PASSIVE COOLING SYSTEMS MAINLY EFFECTIVE IN HOT ARID CLIMATES

There are three major sources of coolness in hot arid regions that can be exploited by the designer: the ambient water vapor which is often very low, the night air which is usually cooler than the indoor space and the internal building surfaces, and the upper atmosphere which is usually characterized by a clear and dark sky. The wind, which is usually used as a source of coolness (by promoting comfort ventilation) in warm humid climates, can also be utilized as an additional source of coolness in hot arid regions during overheating times and when ambient temperatures are low enough to bring comfort.

Due to the variable nature of the weather conditions and the need for a reliable system against any intermittent conditions, storage of coolness might be a necessary option in certain locations. The storage materials that can be used to store coolness are: building structure such as walls, ceilings, and floors, water contained on roof ponds, wall enclosures, or storage tanks, and rocks (or rockbeds). The modes of fluid flow can be as natural flow by buoyancy forces or wind or forced flow by using air fans or water pumps. The effectiveness of any technology that depends on storage of coolness depends on the thermal inertia of the storage material. For thermally thick solids, it is well known (Kreith, 1973) that the thermal inertia of the material governs the rate of rise of the surface temperature and consequently the time to peak. The thermal inertia, $k\rho c$, is the product of the thermal conductivity, k, the density, ρ, and the specific heat, c, of the material.

The strategies that are effective in hot dry climates can be classified according to their coolness mechanism: (i) strategies that have no storage of coolness, (ii) strategies that rely on daily or short term storage of coolness, and (iii) strategies that rely on seasonal storage of coolness. The strategies that rely on evaporative cooling (such as the wind tower and the evaporative coolers) operate without the need for a mechanism to store coolness. Courtyards rely on radiant cooling to the upper atmosphere and storage of coolness in the courtyard's space (as a container of cool air) and in the building's structure as well. Night cooling and roof ponds require a mechanism for storage of coolness such as the use of the building structure or by providing a rockbed or a water tank. Strategies that depend on seasonal storage of coolness utilize the natural cooling of the soil surrounding a building, or earth tubes.

2.5.1 *Courtyards*

2.5.1.1 *Use of courtyards to reduce building heat gains*

There are a number of thermal advantages of using courtyards (Bahadori, 1987). These are as follows:

1. *Solar shading*:
   a. The walls of the courtyard shade each other and help to cast shadow on the courtyard floor surface. This reduces heat gain by solar radiation.
   b. Courtyards are usually planted with trees and other vegetation, which helps to provide shading on the building walls and also on the floor surface.

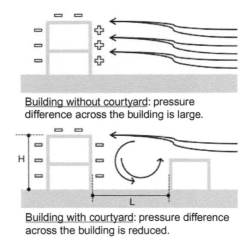

Building without courtyard: pressure
difference across the building is large.

Building with courtyard: pressure difference
across the building is reduced.

Figure 2.9.   Courtyards wind-shadow upstream walls and reduce wind pressure distribution on their
surfaces.

2. *Creating a desirable microclimate – reduction of outdoor temperature and increase of relative humidity*:
   a. During the night the surfaces (walls and floor) of the courtyard lose heat by convection to the cool air and by radiation to the clear sky.
   b. During the day, the air in the courtyard remains cool for several hours, which provides a very desirable and cool environment in the shaded area.
   c. The air in the courtyard is cooled evaporatively when air passes over surfaces of trees, shrubs or other vegetation, as well as water fountains/ponds. The cooled air stays for long time (does not escape quickly) due to the space enclosure of the courtyard.
   d. Also, courtyards help to reduce air conditioning costs indirectly when building occupants stay outside the building in the courtyard for some hours of the day.
3. *Wind shading on upstream walls*:
   a. Courtyards help to wind-shadow upstream walls and reduce wind pressure distribution on their surfaces (Bahadori, 1987; see Fig. 2.9). This also depends on the length to height ratio (L/H) of the courtyard space. For low values of ($L/H < 8$), the wind pressure distribution on the leeward walls becomes negative and the pressure difference across the building is reduced. This can amount to a substantial reduction of heat gains by infiltration.
   b. The reduction of the wind velocity causes reduction of the convective thermal coefficient and hence reduction of building envelope heat gain by convection.

2.5.1.2   *Courtyards radiant cooling effect*

Radiant cooling helps to dissipate the heat absorbed in the building structure components (especially the roof) to the upper atmosphere. This process is more effective when heat is dissipated from an extended roof to a clear night sky; which is usually found in desert climates. After the roof is cooled during the nighttime, it will work the next day as a cold storage medium absorbing the internal heat and cooling the indoor space. According to Anderson (1979), the total cooling effect during a 10-hour night is between 44 and 88 W or 167–335 J/cm^2 (i.e., equivalent to 150–300 Btu/ft^2).

In a courtyard design like the one in the traditional house of Baghdad, the floor and the roof offer a large surface area exposed to the cool night sky (and cool alleys, shaded during the day) to produce an effective radiation loss (Fig. 2.10). During the day, the earth beneath the courtyard and the roof act as a heat sink that receives heat from solar radiation (Fardeheb, 1987). The difference in temperature between these surfaces and the cool air accumulated in the rooms from the previous

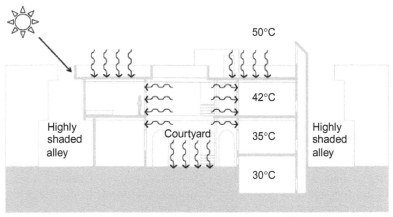

During the day, the earth beneath the courtyard and the roof act as a heat sink that receive heat form solar radiation.

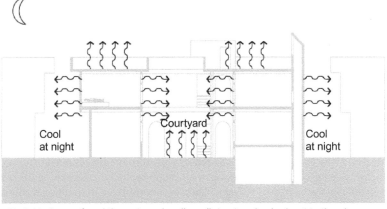

At night, the roof and the external walls radiate stored solar heat to the clear night sky.

Figure 2.10.   Courtyards radiant cooling effect in the traditional house of Baghdad (adapted from Fardeheb, 1987).

night (by the assistance of the wind scoops) create convective currents, which provide comfort during the day. At night, the accumulated heat in the building structure is emitted to the clear night sky causing radiant cooling. The coolness is stored in the thermal mass of the building, which helps to keep the house cool for several hours in the next day.

2.5.1.3   *Courtyards convective and evaporative cooling effects*

Courtyards can also be used to improve convective cooling. The concept of two courtyards can help to improve convective cooling (Cain *et al.*, 1976; Fardeheb, 1987). This requires having two courtyards: one is narrow and deep (with low aspect ratio) that maximizes solar shading most of the time and is cool while the other one is extended and sunny (with high aspect ratio) that promotes heat gain and becomes warm. The two courtyards are connected by a highly shaded outdoor space. The air in the large courtyard is heated by the solar radiation and this results in raising its air up. The air in the small shaded courtyard is still cool and dense from the previous night. This difference in temperature results in a pressure differential (positive pressure in the small courtyard versus negative pressure in the large courtyard); which promotes natural air convection or airflow that goes from the small cool courtyard to the large warm one. The cooling effect can be improved by adding cool porous water jars at the exit openings of the small courtyard to promote evaporative cooling. Other strategies that can help to promote evaporative cooling are to use greenery and water fountains.

2.5.1.4   *Courtyards and thermal comfort issues*

Reynolds (2002) studied courtyards' thermal comfort in his book titled *Courtyards: aesthetic, social and thermal delight*. According to the author, in hot climates many factors affect thermal comfort in the traditional courtyard building such as thermal mass of the building, occupants' behavior and the acceptance of less thermal control compared to air-conditioned buildings. Some key points from the author's analysis and review of other scholars' work with regards to climate and comfort in courtyards can be summarized as follows:

- *Solar shading*: to avoid heat gain in hot climates, shading of courtyard buildings is suggested. According to some case studies from Spain, Mexico, China, and Argentina, shading was done either by plants or some means of translucent or opaque horizontally foldable fabric that covers the courtyard opening at the roof, referred to as a toldo.
- *Catching wind*: in both hot humid and hot arid climates, admitting cool breezes to the courtyard and building is a favorable approach. There are several strategies that can be used to catch wind. Givoni (1998) suggested raising the height of the courtyard wall downwind to catch the wind moving over the courtyard, which will then drop as down draft into the courtyard (Fig. 2.11). This draft can also be enhanced if an outlet is placed at the bottom. Golany (1990) studied the ventilation system used in the Chinese pit-type courtyard houses and cave dwellings; this system incorporates a tall chimney with self-rotating wind catchers constructed above the courtyard level and is dug deep into the earth. It works both ways, at the presence of wind where the outside air is hot, the wind catchers will direct the air downward into the chimney. Here it will be cooled down through the chimney's long dark passage. Heated air in the courtyard will rise causing the air in the cave rooms to be sucked out and replaced by cooler air coming from the chimney, thus establishing effective air circulation within the space. Although weaker, in the case of wind absence, air circulation is induced by the stack effect, where cooler air falls downward and flows into the rooms, and hot air rises through the chimney. This only happens when the outside air near the chimney opening is cooler than the air exiting the space.
- *High thermal mass*: one of the cooling techniques incorporated in traditional courtyard buildings is the use of high thermal mass; this thermal mass is used to delay afternoon heat. To release the heat stored in it during the day, the thermal mass is cooled down by radiation to the night sky, evaporation and ventilation.
- *Evaporative cooling*: the use of evaporative cooling in courtyard buildings in hot dry climates is highly effective in reducing temperature. Mist inducers, water features or a thin layer of water surface on the floor of the courtyard are some evaporative cooling techniques.

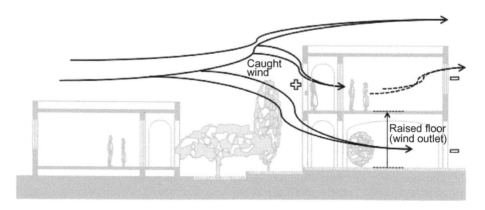

Figure 2.11.   Raising the height of the courtyard wall downwind to catch the wind moving over the courtyard.

2.5.1.5 *Courtyards performance measurement*

Performance of traditional courtyard house in Cairo: in a thermal comparative study between a courtyard traditional house and a modern apartment in Cairo, Nour (1979) reported the interior temperature of the main reception room in the courtyard house was cooler than the temperature recorded in a room of the modern apartment. The room in the traditional house was cooler by 1.5–4.5°C during the early morning and from midday to midnight. The recorded difference in temperature between the air on the roof and the air in the reception room was 11°C between 2:00 and 4:00 pm. The study also found that the courtyard was cooler by 4–5°C than the air temperatures on the roof between noon and 4:00 pm and was generally cooler for a period of 12 hours during the day.

Measuring modern courtyards' bioclimates in Tempe, Arizona: according to a study done by Bagenid (1987) on courtyard bioclimates, two identical midsize courtyards (10 m × 20 m × 7 m) in Tempe Arizona were examined while floor treatments were varied. Results showed that under such climatic conditions when a water feature such as a pool with a fountain was used as a floor treatment in the courtyard, sustainably cooler bioclimatic conditions were observed in contrast to a dry concrete floor. Other findings showed that when soil was placed in the courtyard with 50% grass cover, an increase in the coolness resulted in comparison to a concrete floor that had 18% of its area covered with a running film of water.

Performance of the Andalusian courtyards: the performance of three courtyards was monitored simultaneously with regards to their air temperature range (Reynolds, 2002). The shallowest one (Courtyard A; aspect ratio = 1.64) experienced the highest temperatures nearly identical to those at Cordoba's airport. A much deeper courtyard (Courtyard B with 0.57 aspect ratio) showed a comfortable environment even at the highest temperatures that were almost within the comfort zone in this hot weather. Courtyard C (the deepest one with aspect ratio = 0.53 and also protected by toldo shading) showed the most comfortable environment (least temperature values and limited daily change). Courtyard A was wide and shallow and had a trellis of wisteria and jasmine sitting along one side of the space. There is a second story along two sides (northeast and northwest). The floor was covered with exposed concrete. Courtyards B and C had almost identical aspect ratios and they both have more extensive greenery than courtyard A. They had very different maximum temperatures but their minimum temperatures were somewhat more similar. The difference in performance was attributed to their quite different orientations. Courtyard B was elongated on the east-west axis while courtyard C was elongated at north-south axis. Courtyard C had an advantage of a third floor on its south side and a toldo shading. It was concluded that for these three courtyards, aspect ratio alone would be an inadequate predictor of center courtyard temperature.

The temperatures were also recorded in a room adjacent to each courtyard when Cordoba's official temperature range was between 21 and 38°C. The differences between the rooms during the day were surprisingly small and almost identical at night. The author considered this unexpected, given the temperature difference in the courtyard centers, and the dissimilar conditions (the aspect ratios and the one- versus two-story heights). The room adjacent to courtyard A experienced the greatest temperature variation (26–31°C). This was attributed to the shallow adjacent courtyard and to the roof directly overhead. The difference between the rooms adjacent to courtyards A and C was only 4°C.

Performance of courtyards in Colima, Mexico: Reynolds (2002) also reported the findings of another field measurement of courtyards in which he observed over a hundred courtyards in Colima, Mexico; and took physical measurements of some of them. These observed courtyards serve a wide range of socio-economic groups. Their heights were 1–2 floors, the aspect ratio (floor area/average wall height) ranged from 0.22 to 10.95 and the solar shadow index (S wall height/width N-S of floor) ranged from 0.21 to 3.09.

The study took measurements of two of these courtyard buildings. The first one was the Presidencia Municipal (City Hall) building. It is a one-story building with a sunny and shallow courtyard (a high aspect ratio of 8.22 with a floor area of 296 m^2). On the other hand, the second building, a two-story remodeled hotel called the Tesoreria, has one of the deepest courtyards in Colima (a low aspect ratio of 0.40 with a floor area of 32 m^2).

The measurements showed that the ground-floor offices in the Tesoreria maintained an early June range of 28–31°C, comfortable with air motion by fans which are used widely in Colima. The upper-floor office in the Tesoreria swung from 27°C by night to 33°C in the late afternoon. Its open window helped to provide some breeze. The office in the city hall maintained slightly lower June temperatures than the ground floor in the Tesoreria.

When comparing between the courtyards in Cordoba, Spain with those in Colima, Mexico, one can find that the ones in Colima are shallower (with low aspect-ratio), and hence receive more solar radiation. They are also not protected by top movable shading devices like the Andalusian courtyards, which are shaded by toldo. This initially surprised Reynolds (2002) because Colima's summer is even hotter than those of Cordoba. Yet, the high humidity in Colima was another significant determinant that was probably taken into consideration by the inhabitants there. He then justified this finding by mentioning that if the courtyards in Colima were shaded (like the Andalusian courtyards), this would limit their passive cooling effectiveness by airflow. Also, people in Colima could be more tolerant to withstanding sun heat than the people in Cordoba.

Performance of semi-enclosed courtyards: Meir *et al.* (1995) performed a study on the micro-climatic behavior on two semi-enclosed courtyards identical in shape and surface treatment but differently oriented; one was facing west while the other faced south. The two courtyards were part of dormitories serving a high school campus, located on the Negev highlands that exhibit hot dry climatic conditions. The results showed that orientation, geometry, ground cover and external surface treatment have a huge impact on the comfort level in the courtyards. Both courtyards examined showed an increase in the thermal behavior from that of the surrounding open environment by up to 5°C. What was concluded is that the geometry of the semi-enclosed courtyards influenced their thermal behavior substantially; orientation of such courtyards with respect to solar angles and wind direction have a great effect on the thermal discomfort within its boundaries; microclimatic conditions in courtyards highly rely on controlled ventilation and dynamic shading.

Performance of a courtyard building in Tehran: Safarzadeh and Bahadori (2005) investigated numerically the passive cooling effects of a courtyard in a small building in Tehran, employing an energy-analysis software developed for that purpose. The passive cooling features considered were the shading effects of courtyard walls and two large trees (of various shapes) planted immediately next to the south wall of the building, the presence of a pool, a lawn and flowers in the yard, and the wind shading effects of the walls and trees. The study concluded that these features can reduce the cooling energy requirements of the building to some extent but they cannot maintain thermal comfort by themselves during the hot summer hours in Tehran; and they have an adverse effect of slightly increasing the heating energy requirements of the building.

2.5.2   Wind towers

2.5.2.1   Traditional design issues

There are a number of disadvantages in the traditional wind tower design (Bahadori, 1986). These are as follows:

- A portion of the admitted air can be lost through the leeward opening of the tower.
- Dust can penetrate easily through the tower openings. Also, insects and small birds can enter the building through these openings.
- Since the mass of the tower is relatively small, the stored coolness in the tower mass is limited.
- The evaporative cooling potential of air is not fully utilized even when the air is made to flow over moist surfaces, since these surfaces are generally small.
- When the tower operates as a reversed chimney (i.e., when there is no wind), the operation of the baud-geer is limited to the period that the stored coolness can be retrieved effectively from the storage mass.

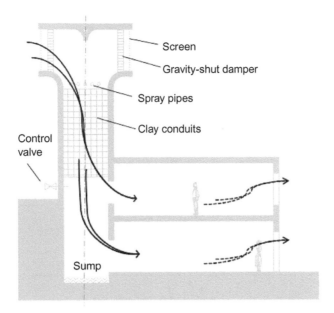

Figure 2.12.   An improved design for the wind tower (adapted from Bahadori, 1986).

2.5.2.2 *Improved design*

An improved design for the wind tower that eliminated the problems mentioned above was developed and hence had the potential to improve its performance (Bahadori, 1978; 1979; 1986). The new design is shown in Figure 2.12. The modifications in the improved design were as follows:

- Adding a 1-way damper on the inlet opening to minimize the escape of air streams from the leeward openings.
- Adding a screen on the inlet openings to prevent insects and birds from entering the tower.
- Building clay conduits on the inner surfaces of the tower walls to increase the surface-to-volume ratio and hence increases the rate of heat transfer stored in the thermal mass.
- Installing a water dripping system to wet the clay conduits and generate evaporative cooling for the incoming air.

2.5.2.3 *Performance of the improved design*

The evaporative cooling potential of the proposed baud-geer design was investigated by Bahadori (1986), using a heat and mass transfer analysis and assuming that nearly 80% of the column was wetted. The study showed the dry-bulb temperature of the air leaving the evaporative cooling column for several inlet conditions and wind velocities (see Fig. 2.13). It also investigated the performance of the proposed design with regards to evaporative cooling potential. Figure 2.14 summarizes the performance of the proposed design and shows the results of the relative humidity of the air as it leaves the column for a wind velocity of 5 m/s. For a wind velocity of this value and a column height of 5 m, air entering the tower at 40°C and 15% relative humidity leaves the column at a velocity of 1.5 m/s, with a temperature of 25°C and a relative humidity of 66%.

Bahadori *et al.* (2008) tested two new designs of wind towers equipped with evaporative cooling systems along with a conventional wind tower. All wind towers were of identical dimensions and arranged side by side. The experiment was carried out during the month of September in the city of Yazd, central region of Iran. One of the new designs had a wetted column consisting of wetted curtains hung in the column, and the other one had wetted surfaces consisting of wetted evaporative cooling pads mounted at its entrance. The wind tower with a wetted column was

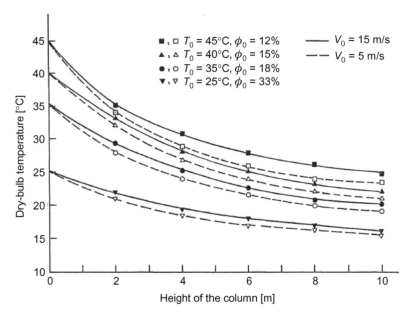

Figure 2.13.   Dry-bulb temperature of the air leaving the evaporative cooling column. V_0 is the wind veloc-ity at the tower height, T_0 and ϕ_0 are the ambient air temperature and relative humidity, respectively. (Source: Bahadori, 1986; with permission of Springer).

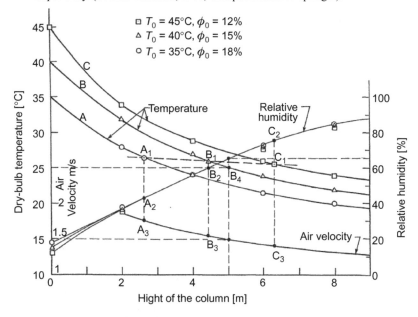

Figure 2.14.   Condition of the air leaving the evaporative cooling column for a wind velocity of 5 m/s. T_0 and ϕ are the ambient air temperature and relative humidity, respectively. (Source: Bahadori, 1986; with permission of Springer).

equipped with cloth curtains, spaced 10 cm from each other and suspended vertically in the wind tower column. They were secured firmly at the bottom to prevent them from fluttering due to the air flowing between them. The wind tower with wetted surfaces was equipped with evaporative cooling pads (similar to those employed in normal evaporative coolers), placed at the wind tower

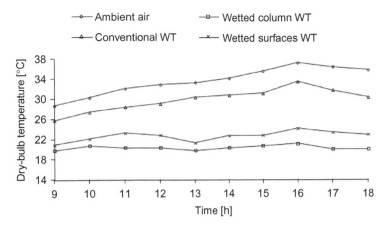

Figure 2.15.   Ambient air temperature and temperature of the air leaving the wind towers (WT) for the conventional, the wetted column, and the wetted surfaces wind towers (produced from data in Bahadori *et al.*, 2008).

openings at the top. The pads were wetted by spraying water on them (similar to the evaporative coolers).

Both new designs performed better than the conventional one, with the air leaving these towers at a much lower temperature and a higher relative humidity than the ambient air (Fig. 2.15). On average, the wetted column is 13°C less than the ambient temperature, while the wetted surfaces is 11°C less than the ambient temperature. The conventional one is 3.74°C less than the ambient temperature. These towers can easily replace the evaporative coolers currently employed in the hot arid regions of Iran, with a great saving in the electrical energy consumed for the summer cooling of buildings. It is recommended that these new designs of wind towers, particularly the design utilizing evaporative cooling pads at the entrance region of the tower, be manufactured in various sizes and incorporated into the designs of new buildings in the hot arid regions of Iran, as well as other countries with similar climates.

2.5.2.4   *Coupling a wind tower with a solar chimney*

The Environmental Research Laboratory at the University of Arizona developed a modified design by coupling a wind tower with a solar chimney (Fig. 2.16), each with a height of 7.6 meters (Cunningham and Thompson, 1986). The solar chimney was intended to provide ventilation air movement within a structure, in combination with a cooled downdraft chimney (i.e., the cooling tower). In the cooling downdraft tower, air is cooled by an evaporative cooler, consisting of wetted packing material, designed as an integral part of the tower. The packing material is kept wet at all times by using a small water pump. The cooled air, heavier than ambient air, falls by gravity through the tower into the structure to be cooled. The system was designed based on an airflow rate of $1.4\,m^3/s$ (2.67 air change per hour). When the wind speed is not enough, the solar chimney provides air updraft; which helps to generate downdraft in the wind tower. The entire roof and the attic are used as a solar collector to assist the solar chimney. All of the return air is exhausted through the ceiling into the attic and this cools the attic and reduces heat gain. This push-pull design provided comfortable conditions in the tested building throughout the summer of Tucson, Arizona.

The building performance was measured in August 22–23, 1985, starting at 6 am the first day and ending at 7 pm the next day (38 h). The results showed that while the outside temperature was ranging from 24.1 to 40.6°C, the tower exit temperature was ranging from 19.7 to 24.9°C. This resulted in comfortable interior dry-bulb temperature ranging from 21.3°C (at 6:00 am, the second day) to 25.4°C (at 4 pm, the first day). The comfort predicted mean vote (*PMV* indicator) ranged from 0.06 to 0.83 and the predicted percentage of dissatisfied (*PPD* indicator) ranged from 5.00 to 19.57.

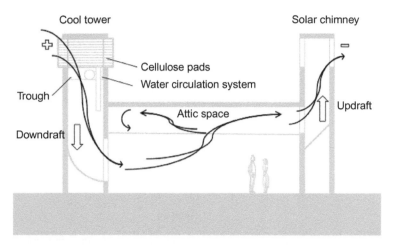

Figure 2.16.   Coupling a wind tower with a solar chimney designed by the Environmental Research Laboratory at the University of Arizona (adapted from Cunningham and Thompson, 1986).

2.5.3   Evaporative coolers

Direct evaporative coolers have been used widely in many parts of the world, especially in desert regions where they are most effective. There are two disadvantages of direct evaporative cooling:

- It requires supplying a large amount of air with a relatively high content of moisture. Consumption of water in large quantities is a problem as water is scarce in the desert regions.
- Sometimes the adiabatically cooled air supplied by the evaporative cooler may not be able to maintain the desirable thermal comfort.

An indirect evaporative cooler can eliminate these two problems (Bahadori, 1986). It employs a heat exchanger to cool the supply air. The exhausted air is cooled evaporatively by passing through one side of the heat exchanger. The outside air that passes through the other side of the heat exchanger is cooled sensibly (i.e., without any change in its moisture content). Before exiting the building, the evaporatively cooled return air may be designed to pass through the attic in order to reduce heat gain.

Gamero-Abarca and Yellott (1983) tested a plate-type indirect evaporative cooler (Australian design, Dricon model PD6) in an unoccupied demonstration house in a suburb of Phoenix, Arizona, during July and August of 1982. The study found that on a typical day, when the outside ambient dry-bulb is 35°C and the wet-bulb temperature is 23.3°C, the indoor dry-bulb and wet-bulb temperatures reached 27.8 and 21.7°C, respectively. During the hottest days, when the mean outdoor air dry-bulb temperature reached 38.9°C in July and August, and the maximum wet-bulb temperature reached 26.1°C in late July, the indoor dry-bulb temperature reached 28.9°C in July and 29.4°C in August. Such extreme temperatures occurred for less than 2% of the July and August hours. During the 2% period that the indoor temperature exceeded the comfort conditions, the study suggested the use of ceiling fans, which could extend the upper limit of thermal comfort zone from 26 to 29°C (Rohles et al., 1983).

2.5.4   Passive cooling with daily or short-term storage of coolness

2.5.4.1   Night ventilation

The International Energy Agency (IEA) produced a report that presented a review of night ventilation among other systems of low energy cooling (IEA, 1995). Night ventilation can be used to cool the building when the outdoor temperature is lower than the indoor temperature. It improves the comfort directly during the night, but also charges the building structure with coolness. The stored coolness helps to reduce the space's temperatures during the following day and hence improve comfort. The efficiency of the system depends on the following:

• The airflow rate,
• The temperature difference between the outdoor and indoor air.
• The effective thermal mass of the building interior.

An integrated design approach is required for the success of this technology. Mechanical or natural ventilation may be used for night cooling, or a combination of the two. However, with natural ventilation there is no direct control of the ventilation rate. With mechanical ventilation, either the normal ventilation system can be used with some modifications, or an additional system may be used.

The thermal mass of the structure plays a major role in the effectiveness of the night ventilation system by reducing temperature swings and hence improving comfort. When designing the thermal mass for the night ventilation system, it is important to make the distinction between the daily inertia (24 hour cycle) and sequential inertia (2 week cycle). In daily inertia, thermal mass helps to reduce the temperature swings between night and day, while in sequential inertia, it helps to reduce the indoor daily mean temperature during warmer days. In most cases, sequences of warmer days do not last more than about five days. One should keep the thermal mass exposed and directly linked to the indoor air. Hence, additional thermal resistances, such as carpets or wall covers, should be avoided or reduced as these materials can reduce the effectiveness of thermal mass in the structure.

Night cooling can be used in either commercial or residential buildings. The design and operation requirements for each of these building types are different due to variations in occupancy patterns, internal heat gains and internal furnishings. One should also take into consideration that commercial buildings typically have small latent cooling loads relative to sensible loads. In residential buildings, domestic activities like cooking, showering and laundry increase the relative importance of latent cooling.

Night ventilation in residential buildings: night ventilation cooling is suited for moderate to moderately hot climates, with night-time temperatures during the warm season less than about 22°C. Large diurnal temperature swings improve the system efficiency. In residential buildings, the airflow is provided through opened windows. Typical air change rates of between 5 and 20 ACH may be achieved in residential buildings with natural ventilation. The designs must allow for cross ventilation, with windows on each side of the building. Windows should enable good control of the airflow rate during the night by having means to control its opened area. Additionally, issues such as privacy, protection against robbery and outdoor noise must also be addressed.

The goal is to precool the building as much as possible during the nighttime in order to prevent overheating during the following day. The air speed must not cause thermal discomfort, especially at night when the outdoor temperature is less than about 15°C. In some cases when there is an air-conditioning system in the building, the designer needs to link its controls to the nighttime ventilation system.

The efficiency of the system will be highly dependent on inhabitant behavior and the potential for opening windows. Depending on the thermal inertia and solar gains, a decrease in the peak temperature of 3 to 7°C can be obtained, according to previous studies that used computer simulations. For higher solar gains, the decrease in the peak is greater, but the absolute temperature is also higher. This technology provides only sensible cooling but care must be taken when applying this technology in hot humid climates, as it may increase the latent cooling loads during the day.

Night ventilation in commercial buildings: night-time ventilation systems have a great effect on the architectural design of a commercial building. The designer must synchronize the design of the spatial layout, partitioning, and thermal mass in order to facilitate the cooling potential of night air. The natural ventilation may be increased through the use of wind towers, passive stack ventilation, ventilation through atria, and solar chimneys. Mechanical ventilation can be used when the flow rates are insufficient or natural ventilation is inappropriate. In any case, controls are needed to monitor the outside and indoor conditions to prevent overcooling and discomfort in the early morning. No additional space is required for simple natural ventilation systems. Some natural ventilation systems (i.e., passive stacks and wind towers) require dedicated space. When solar chimneys are used, they must be thermally decoupled from the rest of the building because of high temperatures that can affect adjacent spaces. Mechanical ventilation systems may require some additional space if more ventilation is provided than required for the standard HVAC system.

Thermal mass plays an important role in the implementation of nighttime ventilation for cooling. The nighttime ventilation may be designed to use different airflow paths from the daytime paths to ensure maximum cooling of the mass. This is done by channeling the nighttime ventilation through voids within the structure, such as hollow floors. A separate ducting system might be necessary for some applications. The mass should not be isolated by floor or wall finishes, such as carpets, dry lining, or suspended ceilings; otherwise its effect will be limited. This can be overcome by using thermally heavy dry lining or a false floor void. In the latter case, the supply air cools the mass underneath the false floor. The most appropriate application of this technology is in buildings which are unoccupied during the night and which have regular cycles of heat gains. Office buildings are a perfect example.

The technology is best suited to hot or moderate climates with large diurnal temperatures over the summer. The humidity ratio of the air should be less than 15 g/kg dry air since the technology provides primarily sensible cooling. In mechanical ventilation systems, nighttime temperatures must be cool enough to make up for the 1 to 2°C heat gain due to the fans.

Night cooling is capable of providing cooling for up to 40 W/m^2 of internal heat gains. Computer simulations indicate that nighttime ventilation can lower the daytime temperatures in heavyweight buildings by about 3°C. When mechanical ventilation is used, electricity is required to drive the fans. For example, energy consumption could be between 1 to 9 kWh for fans to deliver 5 ACH to a 180 m^3 room over a 24 h period.

Slab cooling with air: slab cooling provides sensible cooling only and it is suited best to moderate and moderate-dry climates where the daily outdoor temperature fluctuations are relatively high and where the night temperatures are lower than the indoor temperatures. It offers a convenient method for providing cooling comfort for buildings that tend to overheat in daytime during the intermediate seasons (spring and autumn).

The basic idea behind slab cooling is the exploitation of the thermal inertia of the building mass for the purpose of energy storage. The system takes advantage of the building envelope, which is made of slabs, to store coolness and the channels in the slabs are used as ducts for the ventilation air. When the cool air is passing through the ducts, energy transfer takes place allowing the building mass to be used as a rechargeable energy store. During the summer, the system can run during the night to store cool energy in the building mass. The coolness is then transferred during the day to the supply air, thus decreasing its temperature. Part of the conditioning process takes place in the ductwork, where the heat exchange takes place between the supply air and the building mass. Final conditioning may be left for terminal units.

Slab cooling can be applied to any building type. The technology is not suitable for retrofit applications since it is difficult to install after a building has been completed. Ductwork for the air distribution system is incorporated within the building partitions. Two types of slabs exist for this purpose: hollow core slabs or prefabricated false floor systems.

The energy performance decreases as the nighttime indoor-outdoor temperature differential decreases. The cooling potential of the technology is strongly dependent firstly on the changes of the outdoor temperature and, secondly, on the building mass. An average measured thermal performance of 1.2 W/L/s (including fan pick-up) of supply air was achieved in a small experimental low energy house, using cooling by night ventilation, with the average supply temperature about 1°C below the ambient temperature. Electricity is required to drive the fans of the mechanical ventilation system. The fans can consume 3 to 6 W/L/s of supply air and about 50% of the energy is for overcoming the pressure loss in the ductwork. There is additional pressure loss in a slab cooling system due to the rougher ducts (roughness of 1.0 versus 0.15 for normal ducts) and the connections between the vertical and horizontal ducts. This means that up to 30% more energy may be required for ventilation with slab cooling or up to 1.8 W/L/s of supply air. Assuming a cooling performance of 1.2 W/L/s, this would equate to a minimum seasonal COP of 0.7.

2.5.4.2 *Weekly storage of coolness in heavy walls*

The problem of designing passive cooling systems in moderate climates is relatively easy compared to hot arid or desert climates. This is basically solved by applying night cooling (circulating outside cooler air at nighttime that helps to store coolness in the building structure then retrieve it during the overheated period to maintain thermal comfort). In the harsh conditions of hot climates, employing night cooling strategy is often not sufficient. This could be due to one or more of the following reasons:

- The air motion is weak, and hence the natural convective heat transfer coefficient is small.
- The thermal conductivities of the walls and other structure components are small.
- The thermal capacity of the building mass is small.
- The surface-to-volume area is small.

To overcome these problems, traditional buildings in hot arid regions were constructed with heavy materials such as adobes, bricks, or stones that have high thermal storage capacities. This kind of construction can help to store coolness when there are cool conditions up to several days. Bahadori and Haghighat (1986) numerically investigated weekly storage of coolness in heavy walls (walls with large thermal inertia) by considering one-dimensional heat conduction through the walls. The study used different thicknesses of building materials (i.e., stone, brick, and adobe) for a room exposed to outdoor conditions for all sides, with an area of $16\,m^2$. The ambient air at temperature T_0 was assumed to follow a periodic temperature variation with the maximum and minimum of 35.4 and 20.4°C, occurring at 3 pm and 3 am, respectively. The key findings of the study can be summarized as follows:

- The larger the thermal inertia of the wall (for example adobe as compared with brick), the longer it takes for the wall temperature to reach a steady periodic distribution after the change has occurred. With adobe walls however, the retrieval of the stored coolness is slow, and the mean daily temperature of the room air does not change appreciably (beyond seven days after temperature change) because of the low thermal conductivity of adobe.
- As the wall thicknesses increase, more heat can flow into the wall from the inside (positive slope of temperature distribution at the inside wall). The increase of the wall thickness beyond 50 cm does not improve the thermal performance of the building significantly.
- During the daytime hours, there is always heat flowing into the wall from the outside and inside. During the nighttime hours, the direction is reserved. This illustrates the phenomenon of thermal storage and retrieval.

2.5.4.2.1 Case study of Sana'a vernacular tower house

Previous studies (Al-Sallal, 2013; Al-Sallal *et al.*, 1995; Ayssa, 1995) discussed how the massive construction materials of the vernacular tower house (see Fig. 2.17) of the old city of Sana'a had a very important role in providing a steady internal temperature in both winter and summer when compared with the modern house (see Fig. 2.18). The walls of the vernacular house are constructed with high-density stone material in the first floor and with burnt clay bricks in the upper floors, with an average wall thickness of 40 cm. The modern house is constructed with concrete walls with a wall thickness of 20 cm. The diurnal change of outdoor temperature in Sana'a is relatively large (the average outdoor temperature range is 16–34°C in the summer and 3–24°C in the winter); and hence one can experience some thermal discomfort in the middle of the day in the summer and during late night hours in the winter. In this large variation of external temperatures, the vernacular house provides very steady internal temperatures as seen in Figure 2.18. The same figure also shows the significance of the building orientation in improving

Figure 2.17.   Vernacular tower house in Sana'a (photo by the author).

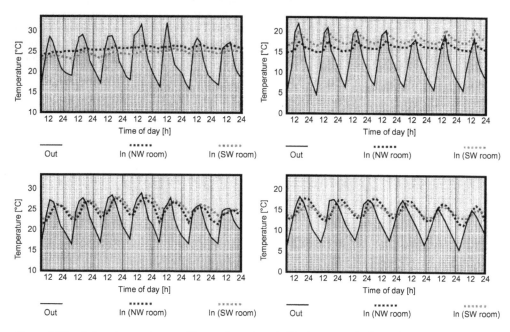

Figure 2.18.   Indoor temperature profile over 1-week period in the vernacular house in July (upper left) and December (upper right); and in the modern house in July (lower left) and December (lower right).

thermal comfort. The study showed that the internal temperature range of the vernacular tower house over a 1-week period was 24–26°C in July, and 15–17°C in December, for a NW room. The room oriented to SW in the vernacular house was a bit cooler in the summer (i.e., approx. 1°C less than the NW room) and a bit warmer in the winter (i.e., approx. 1–2°C higher than the NW room). The internal temperature range of the modern house during the same period was 22–29°C in July, and 12–19°C in December, for a NW room. The room oriented to SW had very similar results.

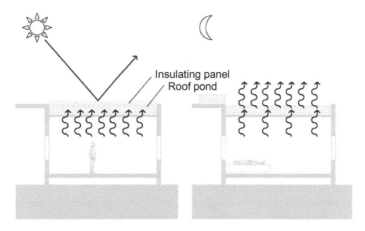

Figure 2.19.   The original roof pond system (Skytherm) introduced by Harold Hay in 1967 (Hay and Yellott, 1969).

2.5.4.3   *Storage of coolness in roof ponds*

The major sources of coolness in roof ponds are the upper atmosphere (sky) and the ambient water vapor. The average monthly sky temperature can be estimated with the Martin and Berdahl Model (1984):

$$T_{sky} = (C_a)\,0.25 \times T_{clear\_sky} \tag{2.7}$$

Effect of cloudiness:

$$C_a = 1.00 + 0.0224 \times CC + 0.0035 \times CC^2 + 0.00028 \times CC^3 \tag{2.8}$$

$$T_{clear\_sky} = T_{air}(e_{clear}^{0.25}) \tag{2.9}$$

Emissivity of clear sky:

$$e_{clear} = 0.711 + 0.56(T_{dp}/100) + 0.73(T_{dp}/100)^2) \tag{2.10}$$

where CC represents cloudiness [0–1; 0 – for clear sky, 1 for totally cloud sky], T_{air} [K] is the air temperature, and T_{dp} [°C] is the Dew point temperature.

In the USA, sky temperature depressions vary between 6–10°F (2.2–5.5°C) in the hot humid southeast region and 14–24°F (7.8–13.3°C) in the arid southwest region. These ranges show the potential of the upper atmosphere in hot arid climates as a source of coolness.

The original roof pond system (Skytherm) was introduced by Harold Hay in 1967 (Fig. 2.19). Hay and Yellott (1969) and Yellott (1982) provided a full review of this passive cooling system. The system has a dynamic thermal insulating panel that expands and covers the roof pond during the day to shade it from the sun and limit heat gain. The heat generated in the building is transferred to the roof pond through a conductive ceiling (no thermal insulation) and stored there until sky temperatures drop at night. At nighttime, the insulating panel is contracted to let the heat stored in the water transfer to the cool dark sky. The water temperature drops down and the night coolness is stored in the water until the sun rises up the following day. The insulating panel then expands again to cover the water and another new cycle is repeated.

There are some disadvantages in this roof pond system:

- It operates only on a flat roof.
- It is a complicated system and its operation method is not practical.
- It is a costly system and requires high maintenance costs.

2.5.4.4   *Water spray on roof with storage of coolness in water*
Without evaporative cooling, radiative cooling alone could not be sufficient in severe hot climates. In a study by Mancini (1983) of a building with a roof pond located in the hot/arid climate of the southwest in the USA, the difference between the mean ambient air temperature and the average temperature of the water in the pond is twice as much as when the pond is not evaporatively cooled.

Another modified design of a roof pond that used both radiative and evaporative cooling was suggested by Brown and Clark (1984). The roof spray system is considered a low energy system (not a completely passive system) as it requires a small pump to circulate the water. The cooling process can be described as follows:

- Water is sprayed on the roof where it is cooled evaporatively and by radiation losses to the sky.
- The cooled water washes over the top surface of the roof before it is collected back into a gutter. This helps to minimize heat gain on the roof.
- A heat pipe is used to transfer the coolness from the water in the roof gutter to a storage tank filled with water inside the building until water in the tank is chilled.
- The chilled water in the tank is circulated through radiant panels that provide cooling to the indoor space by radiation and convection.

The advantages of this system are:

- It consumes 1/3 of the electricity costs required to operate conventional AC.
- It is an easy design to prevent daytime heat gain.
- It requires a compact storage tank with a relatively small exposure area, compared to the original roof pond system (Skytherm) introduced by Harold Hay.
- The cost of the roof is relatively low, compared to Skytherm.
  The disadvantage of the system is the relatively high cost of the radiant panels.

2.5.4.5   *Water spray on roof with storage of coolness in building structure or rock bed*
Another system which combined evaporative cooling with radiative cooling to achieve maximum natural cooling was suggested by Thompson (1983). The cooling process can be described as follows:

- At night, water is sprayed over the metal roof at night, where it is cooled evaporatively and by thermal radiation exchange with the dark sky.
- The cooled water washes over the top surface of the roof and this helps to cool the roof and the air layers beneath it. The water is collected back into a gutter and re-circulated.
- From inside the building, the room air is circulated underneath the conductive roof in order to be cooled by the cool roof surface; then it is channeled through the building structure. The heavy structure of the building functions as a thermal mass medium to store the captured coolness.
- During the day, the coolness is recovered from the building's high thermal mass. Outside fresh air is brought through an opening covered with a damper that controls access of fresh air at nighttime only. It is necessary to close the damper during the daytime.

Another way of storing the coolness is to use rocks (or rock bed) as a storage medium instead of the building structure. The operation of the system is similar to the one discussed above:

- At nighttime, the cooled air that passed under the roof enters a rock bed thus cooling the rocks (charging the rocks with coolness). The air entering the rock bed cools the top layer of the rocks to a temperature nearly equal to the incoming air temperature.
- During the day, the stored coolness in the rock bed is retrieved by circulating the house air through the bed; which is cooled to a temperature nearly equal to the temperature of the bed.

Nayak *et al.* (1982) examined different approaches to the passive cooling of roofs. They found that a roof covered by a film of water and shaded with a vegetable pergola (Fig. 2.20) was the most effective method for thermal load leveling. It also allowed the least average heat flux into a

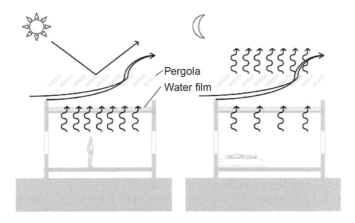

Figure 2.20.   Passive cooling by a roof covered by a film of water and shaded with a vegetable pergola.

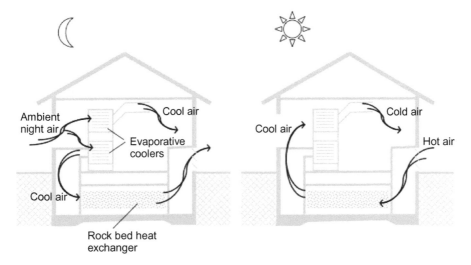

Figure 2.21.   Passive cooling with two evaporative coolers and rockbed storage (adapted from Peck and Kessler, 1981).

room under Indian conditions. Keeping the roof surface mist and evaporatively cooled can help to reduce ceiling temperatures (Bahadori and Haghighat, 1985). This method can be effective especially in low-cost housing in developing countries, where building a roof to support the pond is expensive and operating it properly is not practical.

2.5.4.6   *Evaporative coolers with rock bed storage*

This system was introduced by Peck and Kessler (1981), the Environmental Research Laboratory, University of Arizona. It consists of two evaporative coolers where one is used to supply cooled air to the building space at night, while the other supplies cooled air to the rock bed located underneath the building (Fig. 2.21). Here coolness is stored in the rocks' thermal mass. During the day, the hot dry outdoor air is passed through the rock bed, where it is cooled sensibly (pre-tempered) then flows through the evaporative cooler and cooled further evaporatively. By using this system, the supply air is introduced to the occupants' space at a much lower dry-bulb temperature (and wet bulb temperature) than it would be if the outside air were cooled evaporatively during the day using a direct evaporative cooler.

2.5.5   *Seasonal storage in earth-coupled buildings*

The ground can play an important role in the seasonal storage of thermal energy. The thermal characteristics of soils vary with soil type, compaction, and moisture. In addition, surface conditions (shade, insulating ground cover, sky temperature) affect soil temperatures. At large depths under the ground, the temperature fluctuation is damped out (lower amplitude) and the occurrence of maximum (and minimum) temperature is shifted to a later time. Also, the mean daily soil temperature at any given depth is higher than the daily air temperature in winter and lower than the mean daily air temperature in summer. Hence, this could be an ideal surrounding environment for improving occupants' thermal comfort in both summer and winter, and a source of natural cooling and heating. Yet, this is true only if the building is designed to take advantage of this natural resource, especially with regards to the depth of the soil surrounding the building and the design of the building envelope in contact with the earth.

There are two basic strategies for utilizing earth contact for building cooling: direct contact (where the building envelope is partially or completely buried underground), and indirect contact (where the building is cooled by buried heat-exchangers such as pipes or air tubes). This chapter discusses only the direct contact strategy.

Boyer and Grondzik (1983) showed appreciable savings in the total energy consumption in a study of the energy performance of earth-coupled dwellings in the USA. Other studies showed how the ground temperature could be controlled purposefully (Labs, 1984; Givoni, 1981). These studies recommended shading of the ground by trees and grass, or using large-sized pebbles to shade the ground while it is flooded with water. This helps to reduce the rate of heat flow from the atmosphere to the ground, and alter the normal temperature distribution in the ground. It also helps to increase the time lag between the maximum ambient air temperature and maximum temperature in the ground.

Based on several researches conducted by other investigators, Givoni and Katz (1985) recommended using the following equations to find the relative temperature range for the soil as a function of depth:

$$\Delta t(Z)/\Delta t(0) = \exp(-0.33Z) \tag{2.11}$$

$$\Delta t(Z)/\Delta t(0) = \exp(-0.5Z) \tag{2.12}$$

Equation (2.11) can be used for humid climate regions while the second can be used for dry climate regions, where $\Delta t(Z)$ is the temperature range (in degrees Celsius) at depth Z from the soil surface. Figure 2.22 shows the relative temperature range as a function of the soil thermal diffusivity and Figure 2.23 shows the temperature time lag as determined by many investigators (Givoni and Katz, 1985). These figures can help designers as rules of thumb to predict ground temperature and time lag at a given depth.

Earth coupling works more effectively in areas where the mean annual ambient air temperatures are low. In areas where the mean annual ambient air temperatures are high, reducing the ground surface temperature is needed. Givoni (1981) suggested shading the ground by large pebbles to reduce the solar radiation gains, and evaporatively cool the ground surface by flooding it. These measures can reduce the ground surface temperature significantly.

2.6   PROMOTING AIRFLOW FOR COMFORT OR STRUCTURAL COOLING

2.6.1   *Theoretical background*

2.6.1.1   *Airflow through buildings due to wind effects*
When wind is blowing around a building, it develops regions of positive and negative pressures. The wall surface facing the windward side has a positive pressure $(P_i - P_j) > 0$ while the leeward surface facing the leeward side has a negative pressure $(P_i - P_o) < 0$.

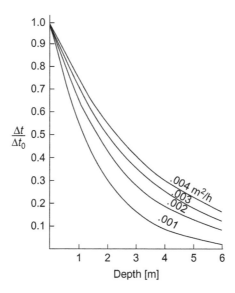

Figure 2.22. Relative temperature range ($\Delta t/\Delta t_0$) as a function of depth for different values of soil diffusivity (Source: Bahadori, 1986; with permission of Springer).

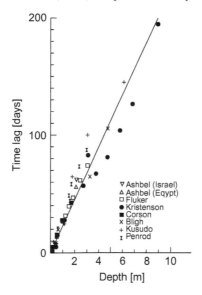

Figure 2.23. Temperature time lag as a function of depth based on experimental studies (source: Bahadori, 1986; with permission of Springer).

The driving potential for airflow through buildings due to wind effects is $P_i - P_j$. The amount of air flowing thru the building, or ϑ, is directly related to $P_i - P_j$, and inversely related to the flow resistance R_{flow} by all objects which may be in the air passage. This can be expressed as follows:

$$\vartheta = \frac{P_i - P_j}{R_{\text{flow}}} \tag{2.13}$$

One can increase ϑ by increasing the pressure difference $P_i - P_j$ and/or reducing R_{flow}. It is customary to present the wind pressure on the building envelope in terms of certain coefficients, called wind pressure coefficients. The wind pressure coefficient on surface "i" can be expressed as follows:

$$Cp_i = \frac{p_i - p_o}{\frac{1}{2}\rho v_o^2} \tag{2.14}$$

where Cp_i is the wind pressure coefficient on the surface i, p_i is the absolute pressure on this surface, p_o is the barometric pressure or the pressure of the ambient air at a distance far away from the building, v_o is the wind velocity at this location, and ρ is the air density. For another surface "j", a similar expression can be written.

The difference between the wind pressure coefficients of these two surfaces is as follows:

$$Cp_i - Cp_j = \frac{p_i - p_o}{\frac{1}{2}\rho v_o^2} - \frac{p_j - p_o}{\frac{1}{2}\rho v_o^2} \tag{2.15}$$

$$Cp_i - Cp_j = \frac{p_i - p_j}{\frac{1}{2}\rho v_o^2} \tag{2.16}$$

And the pressure differential across the surfaces i and j can be expressed as follows:

$$p_i - p_j = (Cp_i - Cp_j)\frac{1}{2}\rho v_o^2 \tag{2.17}$$

The best way to determine wind pressure coefficients is by testing building models in wind tunnels.

2.6.1.2 Air flow through buildings due to buoyancy forces (chimney effect)
If m is the air mass and ρ is the air density, the weight of the air in the stack (chimney) can be calculated using this general equation:

$$W = mg \tag{2.18}$$

The weight of the stack inside the building can also be expressed as follows:

$$W_i = \rho V g \tag{2.19}$$

$$W_i = \rho_i Ah g \tag{2.20}$$

where V is the air volume (or area times height, Ah), and g is the acceleration of gravity. The weight of the air outside the building (i.e., in the imaginary stack outside the building filled with ambient air) can similarly be expressed as follows:

$$W_o = \rho_o Ah g \tag{2.21}$$

When the two stacks are connected together, the difference in the weight of the air columns causes the air to flow in the stack. If $\rho_o > \rho_i$ then air moves from outside into the stack, creating a stack effect, which can be expressed as follows:

$$W_o - W_i = (\rho_o - \rho_i)Ah g \tag{2.22}$$

Pressure is equal to weight over area. So, if we divide both sides of the equation by the area (A), the equation can be expressed in terms of pressure:

$$P_o - P_i = (\rho_o - \rho_i)h g \tag{2.23}$$

$$\Delta P_{\text{stack}} = (\rho_o - \rho_i)h g \tag{2.24}$$

where ΔP is the pressure differential between the outside air column and inside air column. It can also be expressed as follows:

$$\Delta P_{\text{stack}} = hg\left(1 - \frac{\rho_i}{\rho_o}\right)\rho_o \tag{2.25}$$

The density of the air can be related to its temperature and pressure through a simple relation called the ideal-gas equation:

$$\rho = \frac{P}{RT} \tag{2.26}$$

where R is a constant for each gas, P is the absolute pressure, and T is the absolute temperature. When substituting density with these terms, the equation becomes:

$$\Delta P_{\text{stack}} = hg\left(1 - \frac{P_i/RT_i}{P_o/RT_o}\right) P_o/RT_o \tag{2.27}$$

In most cases, $P_i = P_o$, hence the equation can be rewritten as follows:

$$\Delta P_{\text{stack}} = hg\left(1 - \frac{T_o}{T_i}\right) P_o/RT_o \tag{2.28}$$

$$\Delta P_{\text{stack}} = ch\left(1 - \frac{T_o}{T_i}\right) P_o/T_o \tag{2.29}$$

where $c = g/R$ is a constant. The equation form can also be simplified this way:

$$\Delta P_{\text{stack}} = ch\left(\frac{1}{T_o} - \frac{1}{T_i}\right) P_o \tag{2.30}$$

This shows that the pressure differential due to buoyancy forces or the stack effect depends on h, T_o, T_i, and P_o.

2.6.2 *Cross ventilation*

Cross ventilation is an energy-efficient alternative to mechanical cooling under appropriate climate conditions. It works by promoting cooler outdoor air to flow through a space to carry heat out of a building. The design objective may be direct cooling of occupants as a result of increased air speed and lowered air temperature or cooling the building surfaces to provide indirect comfort cooling, as with nighttime flush cooling. The effectiveness of this cooling strategy is a function of the size of the inlet and outlet openings, the capacity of wind speed, the outdoor air temperature, and the indoor spatial and furniture arrangements.

Wind speed and direction is usually changing over the course of the day and throughout the year. One should design cross ventilation for the prevailing desirable wind conditions. The airflow has a driving potential, which is the pressure differential $(P_2 - P_1)$ and there is also a resistance to the airflow (R_{air}). The airflow rate is governed by this equation:

$$V^\circ = \frac{P_2 - P_1}{R_{\text{air}}} \tag{2.31}$$

To increase the airflow, the designer needs to increase $P_2 - P_1$ and/or to reduce R_{air} by proper design such as avoiding wind shadowing and using wind assisting systems such as wing walls, domed roofs, wind towers, or other assisting systems. This is also related to occupants' proper management of the space and windows' operation.

Cross ventilation for occupant comfort requires directing airflow through the building space when the outdoor air temperature is low enough to achieve comfortable conditions. Cross ventilation for nighttime structural cooling requires directing airflow to maximize contact with the building's surfaces. The designer should provide a control system to close the inlet openings when the outdoor relative humidity is at levels that could compromise human comfort. Air speed is critical to direct comfort cooling while airflow rate is critical to structural cooling.

The key architectural issues in designing cross ventilation (Kwok and Grondzik, 2007) are as follows:

- The building form should be designed to maximize exposure to the prevailing wind direction, provide for adequate inlet area, minimize internal obstructions (between inlet and outlet), and provide for adequate outlet area.
- An ideal footprint is the one that is an elongated rectangle with no internal divisions.
- Siting should avoid external obstructions to wind flow (such as trees, bushes, or other buildings). On the other hand, proper placement of vegetation, berms, or wing walls can channel and enhance airflow at windward (inlet) openings.

2.6.3   Stack ventilation

The stack ventilation strategy is derived by a process called natural convection in which indoor warm air rises due to its low density and is evacuated at high level openings, while being replaced by cooler ambient air (higher density) introduced to the space via controlled lower level openings. Stack ventilation should only be allowed when outside air temperature is cooler than the desired inside temperature. Generating a substantial airflow for effective cooling would require a difference between indoor air and ambient outdoor air temperatures to be at least 1.7°C (Grondzik et al., 2010; Kwok and Grondzik, 2007). A greater temperature difference can provide more effective air circulation and cooling. Increasing the height of a stack can help to achieve greater temperature difference, i.e. greater vertical stratification of temperatures. Another way to increase the temperature difference between entering and exiting air is to use solar energy to heat the air.

2.6.3.1   Case study: Queens Building Auditorium, De Montfort University, UK

The Auditorium of the Queens Building in De Montfort University is cooled by stack effect. It is designed as a double height space with a capacity of 150 seats (Thomas, 1996). It is part of a deep floor plan open to outside air on one side only and hence stack ventilation was applied. The spaces have high thermal mass and high ceilings. In summer, the building structure is precooled at night by allowing air movement through the room. This helps to lower the daytime temperature. Air enters at the street side through large openings (Fig. 2.24). It passes through motorized volume

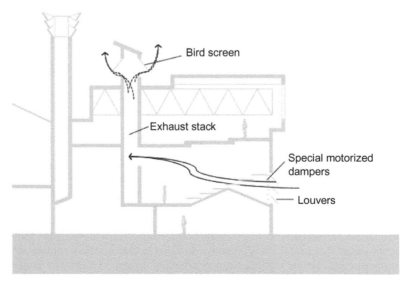

Figure 2.24.   Stack ventilation in the auditorium of the Queens Building in De Montfort University, UK (adapted from Thomas, 1996).

control dampers at the building envelope line and through an acoustically lined plenum. It goes through the void under the seating. Air filters are not used to avoid resistance of airflow. The exist air heated by the occupants and other internal gains rises, and travels into the exhaust stacks. The total cross-sectional area of the two stacks is $4\,m^2$. The area of the vertical exit opening at the top of the stack is $7.9\,m^2$. It was roughly based on a rule of thumb for chimney exit area, which should be as double the cross-sectional area. Some key points were considered in the design:

- The air path from the air intakes, through the space and out through the stacks, is of low resistance, i.e. the pathway is mainly unobstructed.
- The stacks terminate approximately 3 m above the roofline to avoid local turbulence. The air exhausts via automatic opening windows at the stack tops.

2.6.3.2 *BRE Environmental Building in Watford, UK*
BRE Environmental Building in Watford, UK is designed with five vertical Solar Siphons or stacks rectangular in shape covered in glass bricks and topped with circular ducts and cowls (Appleby, 2011). The design of the BRE building helps to increase the temperature difference between entering and exiting air by using solar energy to heat the air (Kwok and Grondzik, 2007). This is achieved by locating the ventilation stacks along the southern facade of the buildings. These stacks are glazed with a translucent material that helps in diffusing solar radiation and heating the air in the stack, causing greater difference in temperature between entering and exiting air, which promotes airflow within the building (Fig. 2.25). These south-oriented stacks are connected by motorized windows for promoting night cooling to the office space positioned behind it (Appleby, 2011). With the aid of the sinusoidal slab, air is channeled through the troughs to maximize thermal mass exposure. When sufficient pressure is not available from wind or stack effect simple axial fans are present at the top of stacks to aid natural ventilation. On the top floor clerestory roof lights are used to provide the draw of natural ventilation since this floor is not connected to the stacks.

2.6.4 *Wing walls*
Wing walls are placed adjacent to windows to increase natural ventilation in buildings and provide comfort cooling. According to research conducted by Chandra *et al.* (1983) on the effect of wing walls on ventilation performance, some important points were found. When wind is blown at

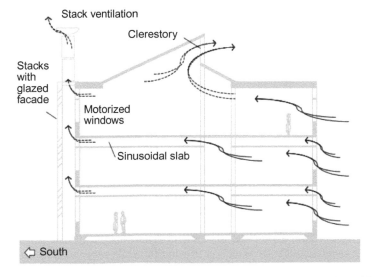

Figure 2.25.　Stack ventilation in the BRE Environmental Building in Watford, UK (adapted from Appleby, 2011).

directions different from normal angle onto the windward wall, wing walls create positive and negative pressure zones at the room openings and hence this helps to improve airflow rate. The authors considered that it is very rare when wind flows exactly normal onto the windward wall and hence the chances of ventilation improvement when using wing walls are very high. They concluded that these simple elements play an important role especially in design cases when no pressure differential is available (or not enough) at different window locations such as when windows exist on the same facade.

Several other natural ventilation designs have also been tested at Florida Solar Energy Center. In another publication, Chandra (1985) presented a brief description of space and window designs that can produce maximum pressure differentials and hence improve natural ventilation. They all take advantage of the fact that when there is a wind blowing over or around a building, there always exist regions of positive and negative pressures on the building envelope.

2.6.5 *Domed roofs with openings in their crowns*

Domed roofs have been used in many countries to cover buildings that require many people to gather in large-span spaces. Also, their strong architectural symbolic expression made them very favorable structures to cover religious, educational, and legislative buildings such as mosques, shrines, churches, universities, schools, and parliament buildings. Moreover, their favorable thermal performance made them attractive structures in other buildings such as bazaars, or marketplaces.

Wind shading can cause great reductions of the airflow. This is because wind shading of one wall on the other can cause the wind pressure coefficients on the shaded wall to be negative. When wind is blowing over a domed roof, a negative pressure or suction is created at the apex of the dome (Bahadori, 1986). This creates a large pressure differential across the building envelope which helps to ensure cross ventilation. This effect can be improved when there are holes at the dome apex. This is the advantage of the domed roof as it helps in maintaining airflow through buildings which may have their openings wind-shaded.

The airflow through a room with a domed roof, with holes in its apex, was investigated through a flow network analysis (Bahadori and Haghighat, 1985). The opening area ratios were 4% and 0.5% of the floor area for the wall and the roof, respectively. The analysis included several locations of the openings, various wind directions, and two cases when the wall apertures were wind-shaded or non-shaded. In this network analysis, no effect of wind turbulence was considered. The results showed the following:

- Ventilation through the building always increased when a domed roof with holes in its apex was employed, and the increase varied from 7.5 to about 70%, depending on the location of the openings in the room.
- The improvement in airflow rate became significant for when all wall apertures were in one wall.
- A room with all apertures in one wall and without holes in the roof had no ventilation as provision of holes in the roof created a wind pressure differential across the building envelope.

In a study conducted by Faghih and Bahadori (2009) on the solar radiation on domed roofs, they found out that a dome with a higher aspect ratio receives more solar radiation than one with a lower aspect ratio at a given location. They also found that covering the domes with glazing tiles could reduce the absorption of solar radiation remarkably. Another study by Faghih and Bahadori (2011) on domed roofs concluded that the thermal performance of the domed roof building was much better to an equivalent building with a flat roof, especially when the dome was covered with glazing tiles on warm days. These domes help in keeping the air relatively cool during the summer. They also found that the direction of wind flow had no importance in decreasing the temperature, and that direction will cause a similar effect. When no wind is present, flat roof buildings had better thermal performance. Faghih and Bahadori (2010) used a numerical method to determine the air pressure distribution over domed roofs by considering a three-dimensional model and laminar inlet airflow. The k-ε RNG method was employed for the turbulent flow simulation method. Simulation was run under three conditions of windows and a hole on top of the dome being open, or closed. The study compared the results with experimental investigations done in similar cases. The results of the study can be used to define the heat transfer coefficient of wind blowing on a domed roof and the passive cooling effect of such structures.

2.7 PASSIVE COOLING EMPLOYED MOSTLY IN HOT HUMID CLIMATES

2.7.1 *Sources of coolness constraints in hot humid climates*

The primary source of discomfort in warm regions like the southeast side of the US and the far east of Asia is the high humidity. Other harsher climates such as the coastal strip of the UAE, Saudi Arabia, Bahrain, and Kuwait, is the combined effect of high humidity and high temperatures. The major parameter distinguishing the hot humid climate and the hot dry climate is the partial pressure of the ambient water vapor, P_v. When P_v is high, the sources of coolness become limited due to the following:

- The dew point temperature of the air (T_{dp}) becomes high and this limits the potential of evaporative cooling.
- The sky temperature T_{sky} becomes high and this limits the potential of radiant cooling.
- The difference between the maximum and minimum of the ambient air temperatures ($T_{a,max} - T_{a,min}$) becomes low and this limits the potential of nocturnal cooling.

Given these conditions, employing the same thermal processes that are used in hot arid regions such as evaporative cooling, convective cooling by night ventilation, radiative cooling to the upper atmosphere, and seasonal storage of coolness in the ground can provide limited benefits in hot humid climates. Hence, the only potential source that is left and could be helpful in hot humid regions is the wind. If the designer manages to promote natural ventilation by bringing cool breezes with sufficient speeds, this can help to provide thermal comfort. The air temperatures of the ambient air in humid climates are usually not as high as in arid regions and the high levels of humidity can be tolerable as long as there is continuous airflow or the air is not still. A main difference between hot arid and hot humid climates with regards to the function of airflow in promoting passive cooling is that in hot arid regions, the nocturnal airflow through the building is primarily used for structural cooling while in hot humid climates, the airflow is used primarily for comfort cooling. When natural airflow is not strong enough, forced airflow by low energy systems such as ceiling fans or whole-house fans can be used to provide thermal comfort in hot humid climates. Dehumidification is another method that depends on chemical and mechanical operations which can help to provide thermal comfort in hot humid climates.

2.7.2 *Dehumidification*

Dehumidification is the removal of water vapor from room air. This can be done by three means: dilution with drier air, condensation, or desiccation. Dilution with drier air is the simplest and cheapest source of dehumidification. It can help in mitigating discomfort caused by high humidity ratios especially in residences that are not air-conditioned due to perspiration, cooking, bathing, and plants. Dilution with drier air occurs naturally in the process of natural ventilation, and is desirable whenever the outdoor dew point is low enough. The design strategies are the same as for comfort cooling by natural ventilation (discussed below). In the case of condensation and desiccation, dehumidification is the exchange of latent heat in air for the sensible heat of water droplets on surfaces; both are the reverse of evaporative cooling and, as such, are adiabatic heating processes. Passive systems in warm humid climates are usually incapable of achieving surface temperatures sufficiently low to cause sufficient condensation.

2.7.3 *Employing natural ventilation*

All the strategies that have the potential to promote airflow for comfort cooling (discussed in Section 2.6 Promoting airflow for comfort or structural cooling) can be applied in hot humid climates such as the following:

- Cross ventilation through windows.
- Stack ventilation through wind towers or the chimney effect.
- Wing walls, which have the potential to improve cross ventilation.
- Domed roofs, which have the potential to improve cross ventilation.

Other approaches that need to be integrated in the architectural design and can provide effective solutions in promoting comfort cooling in hot humid climates were also discussed in the previous sections and in the literature. These are as follows:

- *Proper form configuration of the building*: the spread-out form can be very effective in hot humid climates as it helps to promote airflow.
- *Sparse building arrangements*: sparse buildings' arrangements (as opposed to compact arrangements that are more suitable for hot arid regions) are recommended in hot humid regions for their potential to minimize wind shadowing and thus improving wind flows.
- *Design of the windows*: large operable openings are recommended in hot humid climates since they help to promote cool breezes into the building. The openings have to be shaded by extended porches or shading devices.
- *Adapting the courtyard design to improve comfort cooling*: previous research suggested raising the height of the courtyard wall downwind to catch the wind moving over the courtyard; which will then drop as down draft into the courtyard (Givoni, 1998). Previous research also indicated that top solar shading of the courtyards using elements (such as toldo) could result in limiting passive cooling effectiveness by airflow (Reynolds, 2002).
- *Ventilated basements by wind scoops*: previous research done on the performance of traditional houses of Baghdad showed the significance of wind scoops in reducing high levels of humidity in the designs of serdabs.

2.7.4 *Desiccant cooling*

In a typical desiccant cooling application, moisture (latent load) is absorbed out of the air into the desiccant material while sensible heat is released and the dry-bulb temperature increases. In other words, the air leaves the desiccant dry and warm. This requires additional sensible cooling to counterbalance the sensible gain that is inherent in the desiccation process (e.g., mechanical compression, evaporative cooling).

Desiccant cooling systems have several advantages (IEA, 1995):

- Improvement of indoor air quality – some desiccants act as bactericides as they dehumidify it.
- Production of very low humidity levels.
- Ability to use alternate energy sources and waste heat.
- Potential to minimize electrical consumption.
- Capacity for demand side management by shifting electrical consumption to a thermal source.
- Allowing separate control of humidity and temperature.

There are two types of desiccant cooling systems: solid based and liquid based desiccant systems. In a solid desiccant system (such as the one shown in Fig. 2.26), as air is circulated through porous material, the water molecules physically attach to the surface of the material and permeate into it by diffusion, causing removal of the moisture by adsorption. The material needs to be regenerated by hot air. Solid desiccants are usually applied in the form of a packed bed or a rotating wheel. Examples of organic materials that can be used as solid desiccants are silica gel, lithium bromide, lithium chloride, calcium bromide, molecular sieves and titanium silicate. In liquid desiccant systems, the absorption process is applied by spraying a concentrated solution in

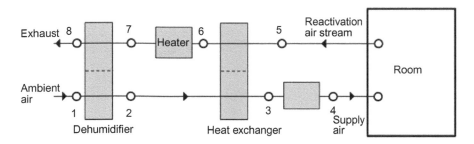

Figure 2.26.   Schematic of a solid desiccant cooling system (adapted from IEA, 1995).

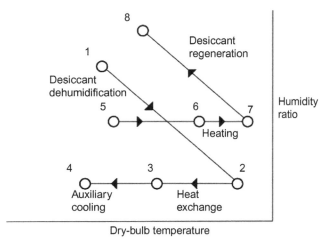

Figure 2.27.   Psychrometric representation of the solid desiccant system (adapted from IEA, 1995).

the air stream to remove the moisture. This results in a diluted solution that needs to be circulated and heated by a hot air stream to re-concentrate the liquid desiccant. Examples of liquid desiccants are glycol and solutions of lithium bromide or calcium chloride in water.

The psychrometric representation of the solid desiccant system described above is shown in Figure 2.27. In this system, a heat exchanger (air-to-air) is used between the reactivation air stream and the dehumidified ambient air to reduce the supply air's dry-bulb temperature. Further reduction of the supply air can be achieved by cooling the reactivation air stream using direct evaporative cooling, prior to the heat exchanger. This would reduce the need for an auxiliary cooling system on the supply airside, yet it would require more heat to reactivate the desiccant.

The desiccant cooling system can be designed to utilize waste heat from sustainable energy resources; for example using excess thermal energy from a photovoltaic (PV) system that can provide thermal energy needed to heat return-air stream required for desiccant regeneration. This concept can be integrated with building façade where the PV system is used as an active building component and connected with the desiccant cooling system, as shown in Figure 2.28. This approach is being investigated by a team at UAE University (including the author of this chapter) in a research project where the excess heat is stored in a phase change material behind a building integrated PV. The stored thermal energy is extracted by water circulation; which can be used to heat the air stream returning from the indoor space required for regenerating the desiccant wheel. The ongoing experimentation work shows high potential of this integrated system in improving the PV efficiency of performance, reducing building cooling load, and recovering energy needed for desiccant regeneration.

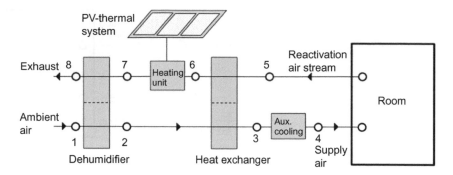

Figure 2.28.   Integration of desiccant cooling system with PV-thermal system.

2.8   CONCLUSIONS

This chapter revisited the classical literature and reviewed recent advances of passive and low energy cooling systems. A review of some basic information was presented in the beginning to help the reader comprehend the technical discussions that come afterwards. This included discussions on problems associated with the extensive use of air conditioning, a theoretical background on thermal comfort and sources of coolness fundamentals, general classifications of passive cooling strategies, methods of minimizing heat gains through architectural design, and a review of passive cooling systems applied in vernacular architecture. After presenting the basic information, the chapter gave deeper and more detailed discussions of systems' governing laws/equations, structures, components, and operations. The chapter depended also on several case studies that were selected from both vernacular and contemporary architecture to illustrate the discussions and exemplify the presented strategies for passive and low energy cooling. The presentation of the cooling methods started with systems that can be more effective in hot and arid regions, followed by airflow systems for comfort cooling, and concluded with systems that have potential to mitigate discomfort problems due to high humidity levels in warm and humid climates.

Passive and low energy cooling systems can help to save cooling energy and operation costs but it will probably add costs for construction. These costs can be minimized if the architect succeeds first in minimizing building heat gains so that the remaining cooling requirements can be covered totally by using passive systems or at least mostly by passive systems with the assistance of another small mechanical cooling system, in case the passive system alone fails to satisfy the cooling requirements. To achieve this, the design stakeholders must set clear objectives from the beginning of the design process and follow a sound architectural design process. Minimizing heat gains through architectural design can be done through choosing/developing a site that helps to improve the microclimate around the building, developing a building form that helps to shade itself and reduce exposure to the sun, arranging the building spaces to receive only the desirable environmental factors (e.g., desirable breezes) while blocking undesirable ones (e.g., intense afternoon west sun), shading the building envelope and openings using porches and shading devices, painting or cladding the building envelope with light colors, using high resistance thermal insulation to minimize heat transfer by conduction, and maximizing reliance on daylighting and using energy-efficient lights and equipment to control internal heat gains.

After scrutinizing sources of heat and applying methods for minimizing heat gains through architectural design, passive and low energy systems can be applied. The process starts by identifying the sources of coolness (or heat sinks) available in the natural environment then choosing the passive systems that have greater potential to utilize these sources.

ACKNOWLEDGEMENTS

This chapter is supported by UAE University through an externally funded research project number 31N103. The author thanks Iman Al-Sallal (Master of Advanced Studies in Architecture, University of British Columbia, Canada) and Maitha Al-Nuaimi (Ph.D. Candidate and Researcher at UAE University, UAE) for their valuable comments and thoughtful and constructive reviews of this chapter.

REFERENCES

Aboul-Naga, M., Al-Sallal, K.A. & El Diasty, R.: Impact of city urban patterns on building energy use: Al-Ain city as a case study for hot-arid climates. *Archit. Sci. Rev.* 43 (2000), pp. 147–158.

Akbari, H., Davis, S., Dorsano, S., Huang, J. & Winett, S.: *Cooling our communities – a guidebook on tree planting and light colored surfacing.* US Environmental Protection Agency, Office of Policy Analysis, Climate Change Division, January, 1992.

Al-Azzawi, S.H.: Oriental houses in Iraq. In: P. Oliver (ed): *Shelter and society.* Barrie and Rockliff, London, UK, 1969, pp. 91–102.

Al-Sallal, K.A.: Vernacular tower architecture of Sana'a: theory and method for deriving sustainable design guidelines. In: A. Sayigh (ed): *Sustainability, energy and architecture: case studies in realizing green buildings.* Elsevier, 2013, pp. 257–288.

Al-Sallal, K.A., Ayssa, A.Z. & Al-Sabahi, H.A.: Thermal performance and energy analysis for Sana'a vernacular house. In: Ben Ghadi, S. (ed): *Proceedings of applications on renewable energy in Yemen.* Workshop, Aden University, Aden, Yemen, 1995.

Anderson, B.: Passive solar design concepts. Los Alamos Scientific Laboratory, Los Alamos, NM, 1979.

Appleby, P.: *Integrated sustainable design of buildings.* Earthscan, London, UK, 2011.

Ayssa, A.: *The thermal performance of vernacular and contemporary houses in Sana'a, Yemen.* PhD Thesis, Open University, UK, 1995.

Bagenid, A.: *Courtyards bioclimates: comparative experiments.* MSc Thesis, Arizona State University, Tempe, AZ, 1987.

Bahadori, M.N.: Passive cooling systems in Iranian architecture. *Sci. Am.* 238:2 (1978), pp. 144–154.

Bahadori, M.N.: Natural cooling in hot arid regions. In A.A.M. Sayigh (ed): *Solar energy applications in buildings.* Academic Press, Inc., 1979, pp. 195–225.

Bahadori, M.N.: Natural air-conditioning systems. Chapter 5 in K.W. Böer (ed): *Advances in solar energy.* 1986, pp. 283–356.

Bahadori, M.N.: Unpublished class notes of Passive Cooling course, School of Architecture, Arizona State University, 1987.

Bahadori, M.N. & Haghighat, F.: Passive cooling in hot arid regions in developing countries by employing domed roofs and reducing the temperature of internal surfaces. *J. Build. Environ.* 20:2 (1985), pp. 103–113.

Bahadori, M.N. & Haghighat, F.: Weekly storage of coolness in heavy brick and adobe walls. *J. Energy Build.* 8:4 (1986), pp. 259–270.

Bahadori, M.N., Mazidi, M. & Dehghani, A.R.: Experimental investigation of new designs of wind towers. *Renew. Energy* 33 (2008), pp. 2273–2281.

Boyer, L.L. & Grondzik, W.T.: Energy performance of earth covered dwellings in the US. *Proceedings of Solar World Congress,* Perth, Australia, August 1983, pp. 46–51.

Brown, P. & Clark, G.: Effect of design changes on performance of a trickle roof cooled residence. *Proceedings of the 9th National Passive Solar Conference,* American Solar Energy Society, Inc., Columbus, OH, September 1984, pp. 50–57.

Cain, A., Ashfar, F., Norton, J. & Daraie, M.R.: Traditional cooling systems in the third world. *The Ecologist* 6:2 (1976), pp. 60–64.

Chandra, S.: Passive cooling for residences in hot humid climates – a review of recent research. *US-India Bi-national Symposium on Solar Energy Research and Applications,* Roorkee, India, August 1985.

Chandra, S., Fairey, P., Houston, M. & Kerestecioglu, A.A.: Wing walls to improve natural ventilation: full-scale results and design strategies. *Proceedings of 8th National Passive Solar Conference,* Santa Fe, NM, September 1983, pp. 855–860.

Cunningham, W.A. & Thompson, T.L.: Passive cooling with natural draft cooling towers in combination with solar chimneys. *PLEA 1986,* 1–5 Sep. 1986, Pecs, Hungary, 1986.

Danby, M.: The internal environmental aspects of the traditional Islamic house and their relevance to modern housing. In A. Bowen (ed): *Proceedings of PLEA 1984*. Pergamon Press, New York, NY, pp. 664–671, 1984.

EIA: *Annual energy outlook* 2012; Early Release. US Energy Information Administration (EIA)/Department of Energy (DoE), 2012.

Faghih, A.K. & Bahadori, M.N.: Solar radiation on domed roofs. *Energy Build.* 41 (2009), pp. 1238–1245.

Faghih, A.K. & Bahadori, M.N.: Three-dimensional numerical investigation of air flow over doomed roof. *J. Wind Eng. Ind. Aerod.* 98 (2010), pp. 161–168.

Faghih, A.K. & Bahadori, M.N.: Thermal performance evaluation of domed roofs. *Energy Build.* 43 (2011), pp. 1254–1263.

Fardeheb, F.: Examination and classification of passive solar cooling strategies in Middle Eastern vernacular architecture. *Passive Solar J.* 4:4 (1987), pp. 377–417.

Fathy, H.: *Natural energy and vernacular architecture: principles and examples with reference to hot arid climates*. University of Chicago Press, 1986.

Gamero-Abarca, J. & Yellott, J.: Operation of a plate-type indirect evaporative cooler in a hot, dry (Arizona) climate. *Solar World Congress*, August 1983, Perth, Australia, pp. 135–139.

Givoni, B.: Earth integrated buildings – an overview. *Archit. Sci. Rev.* 24 (1981), pp. 42–53.

Givoni, B.: *Passive and low energy cooling of buildings*. Van Nostrand Reinhold, New York, NY, 1994a.

Givoni, B.: *Building and urban design guidelines for different climates*. Van Nostrand Reinhold, New York, MY, 1994b.

Givoni, B.: *Climate considerations in building and urban design*. Van Nostrand Reinhold, New York, NY, 1998.

Givoni, B. & Katz, L.: Earth temperatures and underground buildings. *Energy Build.* 8 (1985), pp. 15–25.

Golani, G.: *Design and thermal performance: below-ground dwellings in China*. University of Delaware Press, Newark, 1990.

Grondzik, W., Kwok, A., Stein, B. & Reynolds, J.: *Mechanical and electrical equipment for buildings*. John Wiley & Sons, Inc., Hoboken, NJ, 2010.

Hay, H.R. & Yellott, J.: International aspects of air conditioning with movable insulation. *Solar Energy*, 12 (1969), pp. 427–438.

IEA: Energy Conservation in Buildings and Community Systems Programme: Annex 28 – Low energy cooling: review of low energy technologies: Subtask 1 Report (December 1995), Natural resources Canada, Ottawa, Canada, 1995.

Kreith, F.: *Principles of heat transfer*. 3rd edition, Harper & Row, New York, NY, 1973.

Kwok, A. & Grondzik, W.: *The green studio handbook: environmental strategies for schematic design*. Architectural Press, 2007.

Labs, K.: Experiences and expectations of earth-coupled buildings research. *Proceedings of the Third International PLEA Conference*, Mexico City, Mexico, August 1984, pp. 144–160.

Lezine, A.: La protection contre la chaleur dans l'architecture Musulmane d'Egypte. *Bulletin d'Etudes Orientales* 124, Damascus, Syria, pp. 7–17.

Mancini, T.R.: The performance of a roof-pond solar house: the New Mexico State University experience. *Proceedings of 8th National Passive Solar Conference*, American Solar Energy Society, Inc., Santa Fe, NM, September 1983, pp. 811–815.

Martin, M. & Berdahl, P.: Characteristics of infrared sky radiation in the United States. *Solar Energy* 33 (1984), pp. 321–336.

Meir, I., Pearlmuter, D. & Etzion, Y.: On the microclimatic behavior of two semi-enclosed attached courtyards in a hot dry region. *Build. Environ.* 30:4 (1995), pp. 563–572.

Nayak, J. K., Srivastava, A., Singh, U. & Sodha, M.S.: The relative performance of different approaches to the passive cooling of roofs. *Build. Environ.* 11:2 (1982), pp. 145–161.

Nour, A.: *An analytical study of the traditional Arab domestic architecture*. PhD Thesis, University of New Castle Upon Tyne, UK, 1979.

Peck, J. & Kessler, H.: Evaporative cooling. Chapter 8 in: *Arizona passive design work book*. Western SUN and Arizona Solar Energy Commission, Phoenix, AZ, March 1981.

Reynolds, J.: *Courtyards aesthetic social and thermal delight*. John Wiley and Sons Inc., NY, 2002.

Rohles, F.H., Konz, S.A. & Jones, B.W.: Ceiling fans as extenders of summer comfort envelope. *ASHRAE Transact.* 89, Part 1A, 1983, pp. 245–263.

Safarzadeh, H. & Bahadori, M.N.: Passive cooling effects of courtyards. *Build. Environ.* 40 (2005), pp. 89–104.

Santamouris, M. & Asimakopoulos, D.N. (eds): *Passive cooling of buildings*. James and James Science Publishers, London, UK, 1997.

Thomas, R.: *Environmental design: an introduction for architects and engineers*. E & FN Spon, London, UK, 1996.

Thompson, T.L.: Residential cooling by nocturnal radiation. Report prepared for the US Department of Energy, the Environmental Research Laboratory, University of Arizona, Tucson, AZ, May 1983.

Yellott, J.: Passive and hybrid cooling systems. In: *Advances in solar energy*. American Solar Energy Society, New York, NY, 1982, pp. 241–260.

APPENDIX A – CASE STUDIES

Building Research Establishment (BRE) Office Building

- http://projects.bre.co.uk/envbuild/index.html
- http://www.webpages.uidaho.edu/arch504ukgreenarch/casestudies/bre2.pdf

Breeze Engine Hostel in southern China by Zoka Zola Architecture + Urban Design

- http://www10.aeccafe.com/blogs/arch-showcase/2011/05/25/breeze-engine-in-southern-china-china-by-zoka-zola-architecture-urban-design/
- http://www.zokazola.com/south_china_hostel.html

Masdar City, Abu Dhabi

- http://www.masdar.ae/en/#city/all
- http://www.theecologist.org/News/news_analysis/1879752/masdar_city_a_rising_star.html
- http://peterjphpositivefeedback.wordpress.com/2013/02/05/masdar-city-a-rising-star/

Masdar Institute; Masdar Institute Wind Tower, Abu Dhabi

- https://www.masdar.ac.ae/campus-community/the-campus
- http://archrecord.construction.com/projects/portfolio/2011/05/masdar_institute.asp
- http://www.thenational.ae/news/uae-news/masdar-tower-is-a-big-experiment

The NRP enterprise centre

- https://www.adaptcbe.co.uk/CBE/downloads/consultation/exhibition_panel_11.pdf

Low-energy office building, Belgium

- http://www.europeanconcrete.eu/images/stories/Documents/Office_buildings/BE-SD_Worx_Kortrijk-Full_case.pdf?phpMyAdmin=16bbb563ca43adfed14bd78eb7d8cd8a

Low energy cooling case studies

- IEA, International Energy Agency, Energy conservation in buildings and community systems, Annex 28 – Low energy cooling, case studies of low energy cooling technologies, Zimmermann, M. and Anderson, J. (eds), 1998.
- http://www.ecbcs.org/docs/annex_28_case_study_buildings.pdf

APPENDIX B – OTHER RESOURCE WEBSITES

Natural cooling by Arizona Solar Center

- http://www.azsolarcenter.org/tech-science/solar-for-consumers/passive-solar-energy/passive-solar-design-manual-consumer/passive-solar-design-manual-cooling.html

Natural ventilation, cross ventilation, wing walls

- http://www.wbdg.org/resources/naturalventilation.php
- http://www.architecture.com/RIBA/Aboutus/SustainabilityHub/Designstrategies/Air/1-2-1-3-naturalventilation-crossventilation.aspx
- http://sustainabilityworkshop.autodesk.com/buildings/wind-ventilation#sthash.vfOqnXWg.dpuf

Passive Solar Research Group, University of Nebraska at Omaha

- http://www.ceen.unomaha.edu/solar/solar_fi.php

Passive Cooling, Australian Government

- http://www.yourhome.gov.au/passive-design/passive-cooling
- http://www.yourhome.gov.au/passive-design/shading

Passive cooling, research & design, The Quarterly of the AIA Research Corporation

- http://www.aia.org/aiaucmp/groups/aia/documents/pdf/aiab082771.pdf

Passive and low energy cooling techniques in buildings

- Lain, M. and Hensen, J.L.M.: Passive and low energy cooling techniques in buildings. *Proceedings of the 17th Int. Air-conditioning and Ventilation Conference*, May 2006, pp.1–7. (http://www.bwk.tue.nl/bps/hensen/publications/06_acv_lain.pdf)

Review of Low Energy Cooling Technologies

- IEA: Energy Conservation in Buildings and Community Systems Programme: Annex 28 – Low energy cooling: review of low energy technologies: Subtask 1 Report (December 1995), Natural resources Canada, Ottawa, Canada, 88 p. (http://www.ecbcs.org/docs/annex_28_review.pdf)
- IEA, International Energy Agency Energy Conservation in Buildings and Community Systems, Annex 28 – Low energy cooling, case studies of low energy cooling technologies, Zimmermann, M. and Anderson, J. (eds), 1998. (http://www.ecbcs.org/docs/annex_28_case_study_buildings.pdf)

IEA Energy conservation in buildings and community systems

- http://www.ecbcs.org/docs/

CHAPTER 3

Daylighting

Khaled A. Al-Sallal

3.1 INTRODUCTION

The term daylight usually means any kind of light produced by the sun which reaches the earth directly or indirectly (Steffy, 2008). As described by the IES Lighting Handbook *"Daylighting involves the delivery and distribution of light from the sun and sky to a building interior to provide ambient and/or task lighting to meet the visual and biological needs of the occupants"* (DiLaura *et al.*, 2011). Reinhart (2014) defined daylighting as the "controlled use of natural light in and around buildings". He also described it as "the process by which direct sunlight and diffused daylight are reflected, scattered, admitted and/or blocked to achieve a desired lighting effect". Daylighting can also be defined as the range of brightness and color composition for our vision, it gives us orientation in time and space, and it is a precondition for our perception and evaluation of the built environment (Mueller, 2013). In addition to visual comfort and performance, daylighting can contribute greatly to improving the sustainability level of buildings. It helps to replace artificial lighting by natural lighting; which results in saving a great deal of electricity consumption for lighting due to smaller lighting loads and also for cooling due to smaller cooling loads as a result of reducing internal heat gains caused by electric lighting fixtures.

Daylighting requires integrated architectural designs that harness natural light to illuminate the building and its environment and involves practices such as the placing of windows, gaps and other surfaces that can reflect daylight for the purposes of increasing the performance of daylight.

Approximately half of the solar energy received at the earth's surface is daylight and the majority of what remains is heat while a small portion is ultraviolet radiation. Natural light has two basic components: direct sunlight meaning light that comes directly from the sun's disc and the sky light meaning light that comes indirectly from the sun by reflection and transmission of light through clear or cloudy skies. The third component is produced when natural light (the two mentioned above) is reflected from other surfaces such as the ground, other buildings, or bodies of water.

3.2 HUMAN NECESSITIES AND BENEFITS OF DAYLIGHT

Daylight in buildings plays an essential role besides merely offering enhanced visibility (Baker, 2000). It has various benefits such as substituting high levels of artificial electric lighting energy, providing visual comfort, and promoting human health and user productivity. Previous surveys also proved that it contributes to occupant satisfaction, where more than 60% of office workers favored direct light access into their work place. Furthermore, in terms of health and well-being, occupants expressed preference to operating under natural light conditions due to the active performance of daylight throughout the working hours (Veitch *et al.*, 2003; Vine *et al.*, 1998). Daylighting is crucial for replacing artificial electric lighting energy, visual comfort, human health and user productivity.

3.2.1   Human aspects

Over many years in the history of mankind, human beings have evolved through adaptation to natural light; and through its dynamic changes of day-to-day biorhythms, it became a stimuli influencing human mood and health. Besides offering comfort to the visual senses, research proves that daylight also boosts psychological health. Building occupants expressed that natural light helps provide a deeper connection with the exterior environment, enriching visual value, human emotions and comfort (Edwards and Torcellini, 2002). By allowing more natural light inside the building, the interior environment is transformed into a healthier, livelier and more vibrant one. This in turn positively impacts occupant behavior and productivity.

3.2.1.1   Sleep/wake cycle
As the nature of human rest dictates, it is crucial to provide appropriate light levels throughout the day and night in order to maintain a healthy sleep/wake cycle. That is achieved by high exposure to light during daytime hours for increased attentiveness and productivity, and low exposure at nighttime hours to promote relaxation and lower levels of attentiveness, thus helping the body to sleep. As Edwards and Torcellini (2002) mentioned, sleep/wake cycles are vital for maintaining human health and thus a stable society and economy.

3.2.1.2   Performance and productivity
According to Köster (2013), occupants generally prefer spaces that are well daylit. Robbins (1986) states that well daylit spaces have positive effects on occupants' overall mood and morale, while reducing fatigue and eye strain. Various studies by Edwards and Torcellini (2002) proved that well lit offices in industrial and retail environments are the catalyst for improved performance and efficiency. When moved to a new building with better daylighting conditions, workers productivity increased by 15%. The increase of productivity as a result of lighting satisfaction in the workplace was confirmed by another study (Veitch *et al.*, 2008). The importance of daylight is not limited only to the improvement of productivity in the workplace such as in office spaces and industrial or retail environments. Other studies found that classrooms with high daylighting levels had the potential to improve the score tests of students 7–18% higher than other classrooms that had lower daylighting levels (Heschong, 2002).

3.2.1.3   Health and patient recovery time
Poor daylighting is the cause of many health complications, one of them being seasonal affective disorder (SAD). SAD is a type of mood disorder, which has a seasonal pattern and is characterized by depression due to shorter exposure to daylight (NHS-SAD, 2013). As Boyce *et al.* (2003) stated, this disorder results from a withdrawal of harmonized effects of daylight. Another health complication caused by poor daylighting conditions is the sick building syndrome (SBS), this occurs as a result of inadequately ventilated and electrically lit buildings. Sensory irritation of the eyes and skin, including other infectious diseases are some of its symptoms (Godish, 2010).

According to research, a shortage of vitamin D is caused by a withdrawal of exposure to exterior daylight. One of the many health benefits of sunlight is its ability to heighten the amount of vitamin D in the human body, which is fundamental for the construction of good body tissue.

Studies on hospital design proved that good daylight and views are vital for patients' health improvement and wellbeing. Patients who were placed in well daylit rooms and with views to the outdoor environment showed faster signs of recovery and smaller needs for medication, as opposed to patients placed in rooms with no daylight access (Ulrich, 1984). Another experiment carried out by Beauchemin and Hays (1998) showed that patients who received more sunlight had lower mortality rates. Moreover, those healing from coronary artery bypass graft surgeries had faster recovery times; 7.3 h per 100 lx increase in daylight (Joarder and Price, 2013).

Daylight is essential for the maintenance of metabolic processes, as well as physiological and psychological elements of the human body. Hence, the needs for adequately daylit rooms that promote appropriate task illuminance, comfortable visual environments, savings in electrical energy, and occupant health and wellbeing are crucial in the design of any building.

3.2.2   *Energy savings and environmental benefits*

According to a report conducted by the International Energy Agency (IEA) in 2006 (IEA, 2006), electric power consumption is largely attributed to lighting; this translates to 34% of tertiary-sector electricity consumption, and 14% of residential consumption in OECD (Organization for Economic Co-operation and Development) countries. In non-OECD countries, the rate of electrical power consumption due to lighting is substantially higher. In the USA, energy consumption in buildings constitutes approximately half the source energy consumption and greenhouse gas emissions. Due to buildings' high dependence on electric lighting that is mainly generated from fossil fuels, energy-related greenhouse gas emissions are largely attributed to electric lighting in buildings. Thus, resorting to natural resources such as daylight in buildings is one of the top-priority solutions that need to be executed. Daylighting can play a central role in cutting back these figures; this is the biggest challenge it faces today.

Good daylight design can assist in the overall cutback of energy consumption of a building. This has been proved in numerous studies that showed potential energy savings ranging between 20–60% in office and retail buildings (Galasiu *et al.*, 2007).

When partnered with electric lighting controls in largely daylit spaces, daylighting can reduce electric lighting energy consumption by up to 30–80%. Diminishing lighting energy can also be beneficial in saving cooling load energies, particularly in climates that require heavy cooling. For instance, reducing 100 W of lighting energy translates to 30 W of cooling energy being saved.

3.3   DAYLIGHT SOURCE AND CHARACTERISTICS

3.3.1   *Sunlight and daylight*

The daylighting effect is dependent on the sun's position in the sky and the sky conditions (whether it is clear, somewhat cloudy, or totally cloudy). These two factors are always changing over the course of the day and over the course of the year; thus making daylight exhibit continuous variations in intensity, direction, distribution and color. The two components of daylight are the direct sunlight and the diffuse sky light. Depending on the sky conditions, the effect of each of these two components changes. When the sky is very clear, the direct sunlight becomes more present with crisp lines of shade and shadows and strong impact of reflected light and on visual effects. When the sky is fully cloudy, the sunlight is reflected by and transmits through a dense and diffusing layer of clouds generating diffuse and more uniform light (i.e., sky light). Between these two extreme sky conditions, there are several types of skies and shades of natural lighting.

3.3.2   *Sun position*

The effect of the direct sunlight with regards to its radiation intensity and direction is dependent on the sun's position in the sky. The earth travels one complete revolution around the sun every 365 days along the earth's elliptical orbit. While doing this, it also rotates around itself once per day (every 24 h); that is around its axis that connects between the center of the northern pole and the center of southern pole. The earth's axis is tilted 23.5° away from the earth's orbital plane. It is because of this tilt that the earth experiences seasons as it orbits around the sun. When the north pole of the earth is pointed towards the sun, this is summer for the northern hemisphere (and winter in the southern hemisphere). Then 6 months later, when the earth is at the opposite side of its orbit, the north pole is pointed away from the Sun; that is winter in the north (and summer in the southern hemisphere).

Although in reality the earth moves around the sun, the way we experience it on earth is the sun moving in the sky. In any location on earth this movement starts from east at the sunrise time and ends on west at the sunset time. The sun travels in the sky along an arc called the sun path, and every day has a unique sun path. The solar altitude angle is the vertical angle that the sun makes with the horizontal ground plane. The azimuth angle is the horizontal angle of the sun relative to the true north. Knowing these two angles can identify exactly the sun position in the sky. The longest sun path in the northern hemisphere occurs on June 21st, which is identified as the summer solstice; and the shortest sun path occurs on December 21st, which is identified as the winter solstice. In the middle point between the summer and the winter solstices (i.e., after 3 months from the summer solstice), the fall equinox occurs on September 21st. Similarly, in the middle point between the winter and the summer solstices (i.e., after 3 months from the winter solstice), the spring equinox occurs on March 21st.

The slope of the sun path on the ground plane changes with the geographical latitude. Hence, for the same location, all sun paths have the same slope and they are also parallel to each other. At the equator, the sun paths are perpendicular on the ground plane; while at the center of the northern (or southern) pole, the sun paths are parallel to the ground plane.

In latitudes outside the tropics band (between the tropic of Cancer and the tropic of Capricorn), the sun altitude angle reaches its highest point at the summer solstice, the lowest at the winter solstice and a position mid-way in between during spring and autumn equinox days. For the zone inside the tropics band, it depends on the location. If the location is at the equator, the highest sun position is reached at the equinox. If the location is at one of the tropics, the highest sun position is reached at the summer solstice. If the location is closer to one of the tropics but not on it, the highest sun position is reached on a midway point between that tropic and the equator.

3.3.3   *Sky luminance and classifications*

Skies are classified into three types according to two methods, the percentage of their cloudiness, or the sky ratio of cloud cover. Figure 3.1 shows fish-eye images of the sky taken at different times of the year in Al-Ain city, UAE, showing classifications of the daylight sky. The three types are as follows:

- *Clear sky*: a clear sky type is characterized as having 0–30% cloudiness (or with the sky ratio method: less than, or equal to 0.30 cloud cover). Without the solar component, the clear sky has a 1:3 zenith to horizon brightness ratio
- *Overcast sky*: an overcast sky is one with 80–100% cloudiness (or with the sky ratio method: greater than 0.80 and less than, or equal to, 1.0 cloud cover). The luminance of the overcast sky varies in altitude only with an approximate 3:1 zenith to horizon brightness ratio.

Clear   Partly cloudy   Overcast

Al-Ain, UAE – Lat.: 24° 42' N
Daylighting Simulation Laboratory, UAEU

Figure 3.1.   The three standard classifications of the daylight sky: (left) clear sky, (middle) partly cloudy sky, and (right) overcast sky (photos by the author).

- *Partly cloudy sky*: a partly cloudy sky has 40 to 70% cloudiness (or with the sky ratio method: greater than 0.30, but less than 0.80 cloud cover). The increased percentage of clouds in a partly cloudy sky can lead to a more dynamic range of luminous distributions; and thus its sky contribution cannot be controlled with just a simple solar shading device (as the case in a clear sky conditions).

There is a clear difference in luminous distribution between a *clear sky* and a *standard overcast sky* (see Fig. 3.2). The clear sky is a directional source of light that produces high illuminance levels and varies by orientation. The luminous distribution of the clear sky is determined by two dynamic zones that move simultaneously with solar position: the circumsolar zone which is the brightest area of the sky located around the sun; and the anti-solar zone which is the darkest area of the sky located on the opposite side of the sun. The sky luminance near the sun is ten times greater than the opposite side. In contrast, the standard overcast sky is a diffuse source producing lower illuminance levels that do not vary by orientation. Its brightest zone at the zenith has luminance three times greater than the horizon.

Fenestration design decisions should be made based on predominant sky type. Under an overcast sky, a horizontal opening (e.g., skylight) admits about three times more lumens into the interior than a window of equal size. Therefore the skylight can be about 1/3 the size of the window in order to admit an equal amount of light. Under a clear sky (with the combined effect of direct sunlight and diffuse sky light), the horizontal illuminance is about ten times greater than that of an overcast sky.

3.3.4   *Daylight availability*

The daylight environment in a region depends on the latitude of the location and the amount of cloud cover. It depends also on the atmospheric conditions like pollution and reflected daylight from the surrounding environment. These factors create continuous changes in intensity and color compositions of daylight. Maximum horizontal illuminances exceeding 100 klx under clear sky conditions and mean values of 10 klx under overcast sky conditions are typical of the daylight in our natural environment. The different components of daylight (direct sunlight, diffused sky light, externally reflected light, internally reflected light) are influenced by the climate/environment characteristics. The luminance of the direct sunlight ($1,650,000,000 \, cd/m^2$) is unbearable to the human eye. Clear sky luminance varies between 2000 and 12,000 cd/m^2, depending on the atmospheric scattering effects. The value for the cloudy sky varies between 1000 and 6000 cd/m^2, depending on the altitude of the sun and the density of the cloud cover. The light reflected by external surfaces (such as the ground cover, claddings of the surrounding buildings, and plants) and the internal room surfaces of the room ceiling, walls, floor, and furniture has a great effect on the amount of daylight and its quality. This requires the designer to choose proper reflectivity values, at least for the internal surfaces since the external ones are in most cases not under the designer's control.

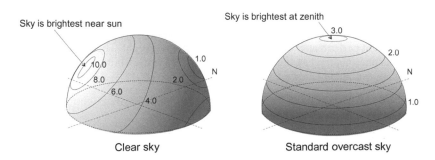

Figure 3.2.   Luminance distribution comparison between clear sky and standard overcast sky.

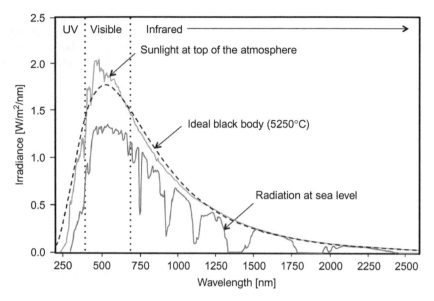

Figure 3.3.   Solar radiation spectrum for direct light at both the top of the earth's atmosphere and the sea level.

3.3.5   *Solar radiation spectrum*

Daylight is terrestrial radiation in the visible range of the electromagnetic spectrum (380–780 nm). This is the band of solar radiation with high intensity that is capable of exciting the human eye retina and producing a visual sensation (Fig. 3.3). Outside this range of the solar spectrum, the radiation is not visible to the human eye. Visible light accounts for about 65% of the total solar radiation reaching the ground. The amount of solar energy received at the earth's surface varies according to the depth and condition of the atmosphere through which light traverses. About half of this received energy is visible radiation and the rest is heat in the form of infrared (40%) and ultraviolet or shorter wavelengths (10%). Once absorbed in any surface, virtually all the radiant energy from the sun is converted to heat. The re-radiation (emission) of absorbed energy occurs in the far-IR region. The distinction between the near-IR and far-IR regions is particularly important architecturally because of the different behavior of glazing materials in these regions.

3.4   PHOTOMETRY

The maximum light sensitivity of the eye peaks at 560 nm. It decreases gradually to the boundary values of the visible light range (380–780 nm; Fig. 3.3). Since the eye has variable response to different wavelengths of light, measuring light is necessary to understand how architectural design affects light performance and occupants' perception of light. The science of measuring of light, known as photometry, has some important principles and definitions that designers need to know. These are explained below:

- *Luminous flux*: the time rate of flow of light; unit is Lumen (lm). For lighting design purposes, "luminous flux" can be considered the equivalent of "light".
- *Luminous intensity*: the luminous flux (light) in a specific direction; unit is candela (cd).
- *Illuminance*: the concentration of the luminous flux (light) striking a surface; unit is lux (or lx). Illuminance describes the light incident on a surface and is usually used as target values in building standards. It is measured by illuminance meters (or lux meters).

Table 3.1. Luminance efficacy for different light sources.

Light source	Luminance efficacy [lm/W]
Incandescent	17
Fluorescent	100
Sun (low-high altitude)	70–140
Sky (diffuse-clear)	125–140
Daylight via glazing	40–200

- *Luminance*: the luminous intensity of a surface in a given direction; the units of luminance are candelas per square meter (cd/m^2). Luminance is the measurable equivalent of "brightness"; which is different from *brightness* – the subjective perception of luminance. It is the characteristic for measuring the image seen of an illuminated surface or space. Luminance meters can be used to give point measurement. Luminance cameras (high dynamic range photographs) are used to give a comprehensive survey.
- *Luminance efficacy*: any light source introduces heat into a building in the process of introducing light. Luminance efficacy is the ratio of lumens emitted to watts of heat introduced into a building by daylighting (or watts consumed by an electric lamp). It is a measure of the luminous efficiency of a light source. It helps to evaluate the energy demand and sustainability of light sources including daylight sources (i.e., sun and sky or diffuse sky – see Table 3.1). For example, incandescent lamps deliver only 10–17 lm/W compared to fluorescent or light emitting diodes (LEDs) that deliver 75–100 lm/W. The sun can deliver 70–140 lm/W depending on the altitude (high altitude has higher efficacy). The sky can deliver 125–140 lm/W depending on the sky conditions (clear sky has higher efficacy compared to diffuse sky). These values show how daylight sources can be much more sustainable compared to the artificial lighting sources; which is attributable to lower heat gains and CO_2 emissions when using daylight sources.

3.5   DAYLIGHTING CHALLENGES

3.5.1   *Urban regulations and daylight access*

The value of a building depends directly on the advantages given by its site. One of the important factors that improve a building's value is its visual connection to natural light and attracting views such as landscape and bodies of water. If a building has limited access to natural lighting due to congested site, this lowers considerably its value. Daylighting design is influenced greatly by the prescribed design guidelines and data standards (such as values, ratios, percentages, or ranges) given in building regulations to legalize urban density and zoning. For instance, the regulations that control the size of urban spaces and building heights and forms affect daylight access to buildings and buildings' overshadowing of neighboring sites.

3.5.2   *Human needs for daylight versus building energy efficiency*

Daylight is necessary for human health, comfort, and performance as discussed in Section 3.2. It also contributes greatly in improving energy efficiency and sustainability levels. Transparency and abundance of daylight access achieved by large glazed facades are attractive features in contemporary architecture. Yet, one just cannot keep increasing the glazing area on the building facade as it can quickly lead to several problems such as visual discomfort caused by too much and direct light, thermal discomfort caused by direct solar radiation transmitted through the glass, or increased winter heat losses or summer heat gains due to high conductance of glass (i.e., poor energy conservation factor). Thus, the daylighting design should not be understood as just a solution of the simple problem of providing the required lux level on the task plane. It is rather an integrated solution of the complex problem involving many factors: daylight availability, visual comfort, light quality, climate, building function and orientation, building materials, and artificial lighting, as well as building mechanical/electrical systems (such as HVAC, BMS, and energy production systems).

3.5.3   Energy efficiency, daylight requirement, and building profits

Another important issue is how the building design affects energy consumption and space efficiency. Improving the thermal conservation of the building envelope might require minimizing the window-to-wall ratio (*WWR*) due to the high U-value of the glazing compared to the solid insulated walls. This can easily lead to increased proportion of energy needed for electric lighting and the total energy consumption. In such cases, considering daylighting as part of the design is a rational choice as it could help to reduce electrical loads as well as construction and operational costs. Yet, designing proper daylit spaces requires smaller depth of floor plans; which might conflict with the design goal of maximizing the rentable floor area to improve building profits that requires deeper floor plans.

3.5.4   Integration with electric lighting

Electric lighting in commercial buildings can be accountable for at least 25%, and can go up to 60%, of the electrical energy used. Reducing energy consumption of electric lighting requires developing designs that rely mostly on daylighting. However, to achieve this, daylighting design and electric lighting design must be integrated. This is a challenge as these two tasks in most projects have always been done by separate design professionals or teams: the architect is in charge of the daylighting design while the electrical engineer or the lighting engineer is in charge of the electric lighting design. The beauty in today's technologies is in the advanced control systems that can synchronize between different building systems and operations (e.g., daylighting and electric lighting systems). For a daylighting system to be most effective in enhancing the sustainability of a building, some kind of coordination with the system controlling the electric lighting is needed. Such a system redirects the natural light into deeper space using movable louvers of a lightshelf, blocks direct sunlight, dims the electric lighting gradually with increasing amounts of daylight, turns off the lights when daylight alone is sufficient or no-one is present in the space, and switches the lights on when the occupant is in the space and electric lighting is needed.

3.5.5   Need for clearly defined design targets

Designers frequently find themselves facing a stringent requirement of "designing a well daylit building" without being given a clear and specific definition of what this statement really means. Without having clear design/performance targets at the outset of the design process against which designers could evaluate their designs and judge to what extent their developed ideas are good or bad, it would be impossible to achieve the required design goals or clients' requirements. Sometimes conflicting targets is the main issue especially when the designers are not real experts in daylighting design. Just to show this problem, some examples of tricky design requirements/goals can be given:

- "A minimum task lighting requirement of 500 lx with no glare". If the requirement for glare is not specified in terms of a clear design target (what value, which position, which angle, which direction, etc.), it would be difficult to evaluate it and thus find a trade-off point between task lighting and glare requirements. Perhaps if the requirement for the minimum task lighting was 300 lx, it would be more realistic to achieve both but this cannot be known unless both targets are well defined. In such a case, the design stakeholders (including the design team with the daylighting expert) can agree on what could be the best and more realistic targets for both, taking into consideration the site factors, the climate changes and sky conditions (clear, partly cloudy, overcast), the occupancy schedules, the energy-savings considerations, and any other influencing factors.

- "Design an atrium in a building in Abu Dhabi with a bright, cheerful, and comfortable interior environment that helps improving the building's energy-efficiency and green building rating (e.g., 2 Pearls by the Abu Dhabi Estidama Pearl Rating System; or Silver green rating by the US Green Building Rating System)". In a desert climate with high intensity of sunlight like Abu Dhabi, such a design goal is not realistic and hence could lead easily to faulty design that has no chance for green building rating. The main problem is in the use of vague expressions and conflicting connotations (bright cheerful atrium versus visually/thermally comfortable environment with high energy efficiency). To avoid such problems, the design stakeholders must agree on clear and realistic definitions of performance requirements based on specific target values/ranges suitable for the environmental conditions of the location and site.

3.5.6   *Daylight for green retrofits and renovation projects*

The proportion of global green retrofits and renovation projects is increasing every year. Between 2012 and 2015, 50% of the sectors with the largest opportunities for green building around the world have plans for renovation projects. In the UK and Singapore, green retrofits and renovation projects are planned by the greatest number of firms at 65% and 69%, respectively (World Green Building Trend, 2006). Considering high daylight performance in renovated projects or building retrofits is more challenging than in new constructions as it usually involves major alterations to building facades and structures. However, there is a lack of design strategies and effective tools for managing and retrofitting the large number of buildings already existing, and predicting their performance. Also, carbon targets for green retrofits and renovation projects are not accessible and great efforts must be made to make it available to designers to meet the increasing green building market requirements.

3.5.7   *Appropriateness of daylight metrics*

Availability of light in spaces has been traditionally evaluated using metrics that clarify daylight performance in buildings. The most widely applied metric is daylight factor (*DF*). *DF* is a simple and widely used metric yet it has several limitations (see Section 3.6.1). A major one is that there is an assumption that the sky type under which the building is evaluated is an overcast sky with uniform light. This means that changing building location and/or orientation does not make any change in the value of *DF*; or in other words *DF* is not sensitive to the change of these design parameters. Over the past decade a set of new daylight assessment metrics have emerged which relies on climatic data of a specific location (see Section 3.6.2). They are referred to as dynamic daylight metrics or climate-based daylight metrics (CBDM) (DiLaura *et al.*, 2011; Reinhart, 2014; Reinhart *et al.*, 2006). These new metrics help to overcome the limitations previously imposed by the traditional metrics.

3.5.8   *Validity of rules of thumb*

Rules of thumb are very useful to designers because they rely on them as the sole quantitative justification for room proportions and the positioning of facade openings. They are appealing to designers for their simplicity as they do not require calculations, and for their relevance to key design decisions.

Despite the common appearance of rules of thumb in design guides, one may not find any documented scientific evidence to support them. Reinhart (2005) reviewed the validity of the ubiquitous rule of thumb that relates window-head-height to the depth of the daylit area adjacent to a facade. The study relied on information taken from prominent daylighting design guides and norms that included rules of thumb exercised in Canada (Enermodal, 2002; Robertson, 2005), Europe (Cofaigh *et al.*, 1999), Germany (DIN V 18599, 2005), North America (Rea, 2000), and the United States (O'Connor *et al.*, 1997; USDOE, 2005). Daylit zone depths of rectangular sidelit spaces were simulated using radiance for a variety of climates, facade orientations, facade geometries, and usage patterns based on the established link between the depth of the daylit zone and daylight autonomy distribution. What was concluded in the study (see Section 3.10.2) came very close to most of the reviewed empirical versions of the rule of thumb.

The most common rules of thumb and design guidelines were produced many years ago in countries such as European countries, US, and Canada. These countries have many locations that are characterized by overcast or highly cloudy skies most of the year; and thus can adopt the simple metric of *DF* to check daylight availability with high or reasonable accuracy. If these established rules of thumb were used in locations with clear skies, this can easily lead to faulty or ineffective designs (e.g., using the rule of thumb that helps to design daylighting in atria based on the *DF* concept developed for overcast skies – presented in Dekay and Brown (2014)). To overcome this problem, gaps in existing research where appropriate rules of thumb do not exist or are not effective must be tackled. The Daylighting Simulation Laboratory of the UAE University has focused on this issue. One of its projects has been investigating daylight availability in adjoining spaces of courtyards under the clear sky conditions of Abu Dhabi (Bin Dalmouk, 2015; Bin Dalmouk and Al-Sallal, 2015), using ranges of different design parameters (sizes of courtyard, building heights, *WWR*s); and another one is investigating the effect of different shading systems in courtyards under the same conditions. These researches relied on computer simulation using climate-based daylight availability metrics (see Section 3.6.2) and annual weather files; and they produced new rules of thumb more applicable to desert climates. The rules of thumb produced by these researches can be easily integrated into building codes, standards, and green building rating systems (such as the Estidama Pearl Rating System of Abu Dhabi Emirate).

3.6 DAYLIGHT PERFORMANCE METRICS

3.6.1 *Daylight factor (DF)*

Due to the fact that the daylight illuminance level at any point is always varying as a result of the continuous change in the sky conditions, specifying absolute values of daylight illuminance are often not a useful metric for design. As a ratio (i.e., a relative measure of the indoor illuminance to the outdoor illuminance), daylight factor (*DF*) is generally stable across time and therefore much more useful and usable as a design metric (Kwok and Grondzik, 2007). The *DF* is a simple and convenient metric of measuring daylighting availability. To calculate daylight factor, we need to measure/find the horizontal illuminance level at two points: one represents a point of interest in a room (E_i) and another one that represents the exterior daylight illuminance under an unobstructed and uniformly overcast sky (E_o). The ratio between the two values (E_i/E_o) is the daylight factor.

As a design metric, the *DF* can be helpful to test and promote design alternatives that help to enhance daylight levels in a space such as window-to-wall ratio (*WWR*), window arrangement, window-head heights, ceiling and wall finishes (proper reflectance values), glazing transmittance, and depth of floor plans.

However, the *DF* is mainly developed for overcast uniform sky and thus has several limitations. It is blind when it comes to the effect of building orientation and/or climate. In other words, changing the building orientation or location would not make any change in the value of the daylight factor, at least theoretically. Another disadvantage of the *DF* metric is due to the fact that it has no limitation for the maximum allowable value of lighting level; it promotes a one-dimensional, "the more the better" approach to daylighting. This could promote a tendency by the designer to increase building glazing surfaces to admit more daylighting; which could accidentally lead to unexpected issues such as creating visual angles of glare or increasing building energy consumption and CO_2 emissions.

3.6.2 *Climate-based daylight-availability metrics (CBDM)*

Daylight is variable over time. The illuminance in a space varies at different times of the day and year and depends on the sky conditions. Using the average illuminance over a period of time provides only limited value. The recently emerged annual metrics (or climate-based metrics) can present a much clearer picture on daylight performance over time including the potential for energy-savings. Annual metrics address the dynamic conditions of daylight at a specific site through representative weather data such as the TMY2 (typical meteorological year) or the EPW (energy plus weather) files that are commonly used for building heating and cooling load analysis (DiLaura *et al.*, 2011).

The basic performance annual metric used to assess the dynamic quality of daylight is daylight autonomy (*DA*). The standard definition of *DA* is the percentage of a defined period during which interior illuminance exceeds a target illumination level (DiLaura *et al.*, 2011). It is a climate-based daylight availability metric because it depends on multiple sky conditions experienced at a site throughout the year. It is used to evaluate the magnitude performance at individual analysis points and can also be used to evaluate the magnitude and general distribution of daylight across space. It is usually used as a measure of how well daylight can replace electric lighting when electric lighting is switched on via a photo sensor. It can be used to assess the number of hours (or percentage of time) when a particular light level is exceeded; this kind of application is useful in the case of illuminance on an artifact in a museum. *DA* calculations may include multiple variables such as: time period; calculation point or area; room geometry, interior finishes, furnishings, etc.; target illuminance; climate data set; and assumptions/algorithms regarding solar control and user behavior.

The concept of *DA* can be applied under different variations that use different methods and limits (DiLaura *et al.*, 2011; Tanteri, 2012). These are as follows:

- The "incremental" method is the standard approach to *DA* where illuminances that meet or exceed the minimum target illuminance count toward the final *DA*.
- The "continuous" method, such as continuous daylight autonomy (*cDA*), awards partial credit to illuminances that are less than target level, which counts toward final *cDA*. Because the continuous method accounts for a range of illuminance levels up to the target level, it is more appropriate for dimming systems.
- Maximum daylight autonomy (*mDA*) is an extension of *DA* that is generally considered to be the percentage of occupied hours that interior daylight levels are ten times or higher than the required illuminance.
- Useful daylight illuminance (*UDI*) is composed of three ranges, including the percentage of occupied hours of the year when daylight illuminance at an interior point is at useful levels (between 100 and 2000 lx), very low levels (below 100 lx), and at high levels indicating the likelihood of glare or discomfort (above 2000 lx).
- Spatial daylight autonomy (*sDA*) is defined as the percentage of space or building area that meets a given *DA* value for a set time period (or a minimum percent of the analysis year). For example, an $sDA_{300/50}$ value indicates the percentage of the space that meets or exceeds 300 lx for 50% of all the annual hours between 8 am and 6 pm. *sDA* assumes that sun control devices (blinds or shades) are operated hourly to prevent direct sun penetration into the space beyond a minimum level. Thus, an annual simulation program used to generate *sDA* needs to be able to model appropriate operation of blinds.

Table 3.2. and Figure 3.4 show daylight availability analysis based on *DA*, *cDA*, *mDA*, and *UDI* distributions for a space adjoining a 12 × 12 m courtyard in a single-floor building located in Abu Dhabi, UAE, assuming that the space is occupied on weekdays between 8 am until 5 pm (Bin Dalmouk, 2015; Bin Dalmouk and Al-Sallal, 2015).

Table 3.2. *DA*, *cDA*, *mDA*, and *UDI* for a space adjoining a 12 × 12 m courtyard in a single-floor building.

DA	cDA	mDA	UDI < 100	UDI (100–2000)	UDI > 2000
76.15%	90.83%	42.67%	3.5%	78.47%	18.06%

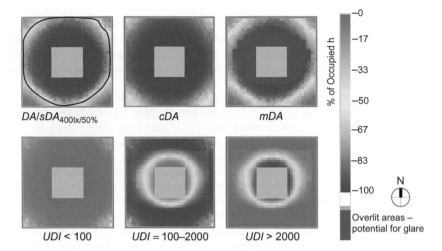

Figure 3.4.   Daylight availability analysis based on climate-based daylight availability metrics for a space adjoining a 12 × 12 m courtyard in a single-floor building located in Abu Dhabi, UAE.

Table 3.3.   Recommended total exposure limits in terms of illuminance hours per year, adapted from IESNA Lighting Handbook (Rea, 1999).

Type of materials Max illuminance	Type of materials Max illuminance [lx]	Type of materials Max illuminance [lx-h/year]
Highly susceptible displayed materials: textiles, cotton, natural fibers, furs, silk, writing inks, paper documents, lace, fugitive dyes, watercolors, wool, some minerals	50	50,000
Moderately susceptible displayed materials: textiles with stable dyes, oil paintings, wood finishes, leather, some plastics	200	480,000
Least susceptible displayed materials: metal, stone, glass, ceramic, most minerals.	Depends on exhibition situation	Max. 300

3.6.3   *Annual light exposure*

Annual light exposure is a metric that reports the cumulative amount of visible light incident on a point of interest, measured in lx-h/year. It is used with another metric (the maximum illuminance level that may fall on the point at any given time) to control light exposure on sensitive artifacts when designing art galleries and museums. CIE includes recommendations for various types of artwork materials. The International Commission on Illumination (CIE) Division 3 TC3.22 "Museum lighting and protection against radiation damage" recommends upper levels for both metrics for a variety of art forms (CIE TC 3-22, 2004). According to IESNA Lighting Handbook, the maximum illuminance levels within the room and the recommended total exposure limits in terms of lx-hours per year should remain within the limits outlined in Table 3.3 (Rea, 1999).

3.6.4   *Glare*

Human eyes cannot tolerate too much light coming from direct sunlight or reflected light. When human eyes experience intolerable amounts of light, such a problem is referred to as glare. Two types of daylight glare that should be avoided in the field of view, are disability glare caused by reflected sunlight and discomfort glare caused by high contrasts. There are several models to evaluate and predict glare from daylight (CIE, 117-1995; DiLaura *et al.*, 2011) such as daylight glare index (*DGI*), unified glare rating (*UGR*), maximum contrast ratio recommended by the Illuminating Engineering Society (IES), and the daylight glare probability (*DGP*). The *DGP* model by Weinold and Christoffersen is based on the vertical eye illuminance in addition to the glare source luminance, its solid angle, and position index (Weinold and Christoffersen, 2006). This model was recommended as it showed strong correlation with occupants' response regarding experience of glare. Creating a design for a daylit space without glare is not an easy task. Tregenza (2011) prescribed practical procedures that help to identify glare and how to know its causes and suggest solutions in workplace lighting.

3.7 MODELING THE DAYLIGHT SKY

Computer simulation programs rely on mathematical models called sky models to calculate the amount of light coming from different parts of the celestial hemisphere that establish the sky luminous distribution. The sky models normally differentiate between direct sunlight coming from the solar disc and diffuse light scattered by the sky ingredients (clouds, air molecules, and water vapor). The ratio of direct sunlight to all daylight falling onto an unshaded horizontal surface over the course of a year depends on the type of climate. In climates with clear and sunny skies such as Phoenix, Arizona, this ratio might be over 70%. In other climates with predominantly overcast skies such as London, England, this ratio can be around 40% (Reinhart, 2011).

3.7.1 *General sky models*

The oldest sky model by Waldram (1950) assumes a uniform sky without direct sunlight; which served as the sky model for the daylight factor metric. Kimball and Hand (1922) noted that cloudy day skies had highest values of luminosity near their zenith, with decreasing luminosity toward the horizon. Twenty years later, Moon and Spencer (1942) formalized the luminance distribution for the overcast sky model. In this model, sky luminance increases by a factor of three from horizon to zenith (see Fig. 3.2), with the distribution of luminance showing radial symmetry with respect to the zenith. This newer model replaced the older uniform one and became an accepted standard for luminance distribution of the overcast sky by the International Commission on Illumination (CIE) in 1955 (Nakamura *et al.*, 1985). In 1967, Kittler came up with a rigorous derivation of the clear day sky model (Kittler, 1967). This was adopted later as a standard sky luminous distribution by the CIE in 1973 CIE, 1973; (Nakamura *et al.*, 1985). Then in 1999, Kittler *et al.* (1997; 1999) introduced a universal sky model based on 15 new sky standards with reference daylight conditions and the model was adopted later by the CIE (CIE, 2003). Another general sky model, called the Perez sky model, was also introduced in the mid-1990s (Perez *et al.*, 1993). Both models attempt to cover the spectrum of intermediate and cloudy skies between the two already standardized clear and overcast sky distributions. The difference between them is that the CIE general sky model defines fifteen standard skies based on discrete parameters used to adopt the sky luminous distribution function to a range of clear, intermediate or overcast sky conditions; whereas the Perez sky model defines the sky based on a continuous change of the luminous distribution parameters. The importance of these models is that they enable characterization of the daylight conditions in any location and thus can be useful in analyzing/comparing measured data and simulating illuminance conditions using the established sky standards. The models were incorporated into standards by the International Organization for Standards (ISO) as well as the CIE (ISO/CIE, 1996).

3.7.2 *Dividing the sky*

There are many ways to divide up the sky into small segments. The CIE (1989) recommended the use of a 145 segment equal-area subdivision based on the early work by Tregenza that suggested that the optimum diameter of a sky zone was approximately 0.2 radians (11.5°) (Tregenza, 1987). The divisions were based on 8 equal altitude bands allowing each zone to be considered as approximating a point source, without noticeable error.

The concept of dividing the sky into segments was originated when researchers were searching for a numeric method that could help speed up calculations in daylighting simulation (Reinhart, 2011). The approach depends on finding the contribution of each sky segment to the total illuminance at various sensor points in a building. The method calculates a quantity called the daylight coefficient that was originally proposed by Tregenza and Waters (1983). It is the illuminance contribution, $E_\alpha(x)$, at a sensor point and orientation, x, for a sky segment, α normalized with the luminance, L_α, and angular size, ΔS_α, of the sky segment; and is governed by the following equation:

$$DC_\alpha(x) = \frac{E_\alpha(x)}{L_\alpha \Delta S_\alpha}$$

3.7.3 High dynamic range (HDR) sky luminance

The potential, limitations, and applicability of the high dynamic range (HDR) photography technique were evaluated as a luminance-mapping tool (Inanici, 2006; Inanici and Galvin, 2004). With the new advanced technologies in digital photography, HDR photographs of the sky luminous distribution can be acquired using a high-end digital camera in combination with a fisheye lens and a luminance meter. The obtained data can be used to light the simulated environment through an image based rendering (IBR) technique. The results of a previous study by Inanici (2009) showed that image based sky models can provide a more accurate and efficient method for defining the sky luminance distributions and the impact of surrounding urban fabric and vegetation as compared to generic CIE sky models and explicit modeling of surrounding urban fabric and forestry. The luminance meter in the HDR photography studies is used for calibration; which is basically improving the accuracy of HDR photographs by comparing the HDR luminance with readings taken in the real scene with a spot luminance meter. The ratio of the two may be used as a calibration factor for successive images (Jacobs, 2007). This method provides an alternative method for acquiring sky luminous distribution data normally produced by sky scanners; yet with much lower costs.

3.7.4 Dynamic simulation

A simulation method is considered static if it simulates interior or exterior lighting levels under merely a particular sky condition such as the CIE overcast sky (Reinhart, 2011). The actual daylighting environment is constantly changing and depends on the local climate conditions. Depending on only a particular sky model such as the CIE overcast cannot give a full picture of the conditions that influence daylighting in buildings. Dynamic simulation overcomes this limitation since it helps to take into account the continuous variations of the climatic conditions over the course of a day (e.g., hour by hour) and throughout an entire year (or a season). In dynamic simulation, the calculations are done based on annual illuminance profiles that use time series with a time step varying from 1 hour down to a minute. This approach is very useful to building designers because the data in these profiles:

- can be translated into climate-based daylighting metrics that consider the quantity and character of daily and seasonal variations of daylight for a given building site, and/or
- can help to predict electric lighting energy use when daylight is not sufficient and can be combined with a thermal simulation program to predict overall energy use for heating, lighting and cooling.

3.8 DAYLIGHT EVALUATION USING PHYSICAL MODELS

Another common method to evaluate daylighting is to rely on physical models that are tested under a real sky or within an artificial daylight environment simulated by the assistance of a simulation machine such as a heliodon or an artificial sky. High caliber cameras are used to take images/videos in the tested model from different view angles for perception analysis, and measurement instruments are operated to measure performance metrics (such as illuminance levels) and log readings for post-test analysis. Conceptually, any physical daylight simulation arrangement should include the following:

- A machine (or a special structure) equipped with a stationary or movable electrical lighting system that simulates the direct sunlight and/or the diffuse sky light with a manual or automatic control system.
- A physical model of the architectural design under investigation made with appropriate (large enough) scale to take photos and fit measurement devices.
- Measurement instruments and/or high caliber cameras.

Traditionally, daylight simulation machines are constructed by lighting experts such as researchers in academic institutions or designers in large firms. Simple versions of these machines can also be made when funding is tight (e.g., students taking a lighting class or conducting research in academic settings). There are also some companies that are specialized in making or constructing these kinds of simulation equipment. The best machines – and usually the most expensive – are the ones that can provide high level of precision and thus are able to produce lighting environments that are very similar to actual environments. To improve accuracy of the results and the quality of the simulated environment, this probably requires increasing the size of the machine as well as using high quality lights (with lamps' emissions similar to the actual sunlight) and advanced technologies that facilitate automation of functions and high definition visualization. However, not all applications will require a high level of sophistication, and thus simpler and lower cost facilities can still be considered. Daylighting simulation machines can be classified into three types, as explained below.

3.8.1 *Heliodon*

The heliodon is the most basic daylighting simulation machine. Its main function is to check/visualize how the beams of direct sunlight penetrate through building openings or masses into architectural indoor or outdoor spaces. Depending on its operation mechanism, the heliodon has either a fixed or a movable platform for holding the tested physical model; and changing the position of the sun is made by either rotating the platform that carries the model or the light (representing the sun). The one with the fixed platform relies on some kind of an artificial parallel light to simulate the direct sunlight and hence it is always used in the indoors (such as in a laboratory); the one with the movable platform can rely on either the real sun (and hence is used in the outdoors) or on an artificial parallel light. Although the heliodon with the movable platform can be used in the outdoors, it is not considered as the best tool to study diffuse daylight that comes from the sky vault. That is because when the model is sloped up/down to receive the sun at a desired angle, the view angle of the horizon will change (giving either less or more view of the horizon); which can lead to an inaccurate ratio of the sky-light to the ground-reflected light. The heliodon can help lighting designers study urban solar access and overshadowing of building masses on sites or on other buildings, conduct sun shading analysis on building facades, and evaluate design of shading devices and optimize their performance.

Figures 3.5–3.8 show examples of different heliodons used in the Daylighting Simulation Laboratory of the Architectural Engineering Department, UAE University. Two of these exemplar heliodons (shown in Fig. 3.5 and 3.6) use similar mechanisms to rotate the model platform around two axes (i.e., horizontal and vertical); so that the model can be positioned with regards to the source of light at the correct solar azimuth and altitude angles, based on a chosen time (date and

Figure 3.5.   Example of a simple movable platform heliodon (in-house-made) used in the Daylighting Simulation Laboratory, UAE University.

Figure 3.6.   Example of a simple movable platform heliodon used in the Daylighting Simulation Laboratory, UAE University.

hour) and location (latitude). The user can use the actual sun or an electric luminaire that produces parallel light (similar to sunlight) as the source of light. To select the desired time of testing (date and hour), the user can depend on the time/latitude dials integrated with the heliodon; or use the Sun Dial chart for the geographical location of the tested building (i.e., the correct chart for the latitude) which can be attached on the model platform (or base). The sun dial has a peg that shows the time by the position of its shadow.

Figure 3.7 shows an example of a heliodon with a fixed-platform and movable lights. This heliodon is easy to understand as it replicates the real-life experience of the sun moving through the

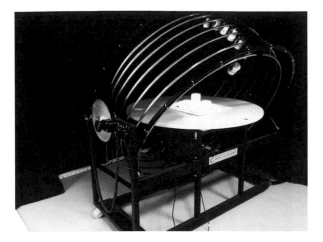

Figure 3.7.   Example of a fixed-platform with movable lights heliodon used in the Daylighting Simulation Laboratory, UAE University.

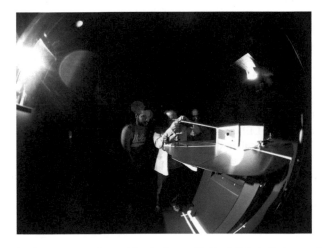

Figure 3.8.   Example of an automated movable-platform with fixed light heliodon used in the Daylighting Simulation Laboratory, UAE University; the author demonstrating to students how to take images inside the model with a borescope.

sky from east to west across a given site (CERES, 2015). It helps to understand the general pattern of sun and shadow movement across a site and thus it is very useful for teaching undergraduate students. It has a 1.2 m support platform for placing the tested model. The sun is simulated with seven lamp hoops (for all 12 monthly solar paths); and the hoops are manually rotated to hourly positions. There is a selector switch to choose the desired month and hence operate its lamp. The user can control the slope of the hoops manually to adjust the latitude, within 0 to 90° positioning.

Figure 3.8 shows an example of a heliodon that has an automated movable-platform with fixed light. This heliodon has a robotic table with a support platform (55 inch, ≈140 cm diameter) for positioning the scale model. It uses a halogen Fresnel light source (1 kW) with a mirror and two stabilizers that can maintain more than 650 lx illuminance on the tested model. The high uniformity of the illumination increases the quality of the video or photo shoots. It can carry a model with a maximum weight of 20 kg and maximum dimensions of $1 \times 1 \times 0.5$ m. It is controlled by software running in a laptop/computer (or iPad) through a serial terminal from a computer with a USB or serial port.

3.8.2 *Artificial sky*

Since the heliodon has a limitation in giving a reliable lighting environment for measuring the diffuse sky light component, one cannot rely on it for accurate measurement of daylight or for complex perception analysis. The concept of the artificial sky was originated to overcome this shortcoming.

The simplest (and least expensive) artificial sky is the *mirror box*. This type can only simulate the standard overcast sky; thus it is useful for daylight factor analysis. A more advanced one is the *full dome artificial sky*, which is capable of reproducing a reliable sky vault for illuminance level measurements, daylight autonomy analysis, and glare studies in any sky. It also helps to give a more realistic perception of the daylight luminous environment inside; and thus it can be useful in evaluating complex interactions of light and glare issues by the naked eyes and cameras. It is relatively costly but very useful and reliable. A third type of the artificial skies is the *virtual dome*. This type depends on a scanning process to simulate the full sky vault. It is very accurate and can reduce cost and consumption of space, compared to the full dome. However, it is not designed to provide direct perception (by the naked eyes) of the daylight environment as the results can only be evaluated through a computer screen.

In 2013, the Daylighting Simulation Laboratory of the UAE University constructed a full dome *artificial sky* with an integrated heliodon. It is a full dome type (Fig. 3.9). Running the simulation and measurement of daylight metrics is controlled through computer programming. It has the ability to reproduce any type of sky with dimmable customized LED luminaires; based on the fifteen standard skies of CIE sky model (CIE, 2003) as well as the continuous Perez sky model (Perez *et al.*, 1993), explained in Section 3.7.1. Each luminous element produces a luminous flux of 6508 lm with optimized LED arrangement that increases the color rendition index (CRI). It also has the capability to simulate direct sunlight at any location on earth using an integrated automatic heliodon on a robotic arm. The sun is simulated with a high quality lamp capable of producing an emission spectrum very similar to the actual sunlight. This advanced simulator is currently assisting students and researchers in optimizing daylighting design in architectural spaces and conducting advanced daylighting studies.

3.8.3 *Solar simulator*

Solar simulators have been used widely for quite some time in industrial applications such as developing coatings that improve performance of photovoltaic panels, plastics and other materials that resist solar radiation adverse effects, and cosmetics creams for sun protection. Nowadays, solar simulators can also be seen in advanced research facilities serving architectural and building technology applications. The great benefit of the solar simulator is its ability to reproduce a near collimated beam of light with a spectrum similar to the direct solar radiation received on earth. This helps researchers to study complex fenestrations/cladding systems and to examine how incident sunlight wavelength can affect the characteristics of a material. With this powerful option, one can examine for example how a particular coating material or thin film can influence the transmission and reflection of sunlight. Another interesting research that employs the power of the solar simulator installed in the Daylighting Simulation Laboratory of the UAE University (Fig. 3.10) is investigating how sun shading by plants affects daylight performance in the indoor space. This simulator, made by BetaNit-Nitter (Bnit, 2014), depends on a luminaire capable of producing a collimated light beam similar to the direct sunlight and a very refined special-made concave mirror. The produced luminance is estimated at around $6,000,000$ cd/m^2 at full aperture.

Figure 3.9.   The full dome artificial sky with an integrated robotic heliodon constructed in the Daylighting
Simulation Laboratory, UAE University; the author is demonstrating to a student how to take
measurements using photometric sensors inside the model.

3.9   DESIGN PROCESS

3.9.1   *Site and weather analysis*

It is very difficult to make a good design for a building that relies on daylight as a main source,
if the area of its skin is limited compared to its volume or if it is highly overshadowed by other
neighboring buildings. Daylight availability is varying all the time. Winter days are short while

Figure 3.10.   The Solar Simulator installed in the Daylighting Simulation Laboratory, UAE University; testing a model using the Solar Simulator and the movable platform in-house made heliodon.

summer days are long. Some days have clear skies while other days have partly cloudy or overcast skies. As a start, a site visit is necessary to experience the context in which a building will be placed.

To guarantee good lighting, the designer needs to survey the site with regards to solar and daylight access. Determining the times when direct sunlight is blocked from reaching any point on the site helps to assess the site's exposure to direct sunlight. A tool called a horizon obstruction diagram is used to do this kind of analysis (Harris *et al.*, 1985). It requires plotting the "skyline" directly onto a sun path diagram for the appropriate latitude with the aid of a transit or theodolite. Anything blocking the sun such as buildings, landforms surrounding the site, trees and other vegetation should be considered. The analysis can be done manually by overlaying a sun path of the location on a fisheye image of the site or by using a simulation software such as Ecotect (Autodesk Ecotect, 2008; Marsh, 2005). Also, there are advanced devices for solar surveying available on the market which can do the solar obstruction analysis with ease such as the Solar Pathfinder and the SunEye 210 Shade Tool (Solar Pathfinder, 2015; Solmetric-SunEye, 2015). The Solar Pathfinder provides a panoramic reflection of a site with overlaid local sun-paths, providing a full year of accurate solar/shade data (see Fig. 3.11). The SunEye 210 is a hand-held electronic tool that assesses the available solar energy by day, month, and year with the press of a button by measuring the shading patterns of a particular site. It provides views of the annual sun path, monthly solar access, and obstruction elevation; and has a live survey mode in which a person can view annual sun paths live as he/she scans the site.

The designer also needs to conduct detailed weather analysis for solar radiation and daylight levels at different dates of the year and on an hourly basis. He/she should depend on reliable information in order to accurately predict exterior and interior surface illuminances on a typical day. He/she can depend on information available in official weather stations such as those found in airports but it is also important to account for differences in weather conditions between the weather station and the building site (caused by other buildings and terrain) by making any required adjustments. The designer should also assess how the topography and context of the site influence other environmental control parameters of the building, e.g., heating and cooling, as well as wind and moisture.

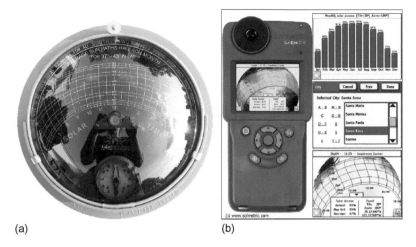

(a)   (b)

Figure 3.11. Examples of solar surveying tools: (a) Solar Pathfinder and (b) SunEye (source: Solar Pathfinder, 2015; Solmetric-SunEye, 2015).

3.9.2 *Programing and project brief*

It is essential that the designer assesses the kinds of tasks to be performed in the building and the requirements associated with these tasks. This might require conducting surveys and gathering information about the building users and client needs, preferences and constraints. Daylighting design in particular, must take into consideration many factors in order to produce a high quality visual environment with the required amount of light and in an architecturally appealing way. It is also important to provide a data sheet for the spaces that will be designed to serve as a reference against which the final design can be judged. The design brief should state clear goals that can tackle one or more of the following issues:

- Visual comfort and satisfaction of occupants.
- Visual and perception needs depending on the age and tasks to be performed.
- Visual access to outside views.
- Performance targets for amount of daylighting.
- Performance targets for energy use and/or carbon emissions.
- Effects of daylight exposure on valuable items such as historical documents or artifacts.
- Ease of use of lighting systems operation or adjustment.
- Safety and security.
- Maintenance of systems.
- Budget constraints.

3.9.3 *Conceptual design*

Once the project stakeholders have materialized a program or project brief, the designers typically engage in schematic design activities during which the design concept is developed. Typical design decisions made during the conceptual design such as the building location in the site, building form and orientation, spatial zoning, and the windows distribution (locations and areas) have great effect on the luminous quality and quantity provided to the interior of a building by daylighting. Daylighting has great potential to significantly improve building sustainability (e.g., energy use, material selection, HVAC system sizing, etc.) and occupants' performance and health, thus integrating it in the early design process and giving priority for the issues that can influence its performance is very important.

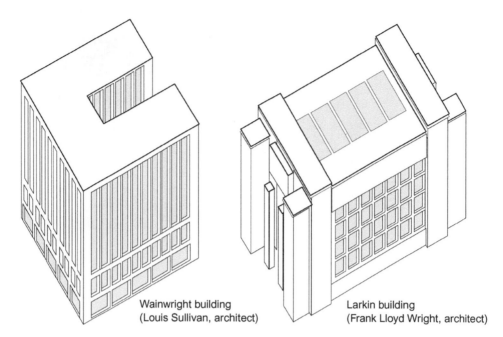

Figure 3.12.   Wainwright Building notched plan allows sidelight and ventilation penetration (left); Larkin Building uses a top-glazed atrium to bring toplight in addition to windows that bring sidelighting.

3.9.4   Building orientation and massing

Building orientation and massing can influence greatly the amount and quality of daylight admitted to the building especially in locations having mostly clear skies. Differences in orientation will affect the amount of daylight available to actually enter the space. When the sky is clear or partly cloudy, a building oriented to north in the northern hemisphere (or south in the southern hemisphere) rarely receives direct sun except for very limited time after sunrise or before sunset in the summer season. If it is oriented to south, east, or west, the direct sun entering through the windows into the building spaces can create several problems such as glare and overheating; thus shading is a necessity especially in hot dry climates. Shading can be provided by the proper choice of building form and massing. Form and massing of a building can influence greatly the amount and quality of daylight admitted to the building especially in locations having mostly clear skies. There are also some methods that can help improve capturing of daylighting such as the use of recessed facades and the use of light-colored room surfaces.

Access to daylighting increases as the surface-to-volume ratio (S/V ratio) increases. Some floor plan configurations such as the finger-elongated or courtyard/atrium floor plans (explained below) lend themselves as effective solutions for obtaining bilateral sidelighting. They can also be useful in promoting natural ventilation and providing visual access to landscape and other views. Such configurations in the past (before the 1970s) were the only available option to naturally illuminate office and other work or commercial spaces before the invention of HVAC systems. Figure 3.12 shows two well-known case studies that represent this era; these are the Wainwright Building by architect Louis Sullivan and the Larkin Building by architect Frank Lloyd Wright. The former building has a notched plan and hence allows side light and ventilation penetration and the latter uses a top-glazed atrium to bring toplighting in addition to windows that bring sidelighting (Moore, 1985; 1993; Ternoey *et al.*, 1985). As a general rule of thumb, a floor plan with increased S/V ratio, such as the finger-elongated floor plan or the courtyard/atrium floor plan, will provide

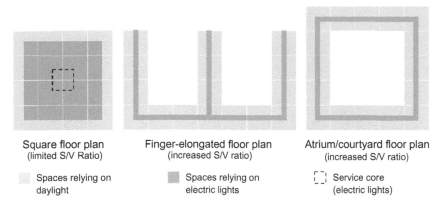

Square floor plan	Finger-elongated floor plan	Atrium/courtyard floor plan
(limited S/V Ratio)	(increased S/V ratio)	(increased S/V ratio)

Spaces relying on daylight

Spaces relying on electric lights

Service core (electric lights)

Figure 3.13. Comparison between three design configurations with equal floor area showing how a floor plan with increased surface-to-volume ratio can provide better access to daylighting.

better access to daylighting than another one with limited *S/V* Ratio. Figure 3.13 demonstrates this fact by comparing between three different design configurations with equal floor area.

Finger elongated floor plans: before the invention of air conditioning technology, finger elongated floor plans used to be a very common practice in designing multistory buildings. The main reasons were to provide natural lighting and ventilation through the operable windows. The depth of the floor plan was limited to avoid dark areas. A narrow floor plan with approximately 14 m facade-to-facade depth seems to be reasonable, as this distance can allow two rooms with a depth of 6 m separated by a 2 m central corridor. This comes from the rule of thumb – room depth is equal to 1.5 room height. Adequate daylight can easily be introduced by up to 4.60 m with conventional height windows. New technologies can redirect natural light to larger depths (up to 9 m) using for example: holographic optical elements, articulated light shelves and light pipes (see Section 3.12: Advanced Daylighting Systems). Narrowing the width of the floor plate to about 14.00 m can help to reduce artificial lighting and optimize natural lighting. In addition, a room with a height to depth ratio of 1:2 with 20% glazing of its external wall area allows good light penetration. The obstruction of fingers onto other parts of the building needs to be taken into account.

Courtyards/atria: courtyards and atria can be employed to provide toplighting, which is very effective as a deep core lighting strategy that helps to transmit daylight into deep floor plans. Courtyards and top-lighted atria can also help to transmit light through many floors to illuminate central spaces in a multistory building. These solutions can be useful in dense urban areas where daylight access is limited.

Staggered building section: a staggered building section combines sidelighting and toplighting methods (Tanteri, 2012). Many well-known daylit buildings use staggered sections to effectively capture and distribute daylighting into indoor spaces. A good case study showing this design approach is the Mount Angel Library by the famous architect Alvar Aalto in St. Benedict, near Mount Angel, Oregon built in 1970 (Fig. 3.14). The use of a staggered building section allows deep penetration of daylight and thus helps to provide greater flexibility in the layout of spaces.

3.9.5 *Design development*

During the design development stage the design team needs to go technically deeper into the design and project details to answer many questions related to daylight and other related systems. The process requires undertaking several actions and relying on a multitude of design and performance evaluation methods including traditional design techniques, computer simulation, testing architectural models under simulated environments (e.g., heliodons, artificial skies, solar

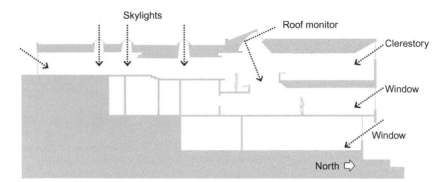

Figure 3.14.   Staggered building section of Mount Angel Library (Mt. Angel, Oregon, architect: Alvar
Aalto) helps to capture daylight and distribute it effectively into indoor spaces.

simulators), measurement and performance improvement. Examples of the performed actions
can be listed as follows:

- Determining the exact location and sizing of fenestration openings.
- Designing the needed shading devices for sun control and diffusion of sunlight.
- Selecting the glazing technologies that will help to control sunlight, glare, sun heat, and divert daylight.
- Selecting color and reflectivity of room surfaces, which depends on the function of the designed space, the activities performed by its occupants, and the atmosphere that needs to be provided.
- Integrating between natural lighting and electrical lighting and implementing devices for lighting control operation (e.g., dimming and/or light switching) based on spatial planning and zoning of the interior spaces.
- Coordinating of light sources with regards to appearance (e.g., color rendering, correlated color and temperature, etc.) and operation actions (e.g., dimming capabilities).
- Analyzing any potential for glare and visual discomfort and taking the required measures to eliminate these problems.
- Coordination between building lighting systems (i.e., daylighting and electrical lights) and other building systems designed for cooling, heating, ventilation, and power generation (including those depending on renewable energy).

3.10   DAYLIGHTING STRATEGIES AND RULES OF THUMB

There are two general strategies to illuminate buildings' indoors with daylight: sidelighting and toplighting. Buildings with a single floor and the top floor in multistory buildings can be illuminated by any one of these two methods. The other floors in a multistory building have less potential solutions as they can be illuminated by sidelighting systems that only cover a limited depth. Optimizing the window design with regards to their location, height, and reflecting methods can help to greatly improve space illumination.

Daylight design strategies have been covered by a great deal of literature such as Ruck *et al.* (2000), Steffy (2008), DiLaura *et al.* (2011), and Tanteri (2012). The discussions below summarize the most important material needed by daylight designers.

3.10.1   *Sidelighting*

Sidelighting is a perimeter daylighting strategy that collects natural light along a building's facade and distributes it inward. After the light is admitted through glazed vertical apertures or windows,

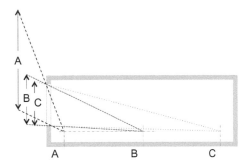

Figure 3.15. Effect of room depth on daylight design. A represents a better design alternative than B and C.

the lateral flow of daylight is directed by means of room geometry, surface reflectance and finish, and interior partitions and furniture. Sidelighting may be admitted from one side (unilateral) or two sides (bilateral). Examples of sidelighting are continuous fenestration and curtain wall construction. It is recommended to limit the distance from the window wall to the inner back wall to twice the window head height with clear glazing to avoid possible visual discomfort that can be caused by large variations in daylight illuminances (i.e., greater than 25:1). Seeing a portion of the sky through the windows is sometimes desirable; which can be achieved by placement of the window head close to the ceiling. Yet, this should be done with careful attention to the potential for glare and direct sun, and how these issues must be addressed by design strategies.

The best way to understand how a room depth can influence daylighting design is to compare between points with different depths in a room, as shown in Figure 3.15. The portion of the sky that can be seen from the deeper location (such as point C or B) is smaller than that of a shallower one (such as point A). Also, when the incident sky light comes from a lower elevation of the sky (which is the case in deep locations), it has a lower luminance since the ratio of the sky luminance from zenith to horizon is 3:1 for continuously overcast conditions. One should consider adding another window wall (i.e., bilateral sidelighting) if the depth of the designed room exceeds the values recommended by the rules of thumb.

Bilateral sidelighting can assist increasing the amount of horizontal daylit area and supply a more uniform level of indoor illuminance. This strategy captures additional lighting from apertures located on a wall opposite to the primary window wall, which can be effective in deep spaces. Since the higher window head height results in increased daylight penetration, the second set of windows can be placed in a very high level of the wall and perhaps with a sloped ceiling. However, this position makes it more exposed to the sun, which necessitates glare control.

The size and location of a window has a great effect on the amount and distribution of daylight inside rooms. The top part of Figure 3.16 shows the extent of the sky that can be seen from a point located in the middle of a room through two windows with different sizes – A is larger than B. The larger window A can allow the designated point in the middle of the room to see a larger portion of the sky and thus can bring more daylight into the room (i.e., coming from a larger area of the sky dome). In a similar way, the lower part of Figure 3.16 compares two windows that have equal areas but are positioned in two different locations – A is higher than B. The higher window A can allow the designated point in the middle of the room to see a brighter portion of an overcast sky and thus can bring more daylight into the room (that is because the ratio of the sky luminance from zenith to horizon is 3:1 for continuously overcast conditions). In clear skies, the ratio of the sky luminance from zenith to horizon is 1:3 (i.e., opposite of overcast sky), and thus position B can bring more daylight into the room than position A; but this is true only if the window is not shaded by other architectural masses or neighboring buildings.

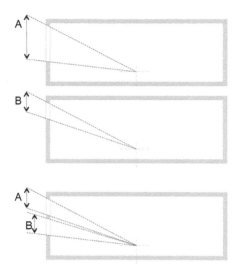

Figure 3.16.   Effect of window size (top) and window location (bottom) on daylight design. A represents a better design alternative than B.

3.10.2   *Sidelighting rule of thumb*

A significant reduction of daylight level results as the distance from the window to a horizontal work plane is increased. One can envisage the space as having two daylit zones (Tanteri, 2012) – a primary and a secondary sidelit zone:

- The average depth of the primary zone is about $1\times$ the window head height, or about 10–15′ ($\approx$3.05–4.57 m).
- The typical depth of the secondary zone starts from $1\times$ and extends to $2\times$ the window head height, continuing from the boundary of the primary daylight zone another 15′ ($\approx$4.57 m).

The division of the space into two zones helps to match the illuminance requirements of various space functions with the spatial distribution of daylight. Hence, the high daylight levels of the primary zone should be more efficiently used by regularly occupied spaces with critical task light levels (e.g., offices and conference rooms). The lower daylight levels of the secondary zone are more suitable for secondary task areas such as circulation and copy rooms. Any area beyond the secondary daylight zone should be considered as a low-lit zone suitable for low use spaces such as storage, mechanical and core elements. To save energy consumed by electrical lighting, which represents a high percentage of the total energy in commercial and office buildings, some building codes require electrical lighting controls to be used in primary, and in some cases secondary, daylight zones.

For adequate daylighting (based on daylight factor $= 1$–2% under overcast conditions) the recommendations (Fig. 3.17) for an effective daylight depth are as follows (Mueller, 2013; Tanteri, 2012):

- For unilateral sidelighting, depth $= 2\times$ window head height above desktop; light shelves can extend depth up to $2.25\times$.
- For bilateral sidelighting, depth $= 5\times$ window head height above desktop.

Reinhart (2005) used simulation to examine the validity of the ubiquitous daylighting rule of thumb (DRT) that related window-head-height to the depth of the daylit area adjacent to a facade

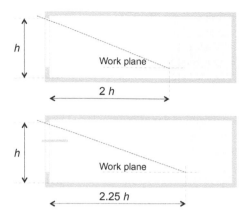

Figure 3.17.   Depth of daylight penetration equals $2\times$ the window head height h (top) or equals $2.25\times$ window head height h if light shelf is used (bottom).

(see Section 3.5.8). The study concluded with a version of the rule of thumb that was very close to most of the empirical versions:

- In a sidelit space with a standard window and venetian blinds, the depth of the daylit area usually lies between 1 and 2 times the size of the window-head-height. The exact number for a particular space is largely influenced by the glazing type and the target illuminance level in the space.
- In case a space does not require the use of a shading device, the ratio range can increase up to 2.5.

3.10.3   *Toplighting*

In toplighting, the daylight is admitted through top light apertures (e.g., skylights) at or near to the plane of the ceiling and goes along a vertical path to reach the task surface. The daylight can also be collected throughout an entire glazed roof (as in top-lit atriums) and can be distributed uniformly using diffuse glass to light a work or gallery space located on a single or top floor building. Toplighting is very effective as a deep core lighting strategy that helps to transmit daylight a large distance from collection to distribution point. Such a strategy can be used to transmit light through many floors to illuminate central spaces in a multistory building when sidelighting might not be possible or perhaps limited. It provides solutions for multistory and high-rise buildings in dense urban areas where daylight access is obstructed from the buildings' sides at varying levels and orientations; but is maintained through the rooftop.

Traditional toplighting devices include several types. These are as follows:

- *Skylights*: skylights are the most widespread system of toplighting. Skylights have many variable glazing geometries such as domes, pyramids, polygons, and other forms. They may use different methods to control heat and glare such as louvers, diffusion panels, or translucent glazing. Some forms of skylights such as the 60°-splayed with matte-finish wells can provide uniform, glare-free daylight distribution.
- *Monitors*: monitors are usually used to daylight large spaces such as airports and industrial buildings. Geometrically, a monitor can be formed by splitting part of a roof (a closed shape, not just a line) and raising it to a higher level with vertical glazing placed in between the two levels to admit daylight. The roofline of the upper level can be extended to shade the glass of the monitor. The interior daylight levels can be increased significantly by adding a high reflectance roof surface to the lower level. The simplest form of the roof monitor has a linear rectangular shape with two sides of glazing across the span of a space.

- *Clerestories*: similar to a monitor, a clerestory can also be formed by splitting a roof into two levels and placing vertical glazing in between the levels to admit daylight. However, the difference between them is in the geometry that each one takes when it is formed. In a clerestory case, the splitting of the roof is done along a line while in a monitor it is done along a closed shape or area. A clerestory can help to overcome the limited daylight penetration of unilateral sidelighting because of its location in the space (i.e., closer to the middle) and its height (i.e., at a higher point of light distribution).
- *Sawtooths*: sawtooths are repetitive clerestory/monitor elements that provide general illumination of large open floor areas such as those often found in industrial type spaces. They usually face north or south and are combined with other features that help to control glare and direct sun access.
- *Atria*: atria are usually used in large public and commercial buildings such as hotels, banks, office buildings, and museums. They can be designed in a variety of geometries and roof forms, e.g., ridge, shed, pyramid and dome. The ratio of glazing to floor area is usually very high in atriums. This requires careful attention to control problems caused by intense direct sunlight by means of shading or minimizing visible transmittance of glazing.

3.10.4   *Toplighting rule of thumb*

Skylight:

The general rule of thumb is to make the daylight area below a skylight equal to the aperture area extended 0.7 ceiling height (*CH*). However, when limitations exist such as a partition or other daylighting systems, some modification of this rule needs to be applied. The rule of thumb as stated in two design standards is described below (ANSI/ASHRAE/IES 90.1-2010, 2010; ASHRAE 189.1-2009, 2009; LCA, 2015). Note that *CH* = ceiling height (floor to ceiling), and *OH* = obstruction height – the height of permanent obstructions (to daylight distribution) such as walls and permanent storage stacks (Fig. 3.18).

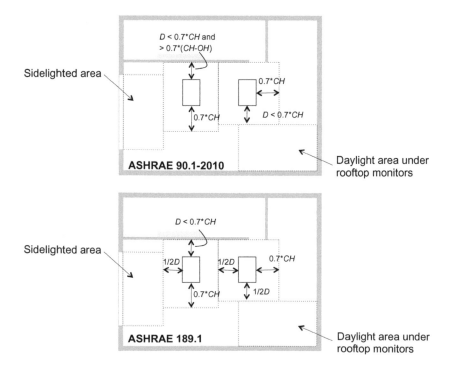

Figure 3.18.   Skylight design rule of thumb.

ASHRAE/IES 90.1-2010:

- Length = skylight length + $(0.7 \times CH)$; width = skylight width + $(0.7 \times CH)$
- If horizontal distance from skylight edge to outermost boundary of primary sidelit areas and daylight areas under rooftop monitors is < $(0.7 \times CH)$, then limit to edge of other daylight zones.
- If horizontal distance from skylight edge to permanent vertical obstruction > $(CH\text{-}OH)$ and < $(0.7 \times CH)$, then to front of obstruction.
- Adjacent zones are considered a single daylight zone, without double counting overlapping areas.

ASHRAE 189.1-2009:

- Length = skylight length + $(0.7 \times CH)$; width = skylight width + $(0.7 \times CH)$
- If horizontal distance from skylight edge to other skylight or vertical fenestration < CH, then limit to 1/2 distance.
- If horizontal distance from skylight edge to 56 inch ($\approx$142 cm) or taller permanent partition < $(0.7 \times CH)$, then limit to front of partition.

Roof monitors:

ASHRAE/IES 90.1-2010:

- Width = same as width of vertical glazing above the ceiling; depth = 1 monitor sill height (MSH) – MSH is the vertical distance from floor to bottom edge of monitor glazing.
- If horizontal distance from glazing to permanent vertical obstruction > $(MSH\text{-}OH)$ and < MSH, then limit to front of obstruction.
- If horizontal distance from glazing to outermost boundary of primary sidelit areas < MSH, then limit to edge of sidelit daylight zone.
- Adjacent zones are considered a single daylight zone, without double counting overlapping areas.

ASHRAE 189.1-2009:

- Length = length of rough opening of the roof monitor + $(0.7 \times CH)$; width = width of rough opening of the roof monitor + $(0.7 \times CH)$.
- If horizontal distance from roof monitor to skylight, roof monitor, clerestory window or vertical fenestration < $(0.7 \times CH)$, then limit to ½ distance.
- If horizontal distance from roof monitor to 56 inch ($\approx$142 cm) or taller permanent partition < $(0.7 \times CH)$, then limit to front of partition.

Clerestory:

ASHRAE/IES 90.1-2010, ASHRAE 189.1-2009:

- Width = width of rough opening of the clerestory + $(0.7 \times CH)$; depth = $0.7 \times CH$.
- Limit to width: if horizontal distance from clerestory to skylight, roof monitor, clerestory window or vertical fenestration < $(0.7 \times CH)$, then limit to ½ distance.

- Limit to depth: if horizontal distance from clerestory to skylight or window $< (0.7 \times CH)$, then limit depth to ½ distance.
- If horizontal distance from clerestory to 56 inch ($\approx$142 cm) or taller permanent partition $< (0.7 \times CH)$, then limit to front of partition.

3.11   DAYLIGHTING APERTURE DESIGN

Different design configurations of size, location, orientation, and geometry of the daylight aper- tures produce very different indoor luminous environments. The more efficient aperture is the one that confronts the largest proportion of the sky as well as receives most of the sky luminance from the zenith direction. Thus, toplighting systems with horizontal glazing (such as skylights) are the most efficient apertures because they face the full sky hemisphere with an exposure angle of 180°, through which a very high proportion of the sky luminance comes from the zenith direction (3 times higher than the horizon direction). In a previous study, Fischer (1985) investigated how different window and skylight systems could vary greatly in aperture size to achieve a daylight factor requirement of 5% in an identical test room. The results showed that to attain a requirement of $DF = 5\%$ in the test room, the skylight aperture with horizontal glazing needs to be 20% of the floor area compared to 50–60% for vertical glazing openings in walls or roofs. However, skylights are not as common as windows because they cannot be used in multistory buildings except for the top floor and they provide very limited view, which is the sky.

For vertical windows, choosing proper orientation is very important. In the northern hemi- sphere, the north direction is best for office and educational activities involving tasks that require uniform and diffuse lighting. The south direction can also be a proper window orientation pro- vided that the window is protected from the direct sun with horizontal shading devices. East and west directions are the least suitable orientations for activities involving tasks that require uniform and diffuse lighting; and they must be shaded by either vertical or egg-crate shading devices. The higher the windows are located, the more illumination can be received toward the middle and deep areas of the room. Windows located below the task level cannot help in contributing proper illumination to the room; yet they could be useful in providing access to views in tall buildings. In thick walls, it is recommended to slant the windows with the inner opening larger than the outer one in order to help reflect more light into the interior. In hot climates with clear skies, one can do the opposite to shade the window and control glare. The window should be sized to provide sufficient daylighting and fully glazed facades should be avoided. If the window to wall area (WWR) is more than 40–60%, openings tend to create glare and solar heat problems.

The light transmittance of the daylight windows and skylights depends on the type and area of the casement frames. It is essential to have high light transmittance and true color rendering for proper daylighting. Other requirements that are important for sustainability are the thermal insulation and solar control of the window.

3.12   ADVANCED DAYLIGHTING SYSTEMS

3.12.1   *Strategies for advanced systems*

Developing an effective advanced daylight system design requires an approach that differentiates between the two known sources of natural light (the sky light and the direct sunlight). Each one of these sources has different characteristics and therefore requires different mechanisms when dealing with them. The sections below present discussions about the needed strategies for the sky light (cloudy and clear skies) and the direct sunlight based on work carried out by the Solar Heating and Cooling (SHC) Programme of the International Energy Agency (IEA) under IEA's Task 21, Energy Conservation in Buildings & Community Systems, Programme Annex 29,

Subtask A: Performance Evaluation of Daylighting Systems; and published by the Lawrence Berkeley National Laboratory (Ruck, 2004; Ruck *et al.*, 2000). The reader can find illustrations of the systems discussed here in Section 3.12.2 and in Ruck *et al.* (2000) and Ruck (2004).

3.12.1.1 *Strategies for sky light*

To effectively employ light coming from the sky (i.e., sky light) for illuminating a space, direct sunlight has to be managed properly since it can cause two different problems (i.e., thermal discomfort and visual discomfort). While the thermal problem is managed by solar shading that protects from direct sunlight, the visual problem is managed by glare protection that moderates high luminances in the field of view.

Strategies for cloudy skies: in predominantly cloudy skies, the daylight openings need to be relatively large with wall openings in high positions in order to illuminate the middle and back areas of a space. Yet, the designer must avoid fully glazed walls and large diffused windows since these openings can cause high glare and visual discomfort. Using large roof openings (such as skylights) can be very effective but when used in climates with direct sunlight, shading must be considered. Advanced daylighting systems such as the anidolic ceiling can enhance daylight penetration in cloudy sky conditions. It can also be designed to control sunlight by adding an external blind device.

Strategies for clear skies: unlike cloudy skies, clear skies have high levels of diffuse skylight combined with direct sunlight. Openings should be designed with respect to the available diffuse daylight levels. Shading systems that protect from direct sunlight, while allowing penetration of diffuse skylight, are applicable such as the anidolic zenithal opening or louvers and venetian blinds.

3.12.1.2 *Strategies for direct sunlight*

Incident sunlight on a building aperture is usually so bright that it can be used to provide adequate daylight levels in large indoor spaces. When there is enough strong sunshine (sunshine probability > 60%), beam daylighting strategies are applicable. Since direct sunlight travels in parallel rays, this allows it to be guided or piped relatively easily. This can be done by using optical systems for direct light guiding and systems for light-transport. Apertures designed for the redirection of sunlight do not usually provide visual connection to outdoor views; therefore such systems need to be combined with other openings providing connection to external views if this is needed.

3.12.2 *Advanced daylighting systems categories*

There are two main groups of daylighting systems: those with and without shading. Daylighting systems with shading use advanced functions developed to protect the area near the window from direct sunlight while sending direct and/or diffuse daylight into the interior of the room. They may tackle other daylighting issues too, such as protection from glare and redirection of direct or diffuse daylight. The systems with shading have two types:

- Systems that reject direct sunlight and rely mainly on diffuse sky light.
- Systems that use direct sunlight by sending it onto the ceiling or to locations above eye height.

Daylighting systems without shading are designed primarily to redirect daylight to areas away from a window or skylight opening. These may or may not block direct sunlight. These systems include four categories:

- Diffuse light-guiding systems redirect daylight from specific areas of the sky vault to the interior of the room. They are designed to utilize the area around the sky zenith since it is much brighter than the area close to the horizon. They can be useful in dense urban environments where tall external obstructions exist and the upper portion of the sky may be the only source of daylight.
- Direct light-guiding systems send direct sunlight to the interior of the room with proper measures to prevent or minimize effects of glare and overheating.

- Light-scattering or diffusing systems are used in toplighting apertures to produce uniform daylight distribution. They should not be used in vertical window apertures, as serious glare could result.
- Light transport systems collect and transport sunlight over long distances to the core of a building via fiber-optics or light pipes.

3.13　CONCLUSIONS

This chapter presented a review of daylight/daylighting design literature with focus on recent advances in sustainable daylight performance evaluation and design methods and technologies. Some necessary fundamentals were presented at the outset; then detailed concepts were added in gradually to build up a knowledge base useful for comprehending deeper discussions that came later in the chapter about daylight metrics, daylight performance modeling and evaluation, and daylighting strategies and advanced systems.

The chapter also devoted a section to discussing current daylighting challenges. This included challenges at the application level with focus on design sustainability topics such as energy efficiency, integration of daylight with electric light, and daylight for green retrofits and renovation projects. Other important challenges discussed in the chapter were mainly derived from lack of research in developing daylight metrics and rules of thumb (or lack of applying newly developed research) that can work reasonably accurately for locations under clear and partly cloudy skies.

The point-in-time metrics, such as *DF*, are greatly used by architects and designers to size building spaces and design apertures for effective daylight admittance. These metrics work correctly under certain conditions (i.e., in overcast sky climates); and when used in clear sky conditions they can lead to inaccurate results. Due to its ability to simulate different climatic conditions through the whole year, CBDM is highly recommended for daylight studies especially in clear sky locations.

The chapter also briefly reviewed the established general models for modeling the daylight sky and highlighted the importance of dynamic daylighting simulation. Validated daylight simulation programs are currently capable of correctly simulating annual daylighting levels in daylit spaces equipped with advanced glazing technologies and shading systems. The simulation results can be combined with occupant behavior models that simulate occupant use of movable shading devices and lighting controls; and predict the likelihood of occupants experiencing discomfort glare and the consequences on energy consumption due to automated controls (the human factor in sustainable architecture is discussed thoroughly in Chapter 5). Further research is needed to effectively convert the results from daylight simulations into suitable design metrics and increase the number of complex fenestration systems that can be reliably modeled.

Lastly the closing sections of the chapter presented topics directly useful to designers such as daylighting strategies, daylit space rules of thumb, daylighting aperture design, and advanced daylighting systems.

ACKNOWLEDGEMENTS

The author thanks Maitha Bin Dalmouk Al-Nuaimi (PhD Candidate and Researcher at UAE University, UAE) and Iman Al-Sallal (Master of Advanced Studies in Architecture, University of British Columbia, Canada) for their valuable comments and thoughtful and constructive reviews of this chapter.

REFERENCES

ANSI/ASHRAE/IES 90.1-2010: Energy standard for buildings except low-rise residential buildings. American Society of Heating, Refrigerating and Air-Conditioning Engineers, Atlanta, GA, 2010.

ASHRAE 189.1-2009: Standard for the design of high-performance green buildings except low-rise residential buildings, 2009.

Autodesk. Ecotect. 2008. http://ecotect.com (accessed May 2015).

Baker, N.: We are all outdoor animals. In: K. Steemers, and S. Yannas (eds): *Proceedings of PLEA, 2000, Architecture City Environment*. James and James, London, UK, 2000, pp. 553–555.

Beauchemin, K.M. & Hays, P.: Dying in the dark: sunshine, gender and outcomes in myocardial infarction. *J. Roy. Soc. Med.* 91:7 (1998), pp. 352–354.

Bin Dalmouk, M.: *Tools developed for courtyard design to enhance daylight performance in adjacent spaces under desert clear sky conditions*. PhD Thesis, UAE University, Al-Ain, UAE, 2015.

Bin Dalmouk, M. & Al-Sallal, K. A.: Chapter 67: Courtyards – optimum use as means of providing daylight into adjacent zones. In: A. Sayigh (ed): Renewable energy in the service of mankind, Volume I. Springer International Publishing Switzerland, 2015, pp. 751–760.

Boyce, P., Hunter, C. & Howlett, O.: The benefits of daylight through windows. Rensselaer Polytechnic Institute, Troy, NY, 2003. http://thedaylightsite.com/wp-content/uploads/papers/DaylightBenefits.pdf (accessed September 2015).

CERES: CERES Lighting Laboratory, Heliodon Facility. http://ceres.iweb.bsu.edu/heliodon/maininterface.html, 2015 (accessed February 2015).

CIE Standard S 011/E-2003: Spatial distribution of daylight – CIE Standard General Sky. Commission Internationale de l'Eclairage (CIE), Vienna, Austria, 2003.

CIE TC-3-22: Control of damage to museum objects by optical radiation. Technical Report, CIE, Vienna, Austria, 2004.

CIE Technical Committee 4.2-1973: Standardization of the luminance distribution on clear Skies. CIE Publication 22, Commission Internationale d'Eclairage, Paris, France, 1973.

CIE Technical Report 117-1995. Discomfort glare in interior lighting. CIE, Commission Internationale de l'Éclairage, Vienna, Austria, 1995.

CIE-1989: *Guide to recommended practice of daylight measurement:* J.D. Kendrick (ed). Commission Internationale de l'Eclairage (CIE), 1989. http://www.cie.co.at/publ/abst/108-94.html (accessed September 2015).

Cofaigh, E.O., Fitzgerald, E., Alcock, R., Lewis J.O., Peltonen, V. & Marucco, A.: *A green vitruvius – principles and practice of sustainable architectural design*. James & James, London, UK, 1999.

DeKay, M. & Brown, G.Z.: *Sun, wind & light: architectural design strategies*. 3rd edition, Wiley & Sons, 2014.

DiLaura, D.L., Houser, K.W., Mistrick, R.G. & Steffy, G.R. (eds): *The lighting handbook*. Illuminating Engineering Society, 2011.

DIN V 18599-4. *Energy performance of buildings – Part 4: Lighting*, 2005.

Edwards, L. & Torcellini, P.A.: A literature review of the effects of natural light on building occupants. Technical Report NREL/TP-550-30769, National Renewable Energy Laboratory, Golden, CO, July 2002. http://www.parans.com/swe/lightacademy/pdf/A%20literature%20review%20of%20the%20effects%20of%20Naural%20Light%20on%20Building%20occupants.pdf (accessed September 2015).

Enermodal Engineering Ltd for Public Works & Government Services Canada 2002. Daylighting guide for Canadian commercial buildings.

Fischer U.: *Tageslichttechnik*. Verlagsgesellschaft Rudolf Müller, Köln, Germany, 1982

Galasiu, A.D., Newsham, G.R., Suvagau, C. & Sander, D.M.: Energy saving lighting control systems for open-plan offices: a field study. *Leukos* 4:1 (2007), pp. 7–29.

Godish, T.: *Indoor environmental quality*. CRC Press, Boca Raton, FL, 2010.

Harris, N.C., Miller, C.E. & Thomas, I.E.: *Solar energy systems. Design*. John Wiley and Sons, New York, NY, 1985.

Heschong, L.: Daylighting and human perfomance. *ASHRAE J.* 44:6 (2002), pp. 65–67.

IEA-2006: Light's labour's lost: policies for energy-efficient lighting. IEA Publications, Paris, France, 2006.

Inanici, M.: Evaluation of high dynamic range photography as a luminance data acquisition system. *Lighting Res. Technol.* 38:2 (2006), pp. 123–136.

Inanici, M.: Applications of image based rendering in lighting simulation: development and evaluation of image based sky models: *Proceedings of Building Simulation 2009*, Glasgow, UK, 2009, pp. 264–271.

Inanici, M. & Galvin, J.: Evaluation of high dynamic range photography as a luminance mapping technique. Lighting Research Group, Building Technologies Department, Environmental Energy Technologies Division, Lawrence Berkeley National Laboratory, Berkeley, CA, December 2004.

ISO 15469/CIE S003-1996: Spatial distribution of daylight – CIE standard overcast sky and clear sky, 1996.

Jacobs, A.: High dynamic range imaging and its application in building research. *Adv. Build. Energy Res.* 1:1 (2007), pp. 77–202.

Joarder, A.R. & Price, A.D.: Impact of daylight illumination on reducing patient length of stay in hospitals after coronary artery bypass graft surgery. *Lighting Res. Technol.* 45 (2013), pp. 435–449.

Kimball, H.H. & Hand, I.F.: Daylight illumination on horizontal, vertical and sloping surfaces. *Mon. Weather Rev.* 50:12 (1922), pp. 615–628.

Kittler, R.: Standardisation of the outdoor conditions for the calculation of the daylight factor with clear skies. *Proceedings, Conference Sunlight in Buildings*, Bouwcentrum, Rotterdam, The Netherlands, 1967, pp. 273–286.

Kittler, R., Darula, S. & Perez, R.: A new generation of sky standards. *Proceedings, LuxEuropa Conference*, Amsterdam, The Netherlands 1997, pp. 359–373.

Kittler, R., Perez, R. & Darula, S.: Universal models of reference daylight conditions based on new sky standards. *Proceedings of the 24th Session of the CIE*, 1999, pp. 243–248.

Köster, H.: Daylighting controls, performance and global impacts. In: V. Loftness & D. Haase (eds): *Sustainable built environments*. Springer, New York, NY, 2013, pp. 227–255. http://link.springer.com/content/pdf/10.1007/978-1-4419-0851-3_198.pdf (accessed September 2015).

Kwok, A. & Grondzik, W.: *The green studio handbook: environmental strategies for schematic design*. Architectural Press, 2007.

LCA: Lighting Controls Association. http://lightingcontrolsassociation.org/lca/topics/daylight-harvesting/, 2015 (accessed February 2015).

Marsh, A.J.: The application of shading masks in building simulation. *Proceedings, Building Simulation 2005, 9th International IBPSA Conference*, Montreal, Canada, 2005.

Moon, P. & Spencer, D.: Illumination from a non-uniform sky. *Illumin. Eng.* 37:10 (1942), pp. 707–726.

Moore, F.: *Concepts and practice of architectural lighting*. VNR, New York, NY, 1985.

Moore, F.: *Environmental control systems: heating cooling lighting*. McGraw-Hill, New York, NY, 1993.

Mueller, H.F.O.: Daylighting. In: A. Sayigh (ed): *Sustainability, energy and architecture: case studies in realizing green buildings*. Elsevier, 2013, pp. 227–255.

Nakamura, H., Oki, M. & Hayashi, Y.: Luminance distribution of intermediate sky. *JLVE* 9:1 (1985), pp. 6–13.

NHS-SAD, 2013: http://www.nhs.uk/conditions/Seasonal-affective-disorder/Pages/Introduction.aspx (accessed June 2015), 2013.

O'Connor, J., Lee, E., Rubinstein, F. & Selkowitz, S.: Tips for daylighting with windows. Lawrence Berkeley National Laboratory, LBNL report 39945, 1997.

Perez, R., Seals, R. & Michalsky, J.: An all-weather model for sky luminance distribution – a preliminary configuration and validation. *Solar Energy* 50:3 (1993), pp. 235–245.

Rea, M. S. (ed): *Lighting handbook*. Illuminating Engineering Society of North America, New York, NY, 1999.

Rea, M. S. (ed.): *Lighting handbook*. Illuminating Engineering Society of North America, New York, NY, 2000.

Reinhart, C.F.: A simulation-based review of the ubiquitous window-head-height to daylit zone depth rule-of-thumb. *Proceedings, Ninth International IBPSA Conference*, Montréal, Canada August 15–18, 2005, Volume 9, 2005, pp. 1011–1018.

Reinhart, C.F.: Daylight performance predictions. In: J. Hensen & R. Lamberts (eds): *Building performance simulation for design and operation*. Spon Press, New York, NY, 2011, pp. 235–276.

Reinhart, C.: *Daylighting handbook I*. Cambridge, MA: Handbook Publishers, 2014.

Reinhart, C., Mardaljevic, J. & Rogers, Z.: Dynamic daylight performance metrics for sustainable building design. *Leukos* 3:1 (2006), pp. 7–31.

Robbins, C.L.: *Daylighting: design and analysis*. Van Nostrand Reinhold, New York, NY, 1986.

Robertson, K.: Daylighting guide for buildings, distributed through the Canadian Mortgage & Housing Corporation, 2005.

Ruck, N.: *BDP Environment Design Guide: DES 6*. Australian Council of Building Design Professions (BDP), The Royal Australian Institute of Architects, August 2004.

Ruck, N., Aschehoug, O., Aydinli, S., Christoffersen, J., Edmonds, I., Jakobiak, R., Kischkoweit-Lopin, M., Klinger, M., Lee, E., Michel, M., Scartezzini, J. & Selkowitz, S.: *Daylight in buildings – a source book on daylighting systems and components* (No. LESO-PB-BOOK-2000-004). Lawrence Berkeley National Laboratory, Berkeley, CA, 2000.

Solar Pathfinder. 2015, http://www.solarpathfinder.com (accessed February 2015).

Solmetric-SunEye, 2015, http://www.solmetric.com/buy210.html (accessed February 2015).

Steffy, G.: *Architecture lighting design*. John Wiley & Son Inc, Hoboken, NJ, 2008.

Tanteri, M.: *IES Daylighting Seminar*, 2012.

Ternoey, S., Bickle, L., Robbins, C., Busch, R. & Mc Cord, K: *The design of energy-responsive commercial buildings*. John Wiley and Sons, New York, NY, 1985.

Tregenza, P.R. & Waters, I.M.: Daylight coefficients. *Lighting Res. Technol.* 15:2 (1983), pp. 65–71.

Tregenza, P.: Subdivision of the sky hemisphere for luminance measurements. *Lighting Res. Technol.* 19:1 (1987), pp. 13–14.

Tregrenza, P. & Wilson, M.: *Daylighting: architecture and lighting design*. Routledge, New York, NY, 2011.

Ulrich, R.: View through a window may influence recovery. *Science* 224:4647 (1984), pp. 224–225.

US-DOE 2005: Energy Efficiency & Renewable Energy, Building Toolbox. Department of Energy (DOE), http://www.eere.energy.gov/buildings/info/design/ (accessed September 2015).

Veitch, J.A., Charles, K.E., Newsham, G.R., Marquardt, C.J.G. & Geerts, J.: Environmental satisfaction in open-plan environments: 5, Workstation and Physical Condition Effects, IRC-RR-154, Oct. 20, 2003, http://nparc.cisti-icist.nrc-cnrc.gc.ca/npsi/ctrl?action=rtdoc&an=20378817 (accessed September 2015).

Veitch, J.A., Newsham, G.R., Boyce, P.R. & Jones, C.C. Lighting appraisal, well-being and performance in open-plan offices: a linked mechanisms approach. *Lighting Res. Technol.* 40:2 (2008), pp. 133–151.

Vine, E., Lee, E., Clear, R., DiBartolomeo, D. & Selkowitz, S.: Office worker response to an automated venetian blind and electric lighting system: a pilot study. *Energy Buildings* 28:2 (1998), pp. 205–218.

Waldram, P.G.: *A measuring diagram for daylight illumination*. Batsford, London, UK, 1950.

Wienhold, J. & Christoffersen, J.: Evaluation methods and development of a new glare prediction model for 152 daylight environments with the use of CCD cameras. *Energy Buildings* 38 (2006), pp. 743–757.

World Green Building Trend: *Business benefits driving new and retrofit market opportunities in over 60 countries – smart market report*. McGraw Hill Construction, 2006.

CHAPTER 4

Designing passive solar-heated spaces

Ulrike Passe & Timothy Lentz

4.1 INTRODUCTION: PASSIVE SOLAR TRADITION, PRESENT AND FUTURE

Passive solar design techniques have been an intrinsic part of architectural conceptual thought for at least two millennia. References to the powerful relationship between orientation, spatial composition, apertures and materials can be found in major architecture treatises from Vitruvius to Alberti to Le Corbusier. Those strategies are still very valid today as an efficient use of onsite energy is becoming a societal necessity.

The letters of the Younger Pliny (61AD–112AD) provide a fascinating example how passive solar concepts were already well understood in the architecture of antiquity:

> *"In the angle of this room is the dining room and the dining room has a corner, which retains and intensifies the concentrated warmth of the sun, and this is the winter quarters and gymnasium of my household for no winds can be heard there...."* (Pliny, 1969).

> *"As the sun beats down, the arcade increases its heat by reflection and not only retains the sun but keeps off the northeast wind so that it is as hot in front as it is cool behind"* (Pliny, 1969).

> *"Here is a sun-parlor facing the terrace on one side, the sea on the other, and the sun on both. There is also a room, which has folding doors opening on to the arcade and a window looking out on the sea. Then there is an ante-room and a second bedroom, built out to face the sun and catch its rays the moment it rises, and retains them until after midday, but then at an angle"* (Pliny, 1969).

Vitruvius, the important Roman architectural writer of the 1st century BC, who founded the principles of Western architecture as Firmitas, Utilitas, Venustas, advised builders in the Italian peninsula, *"Buildings should be thoroughly shut in rather than exposed toward the north, and the main portion should face the warmer south side."* (Smith, 2003).

Leon Battista Alberti wrote in his major treatise of the Renaissance, that

> *"The useful practice of the 'ancients' should be employed on the site so that loggias should be filled with winter sun, but shaded in the summer."* (Alberti, 1988).

This statement indicates the importance of seasonal variation and the necessity to protect from the sun at certain times while giving access to it at other times.

During the 17th and 18th centuries northern European noble families explored the use of passive solar strategies for luxury recreational usage and stored precious exotic plants behind glass in their elaborate greenhouses to endure the cool or cold winter months. Frederick the Great (German: Friedrich II; 24 January 1712–17 August 1786; King of Prussia (1740–1786)) envisioned Sanssouci (Fig. 4.1) (built by Georg Wenzeslaus von Knobelsdorff (1745–1747) (SPSG, 2014)), where the vineyard as a glass covered mountain faces straight south and vine and fig trees survive the winter behind glass, while the glass doors swing open during the summer. This project already combines the essential use of thermal mass behind the glass to store the heat for cooler evenings and nights, when the heat re-radiates back out and prevents those small glass house spaces from freezing.

Figure 4.1.   Passive solar architecture in history: Sanssouci palace in Potsdam, Germany (image courtesy of Stiftung Preußische Schlösser und Gärten Berlin-Brandenburg).

Besides these documents archaeological evidence exists, that passive solar architecture was understood in all continents and throughout all times.

In the US the first consciously designed passive solar house is usually attributed to George F. Keck, who in 1940 developed a passive solar home for a Chicago area real estate developer named Howard Sloan in Glenview, Illinois (Collins and Nash, 2002). But also Frank Lloyd Wright developed a prototype with the Hemicycle house or Jacobs II in Middleton, Wisconsin in the 1940s (Figs. 4.5, 4.7, 4.8, 4.9) (Storrer, 1993).

Most likely design guidance for scale and size of openings, overhangs and other architectural elements derived from observation and experience, but often geometric proportions might have been favored over solar geometric relationships. Thus Ed Mazria's extensive and comprehensive design and calculation guidelines for passive solar in his 1979 publications (Mazria, 1978; 1979) as well as performance evaluation research conducted by Balcomb (1992) and others were extremely valuable in understanding the scale of the necessary relationships to solar geometry in order to maximize the output of solar thermal energy. In addition architects in the past had only very few material options to choose from, the choices for glass today are endless, but the need for proportional design relationships remains. This chapter will analyze the proportional relationships implemented in six 2009 Solar Decathlon homes and a competition entry to the 2011 Living Building Challenge competition and thus highlight the most successful passive solar design strategies and showcase the balance between those strategies' principles.

4.2   PRINCIPLES OF PASSIVE SOLAR DESIGN: BALANCING SUN, MASS AND AIR

Passive solar design strategies are essentially architectural strategies which manage, balance and support solar thermal energy inside building enclosures in order to provide thermal comfort.

As noted earlier the ingredients of passive solar architecture have been known for centuries, but it seems a contemporary revival could well be enhanced to new levels. The major principles of passive solar design are utilizing spatial design elements to capture solar thermal energy:

- Direct penetration of solar radiation through a transparent or translucent media (As close to direct normal as possible).
- Transformation of radiation into long wave infrared heat.
- Distribution of the thermal energy through convective air currents.

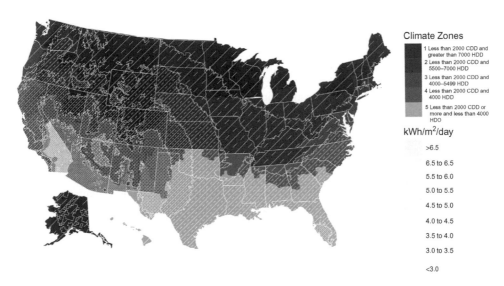

Climate Zones

1 Less than 2000 CDD and greater than 7000 HDD
2 Less than 2000 CDD and 5500–7000 HDD
3 Less than 2000 CDD and 4000–5499 HDD
4 Less than 2000 CDD and 4000 HDD
5 Less than 2000 CDD or more and less than 4000 HDO

kWh/m²/day

>6.5
6.5 to 6.5
5.5 to 6.0
5.0 to 5.5
4.5 to 5.0
4.0 to 4.5
3.5 to 4.0
3.0 to 3.5
<3.0

Figure 4.2.   Map of the regions with cold climates and good solar radiation Source: US Energy Information Administration (02/11/04) (http://www.eia.gov/emeu/recs/climate_zone.html).

- Trapping of this thermal heating energy inside the space through well insulated air tight envelopes.
- Storage of heat in appropriate materials to reduce warming of air and distributing heat over a longer period of time.

The proportional relationship of these five key elements is important for the success of the architectural design strategy and is dependent on the climate and location (latitude) as these proportions will balance between excessive gains and heat retention and thermal comfort. If done well, there is no need for added technology. Technological advances can enhance the performance of those strategies, but a detrimental orientation cannot be offset with better glass. In order to achieve these relationships various design rules of thumb have been established, which will be demonstrated here (Williams, 1983).

4.3   IMPACT OF COLD CLIMATE

On the northern hemisphere passive solar strategies are used from the West Coast of the US (for example the Thesen Home in Palo Alto, California constructed in 2011 (Project Green Home, 2014) and Canada to the high altitudes of the Himalaya (Chandel and Aggarwal, 2008), where passive solar architecture enjoys a long lived tradition as highlighted by many scholars who visited these remote regions. Human settlements with resource scarcity have always looked to the sun for warmth and light. The ideal climate for passive solar architecture encounters lots of direct solar radiation in the cold season (Fig. 4.2), but diffuse radiation as known in northern Europe is not completely detrimental to passive solar architecture. Just the gathered energy is reduced and other measures might need to be increased for an enhanced balance.

4.4   PASSIVE SOLAR DESIGN TECHNIQUES AND THEIR DESIGN RULES OF THUMB

Over the course of the past 30–50 years rules of thumb have been developed and implemented to quantify and maximize the outcome of direct-gain, indirect-gain, shading, thermal mass, insulation and window sizing (Lentz, 2010). Basic daylighting strategies have been established and are widely published. Of course passive solar design strategies can now also be tested with elaborate energy performance simulation software packages that model heat transfer and energy usage in order to optimize the design, but this undirected parametric design might take a long time to arrive at an acceptable design strategy. Thus a common design alternative is still the use of rules of thumb. These quick calculations are used in the design phase to estimate proper sizing and placement of passive solar features.

4.5   THE 2009 US DOE SOLAR DECATHLON COMPETITION

In order to promote energy efficiency and building integrated photovoltaic systems and to show-case new and effective design practices, the United States Department of Energy (DOE) has created the Solar Decathlon competition (NREL, 2010). The Solar Decathlon is an international competition that challenges 20 university teams every two years to design and build an energy-efficient, single-family 74.3 m^2 (800 square feet) home that is completely solar powered. The 2009 competition culminated at the end of two years of planning and building with a contest on the National Mall in Washington, DC. The teams transport their houses to Washington DC for several days of measured and subjective contests. There have been five Solar Decathlon competitions to date: in 2002, 2005, 2007, 2009 and 2011.

Although the focus of the competition was mostly driven by active solar energy conversion several passive solar design techniques were implemented to make more effective use of the space and design of the homes in order to reduce the need for active energy production. In order to explain the most common rules of thumb the homes built for the 2009 US DOE Solar Decathlon were used as case study. This chapter is based on a master's thesis (Lentz, 2010) which analyzed the effectiveness of these common rules of thumb by comparing how well the houses conformed to the rules. This thesis used actual data collected during the competition, especially how well the

Table 4.1.   2009 Solar Decathlon teams, their house names and competition place[1].

2009 Solar Decathlon teams	House name	Overall competition result
Cornell University	Silo House	7
Iowa State University	*Interlock House*	*12*
Ohio State University	Ohio-centric	10
Penn State University	Natural Fusion	16
Rice University	Zerow House	8
Team Alberta	*SolAbode*	*6*
Team Boston	Curio.House	15
Team California	Refract House	3
Team Germany	*SolarHome*	*1*
Team Missouri	Show-Me House	11
Team Ontario-BC	*North House*	*4*
Team Spain	Sigue el Sol	14
Universidad de Puerto Rico	Casa Solar	17
University of Arizona	SEED [pod]	18
University of Illinois	*Gable Home*	*2*
University Kentucky	sky blue house	9
University of Louisiana	BeauSoleil	19
University of Minnesota	*ICON House*	*5*
University of Wisconsin	Meltwater	20
Virginia Tech	Lumenhaus	13

[1]Houses in *Italics* are analyzed later in the chapter

teams were able to keep their houses comfortable. The thesis did not find many correlations in predicting the competition outcome by analyzing the adherence to rules of thumbs, but concluded that the rules of thumb are still very valid indicators for early design decisions (NREL, 2010).

The 20 university teams competing in the 2009 event are listed with the name of their houses and overall placement in the competition in Table 4.1.

The 2009 Solar Decathlon competition (Figs. 4.3 and 4.4) measured the houses in ten subjective and objective contests. The subjective contests (architecture, engineering, market viability,

Figure 4.3.   Solar Decathlon 2009: Construction sequence of Iowa State University's Interlock House.

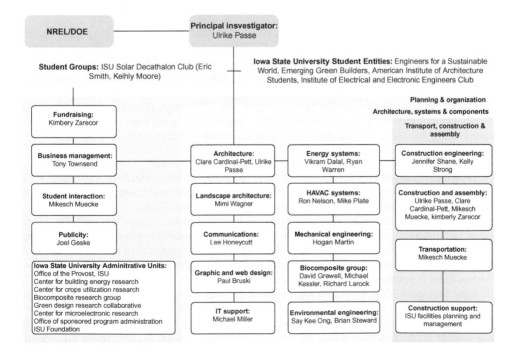

Figure 4.4.   Solar Decathlon 2009: Iowa State University's competition team organization sequence.

communications, lighting design) were judged by juries composed of industry professionals. The objective contests (comfort zone, hot water, net metering/energy balance, appliances, home entertainment) were directly measured. Some contests were task based such as hot water and appliances. Other contests such as comfort zone and net metering were directly measured throughout the course of the competition time. Each house was equipped with shielded temperature and humidity sensors to measure interior air characteristics. Bidirectional watt nodes and current transformers were used in each house to determine the overall energy balance. Sensors were also placed in the center of the exterior competition site to measure environmental temperature, humidity and insolation on a horizontal surface. The competition assigned points based on how well a team met the criteria for each of the ten competitions. Specifically, full points were awarded in the comfort zone competition for maintaining an internal temperature between 22.2 and 24.4°C (72 and 76°F) and maintaining an internal relative humidity level between 40 and 55% during all scoring periods.

4.6   DETAILED ANALYSIS OF PASSIVE SOLAR DESIGN RULES OF THUMB

4.6.1   *Direct penetration of solar radiation through a transparent or translucent media: direct-gain*

Solar radiation falling onto a surface is dependent on the angle at which it falls onto the surface and on the property of glass. The optimum orientation of the solar radiation collecting surfaces on the northern hemisphere is south facing.

The first common rule of thumb (Chiras, 2002) evaluates the total area of solar glazing (south facing windows) in relationship to total floor space and most literature assume this ratio in situations without thermal mass to be between 7 and 12%. The variation depends greatly upon the local climate and the amount of thermal mass available, which offsets excess gains. Direct gain works in collaboration with thermal mass, which balances temperature swings and stores thermal energy for later use. Increasing the amount of thermal mass allows for an increase in provided south-facing window to floor-space ratio.

The mentioned (Lentz, 2010) calculated results for direct-gain solar glazing ratio, thermal mass and unaccounted for solar glazing of all 2009 Solar Decathlon houses. Unaccounted for solar glazing is the amount of glazing area that is not compensated for with incidental thermal mass or with additional dedicated thermal mass. According to these recalculations only Cornell managed to design a house that stayed below the 7% solar glazing ratio. Additionally, Iowa State, Louisiana and Rice all managed to be close to the 7 to 12% range. Based on the design rule of thumb, the majority of the houses in the competition were designed with excessive amounts of solar glazing, but the ratio itself did not indicate any ranking in the actual competition.

4.6.1.1   *Storing energy – thermal mass as temporary energy storage*
Thermal mass is usually a solid material used in the construction or finish of a building, which has a high heat capacity and can thus store a large amount of thermal energy without raising the temperature of the surrounding air. In conventional construction thermal mass is usually exposed brick or exposed concrete. Innovative new materials called phase change materials (PCM) use the energy absorbed or stored during the phase change of a material to store excess solar energy.

When sizing thermal mass for a direct-gain system, it is appropriate to take into account the amount of solar glazing. If solar glazing is less than 7% of the total floor space, no additional thermal mass is needed because the "incidental" thermal mass of the building will account for the solar heat gain. If the solar glazing area is larger than 7% of the total floor area, additional thermal mass will be necessary. Depending on the location of the mass certain ratios compared to glazing area can be used. Thermal mass that is placed in directly lit floor areas should be designed at a ratio of 1:5.5. This means that for every square-meter of solar glazing above the 7% limit there should be 5.5 m^2 of sunlit thermal mass. For unlit floor areas the ratio is 1:40 and for unlit mass walls the ratio is 1:8.3 (Chiras, 2002).

The addition of thermal mass played a strong role in the ability of the solar houses to control temperature and maintaining high temperatures during cooler times. Only two houses had enough thermal mass to compensate for all of the solar glazing, but no strong correlation could be drawn between temperature controls and fully accounted for solar glazing.

Since thermal mass can even out temperature swings and prevent overheating, it plays an important role in designing with the direct-gain passive solar heating strategy. Seven teams in the 2009 Solar Decathlon had very deliberate uses of thermal mass while the remaining teams had thermal mass included in furnishings and counters. Of seven teams, which best designed their thermal mass according to the rules of thumb, four used phase change materials (PCM) (Penn State, Ontario/BC, Germany, and Spain), two used water (Ohio State and Arizona), and one used concrete (Virginia Tech) as thermal mass. For Spain and Ontario/BC the total capacity for the PCM was given for a defined temperature range which allowed the storage capacity to be presented in Joule (kWh)/°C (Btu/°F) units. The PCM used by Penn State and Germany did not have a defined temperature range and thus could not be calculated on a per degree basis. In order to make these calculations comparable to other forms of thermal mass, the total storage capacity was divided by 11.2°C (20°F). This temperature range was chosen based on the daily temperature swings measured in several of the houses.

Factoring in thermal mass, the excessiveness of some teams' solar glazing is reduced. Team Germany's house, which is ranked 12th in the amount of solar glazing, has enough thermal mass to compensate for all of the window area above the 7% threshold. This puts Germany alongside Cornell as the two teams that best designed their houses for direct-gain. Germany's design most likely had a greater effect on the performance of the house due to the greater thermal collection and storage capability. However, Cornell's design allows for less thermal loss through windows due to the reduced solar glazing area.

4.6.1.2 *Direct-gain – final discussion of calculated results*

Direct-gain windows oriented south (Fig. 4.5) proved to be a useful design feature for many teams. The majority of teams added more solar glazing than was recommended by the rule of thumb but the correlations showed that the extra glazing was advantageous. The collected data suggests that the solar glazing limit suggested by the rule of thumb could be increased from 12 to 40%.

4.6.1.3 *Indirect-gain – rules and results*

Sun spaces can be added internally or externally to the floor plan of the house. Generally an enveloped sunspace has better thermal performance than an added sunspace because most of the

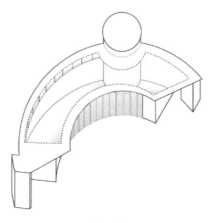

Figure 4.5.   The Hemicycle House by Frank Lloyd Wright (1943) contains most necessary design features for a passive solar heated house: Southern fenestration.

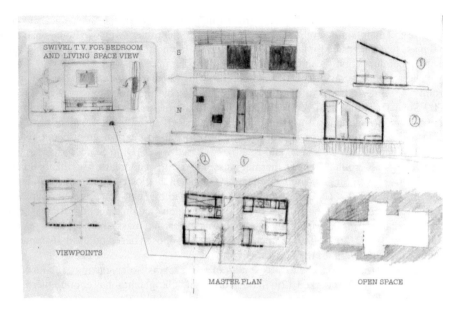

Figure 4.6.   Principle diagram for Iowa State University's 2009 Solar Decathlon entry.

heat is lost to the internal living spaces as opposed to the outdoors. As with the direct-gain design approach, heat is gained through south facing windows in the sunspace. Windows can also be incorporated into the roof and walls of the sunspace to increase the solar collection area. However, the decreased thermal resistance of windows causes the extra glazing to decrease the performance of the space if not properly insulated. In northern heating dominated climates it is recommended that double or even triple pane windows be used in a sunspace to increase performance (Williams, 1983). With a properly sized sunspace, a house can achieve a solar heating fraction between 55 and 84% (Chiras, 2002).

Only two 2009 Solar Decathlon teams designed houses with indirect-gain sunspaces: Iowa State (Fig. 4.6) and Spain. Iowa State's ratio of sun space to living space was about 10%, while Team Spain's was about 5% of the living space. The sun space in the Iowa State House makes up most of the solar gaining south facing glass.

Spain has a much larger solar gain ratio (400%) but no thermal mass and a smaller ratio of sunspace to living space. Theoretically this would mean that the air in the sunspace would get very hot but there would only be a short period of heat gain in the living space. Since the Iowa State House has thermal mass, more heat can be stored and released gradually into the living space, when radiation diminishes.

4.6.1.4   *Shading – modulating the gain*
With more diverse seasonal conditions it can be beneficial, if not essential to be able to reduce solar radiation gain occasionally. Two major differences can be detected amongst shading devices: dynamic shading strategies, which can adapt to outside conditions and seasons and static shading strategies, which have to be carefully designed to provide the best options for all seasons.

No matter what type of shading device is used, there are several goals that the design should accomplish:

- Shades should be on the outside of the opening, because this reduces the intake of heat during summer.
- Shades should be made of light and reflective materials to avoid absorption and re-radiation. Materials should have low heat storage capacity for rapid cooling.

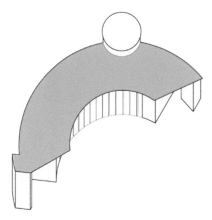

Figure 4.7. The Hemicycle House by Frank Lloyd Wright (1943) contains most necessary design features for a passive solar heated house: Shading the southern exposure.

- The design should prevent reflection onto any part of the building or openings or neighbors.
- Hot air should not be trapped against the building (Chiras, 2002).

Shading (Fig. 4.7) is one of the most architecturally visible passive solar elements and thus either greatly taken care of or dismissed. The homes discussed later use a large variation of shading devices.

4.6.2 *Heat retention through insulation*

Many solar designers in the US recommend in their rules of thumb and guidelines to insulate walls with thermal resistance between *R*-22 and *R*-30 and ceilings between *R*-40 and *R*-50 in US units (ft^2°Fh/Btu), which would equal approximately 3.5 and 5.3 m^{2}K/W and between 7 and 9 m^{2}K/W in SI units. The actual value will depend on the severity of winter as indicated by Heating Degree Days. Specific values can be evaluated by calculating the basic heat loss through the envelope per area of heated floor (Lentz, 2010).

Increasing insulation beyond these points helps reduction of heat transfer, but there is a point where increasing insulation might not be cost effective (Chiras, 2002). The Passive House Standard initiated in Germany increases insulation up to the point where furnaces become obsolete and reduces costs by eliminating the furnace (Passiv Haus Institut Germany, 2012). Thus most solar houses should have little to no openings other than facing south (Fig. 4.7).

The resistance of a wall is determined by the paths that heat can take to transfer out of the wall. Similar to circuits, paths are either parallel or in series. Walls are typically a combination of both parallel and series paths. The effective *R*-value, R_e, for a series path is:

$$R_\mathrm{e} = R_\mathrm{si} + R_\mathrm{so} + \Sigma R_i X_i$$

where R_si = inside surface resistance
R_so = outside surface resistance
R_i = *R*-value of the *i*th material
X_i = thickness of material *i*

The effective heat loss coefficient through *j* parallel paths is:

$$U_\mathrm{e} = \frac{\Sigma U_j A_j}{\Sigma A_j}$$

where A_j is the cross sectional area of the *j*th material (Williams 1983)
U_j is the conductance ($1/R_j$)

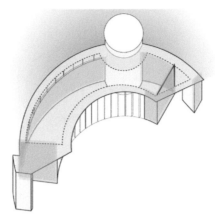

Figure 4.8.   The Hemicycle House by Frank Lloyd Wright (1943) contains most necessary design features for a passive solar heated house: Outer wall insulation impact on passive solar architecture.

4.6.2.1   *Insulation calculations*

"Analysis of the passive design and solar collection techniques of the houses in the 2009 US Department of Energy's Solar Decathlon competition" (Lentz, 2010) recalculated the R-values of all 20 houses. The R-values for walls alone, walls with windows and doors, roof alone and the entire envelope are indeed very diverse ranging from single digits to R-40 for the whole building envelope. Illinois had the highest insulated house with the whole envelope R-value doubling that of the third most insulated house, Iowa State. The top two most insulated houses, Illinois and Germany also placed in the top two spots of the comfort zone competition. Germany won the competition with Illinois taking second. While the R-value is not a direct correlation to performance in the comfort zone competition, it surely has a strong effect on the outcome, because a high R-value reduces temperature swings induced by climatic changes. Data actually showed, that good solid R-values with reduced thermal bridges were the single best indicator for good thermal comfort performance.

Some teams managed to design for the recommended R-values in the opaque walls but still ended up with an extremely low whole envelope R-value. This was usually due to large window and door area or low quality windows and doors. Most houses with higher insulation values were from northern colder climates while those with lower insulation values were from southern climates, which indicates higher awareness in those teams of the correlation between envelope heat retention, energy and comfort in colder climates.

4.6.2.2   *Insulation – results*

Data clearly shows a strong correlation between increasing R-values in all parts of the envelope and increased internal temperature and improved temperature control. The rule of thumb specifying a minimum of R-22 ($\sim$3.9 m^{2}K/W) for walls and R-40 ($\sim$5.3 m^{2}K/W) for roofs is also supported strongly by the data. Increasing R-values beyond these points leads to further improved temperature control and higher average temperatures, thus it can truly be recommended to provide as much insulation as possible for good thermal comfort in a passive solar heated home.

4.6.3   *Window assembly properties*

Windows are the main means for solar gain but they must also be designed for efficient retention of thermal energy, which is usually indicated by the U-value of the assembly (glass and frame). Compared to walls and roofs, windows are typically the big energy losers of a building. There are several factors to consider when determining the thermal efficiency of a window. The type of window is a big factor. Non-operable windows are the most efficient type because there are no moving parts and no gaskets. The whole window is built to be permanently sealed. But if natural ventilation is supposed to be an integral strategy of the design, at least some of the windows need to be operable. Thus wise placement of windows can increase the efficiency of the house tremendously. If an operable window is needed for ventilation, it should be located to maximize ventilation reducing the need for excessive numbers of operable windows.

Secondly, choosing windows with the proper solar heat gain coefficient (*SHGC*) for the region is important. It is recommended that for hot climates *SHGC* should be less than 0.4; between 0.4 and 0.55 for intermediate climates; greater than 0.55 for cold climates. Also, certain coatings could be chosen in accordance with the direction a window faces. East and north facing windows should have low-e coatings while west windows should have heat rejecting coatings to prevent excessive heat gain. Additionally, the north and east windows should account for no more than 4% of the floor space and the west windows should not account for more than 2% of floor space to limit thermal losses and unwanted gains. South facing windows can be either uncoated or use low-e coatings for solar heat gain (Chiras, 2002).

4.6.3.1   *Windows – calculation*
The calculations weigh solar heat gain coefficient of southern windows, *R*-value, and glazing as a percent of conditioned floor space based on orientation. Several teams managed to create a design that either had the proper *SHGC* or a high *R*-value or windows in the proper proportions for energy efficiency but only Illinois managed to fulfill all of these criteria. This helps explain why Illinois has such a high envelope *R*-value.

Several teams had extremely high percentages of windows, especially on the north wall, which is detrimental to heat retention and does not benefit any solar gains. These high amounts of windows were probably designed with daylighting in mind, while less would have been sufficient. Many of the designs employed excessive amounts of windows even for recommended daylighting standards, which indicate, that a lot of misconceptions exist regarding daylighting. Also notable, along with the large amounts of windows, is that the median window *R*-value is 3.7 ($\sim$2.06 m^{2}K/W) to 3.75 (2.08 m^{2}K/W). While this is not a terribly low *R*-value, it is also not outstanding considering the importance of heat retention. The expectation for an efficient housing design competition would be that the windows would be far better than those in an average house, especially since windows are typically the weak point in the building envelope. Costs could have played a role here, but also frankly lack of experience and understanding.

4.6.3.2   *Windows – results*
The rule of thumb governing solar heat gain coefficient was supported by the data. Houses with a *SHGC* value between 0.4 and 0.55 typically retained higher average temperatures. The rule of thumb for sizing north, east and west windows was partially supported by the data. The trends indicated that decreasing north windows helped maintain a higher indoor temperature but there were not enough data points that fell below the rule of thumb value to judge the effectiveness of the rule. The data points for east windows that fell below the rule of thumb value suggested that following the rule resulted in a cooler indoor temperature which is not the goal for this rule. For west windows the goal is to keep out unwanted heat gain and the data supports this. Data points falling below the 2% rule of thumb value were typically at a lower temperature. The trends also suggested that decreasing window areas on all sides helps improve temperature control.

4.6.4   *Daylighting*

The third major function of windows with respect to passive solar design of course is daylighting and two basic, but effective rules of thumbs have been tested for the solar homes. The first rule of thumb for daylighting involves effective aperture (*EA*) which is used to determine the ideal

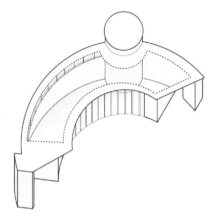

Figure 4.9.   The Hemicycle House by Frank Lloyd Wright (1943) contains most necessary design features for a passive solar heated house: Northern clerestory: Impact of daylighting on passive solar energy.

amount of glazing. The effective aperture is the product of the window-to-wall ratio, *WWR*, and the visible transmittance, τ, of the windows.

$$EA = WWR \times \tau$$

The target value for effective aperture is 0.18 (Ander, 1995). Any value above this target increases cooling loads more than it relieves lighting loads. This is the point considered to achieve daylighting saturation.

Daylighting and solar access need to be brought into a proportional balance. For many tasks control of daylight is very important due to the dynamic changes in the external light levels between summer and winter and direct sunlight as well as overcast sunlight. Small northern clerestory windows provide daylight without the added heat gains and losses (Fig. 4.9).

Calculated values for the effective aperture of each house were also calculated. Team Alberta and Louisiana best designed for the 0.18 effective aperture value, differing by only 0.005 and 0.006, respectively. Most teams had an *EA* value higher than 0.18. This is more likely due to a high window-to-wall ratio than to a high visible transmittance value. As a result, these houses lose more energy through windows than is saved through daylighting.

The data suggested that increasing the effective aperture for daylighting beyond 0.18 also increased average temperature and improved temperature control. The rule of thumb is based on losing energy due to excessive window area which would result in lower average temperature and poor temperature control. It is likely that the trend visible in the daylighting graphs is due to the solar glazing which also caused a similar trend in the direct-gain graphs.

4.7   METHODOLOGY

"Analysis of the passive design and solar collection techniques of the houses in the 2009 US Department of Energy's Solar Decathlon competition" (Lentz, 2010) evaluates the 2009 Solar Decathlon houses based on the performance of their passive building techniques. Quantitative values were calculated for the direct-gain rule of thumb comparing solar glazing as a percentage of floor space. For indirect-gain, the total area of solar glazing was calculated as was the thermal mass and volume of the sunspace. The sunspace volume was also compared to the interior volume of the living space. Daylighting values were calculated using the effective aperture rule of thumb. Thermal mass was analyzed both for total thermal storage and as a ratio to extra solar glazing area. The building envelope for each house was analyzed and building average *R*-values for walls, windows, doors and roofs were calculated. The windows chosen for each house were analyzed for *R*-value and *SHGC*. North, east and west window areas were also analyzed as a percentage of the conditioned floor space. Shading was analyzed on a qualitative basis using the five design points listed at the end of the shading discussion in the literature review.

Full measurement data for all of the houses has been provided by the National Renewable Energy Laboratory (NREL) as part of the competition management (Solar Decathlon, 2014). Temperature and humidity readings were recorded in 15 minute intervals for the entirety of the competition week. For each house the average temperature and humidity levels were calculated over the entire competition period and over the comfort zone scoring period only. Additionally, the sample standard deviation for temperature and humidity was calculated. (Standard deviation is a measure of how much variation from the average value exists in the data set. In the case of temperature and humidity measurements, standard deviation is also a measure of how well the house controls these parameters with smaller values meaning better control).

4.8   TEMPERATURE AS PERFORMANCE INDICATOR

The competition week was comprised of several rainy and cloudy days interspersed with one or two sunny days. The average outdoor temperature for the competition was 15.67°C (60.2°F) with a maximum temperature of 30.33°C (86.6°F) and a minimum temperature of 7.28°C (45.1°F). With the average temperature being well below the comfort zone competition temperature range, the goal was to retain as much heat as possible in order to keep the internal temperature close to the 22.22 to 24.44°C (72 to 76°F) range. As Figures 4.10 and 4.11 show, few teams were able to maintain a high enough temperature over the course of the competition week. This is most likely due to public tours which frequently require leaving doors open for long periods of time. After removing the public tour hours from the calculation, eight teams have average temperatures within the required range.

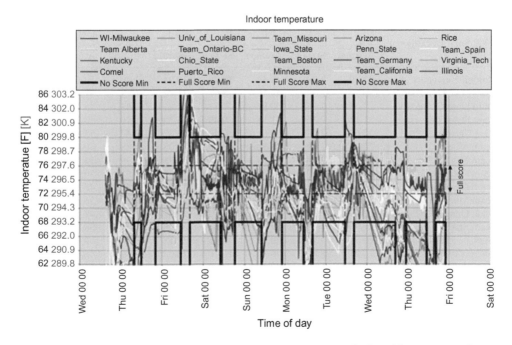

Figure 4.10.   Measured indoor temperatures during the 2009 US DOE Solar Decathlon contest week.

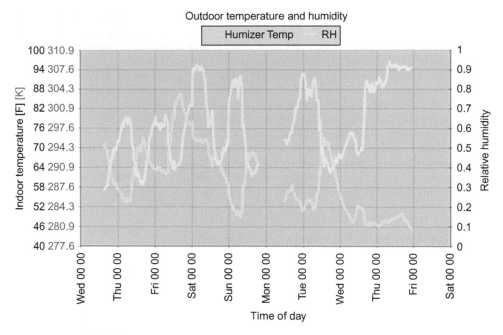

Figure 4.11.   Measured outdoor temperature and humidity during the 2009 US DOE Solar Decathlon contest week.

Table 4.2.   Average indoor temperature and relative humidity measurements in the competition homes.

2009 Solar Decathlon teams	Whole event period				Competition contest hours only			
	Temperature [°C]		Relative humidity [%]		Temperature [°C]		Relative humidity [%]	
	Average	St. dev.	Average	St. dev.	Average	St. dev.	Average	St. dev.
Cornell University	21.10	3.29	48.56	10.78	29.91	2.40	52.67	9.10
Iowa State University	20.74	2.35	49.10	8.08	21.22	1.73	51.60	6.83
Ohio State University	22.15	2.07	47.87	7.71	22.41	1.43	48.72	6.88
Penn State University	21.14	2.36	51.58	9.49	23.29	1.70	53.93	7.97
Rice University	20.62	3.58	52.01	10.40	20.77	3.42	53.81	8.26
Team Alberta	22.79	2.61	47.32	8.80	23.03	1.88	49.40	7.38
Team Boston	21.48	2.74	51.53	7.53	22.38	2.39	54.50	7.20
Team California	21.04	3.39	48.66	9.49	21.36	3.61	50.05	9.87
Team Germany	23.54	1.91	44.36	7.17	23.42	1.56	46.93	6.47
Team Missouri	21.43	3.44	52.64	9.19	21.32	3.37	56.34	6.41
Team Ontario-BC	22.89	1.38	46.51	6.73	23.07	1.16	47.77	6.22
Team Spain	20.70	4.31	54.84	13.01	21.28	3.34	58.47	10.78
Univ. de Puerto Rico	20.34	3.75	50.91	8.85	21.18	3.56	52.01	7.59
University of Arizona	20.21	3.25	46.36	11.20	20.63	2.99	46.83	11.77
University of Illinois	23.47	1.47	48.39	8.66	23.57	0.80	51.79	5.29
University Kentucky	21.32	3.24	52.40	10.76	21.88	2.80	66.56	8.98
University of Louisiana	21.65	3.02	45.40	7.66	21.89	2.60	45.45	5.14
University of Minnesota	21.72	2.35	51.28	8.45	22.11	1.85	53.22	6.68
University of Wisconsin	21.90	3.06	48.02	7.73	23.42	2.07	47.42	6.28
Virginia Tech	21.79	2.86	48.53	10.79	21.64	2.90	50.64	10.28

Standard deviation of the temperature data set is also included in Figure 4.10. Standard deviation allows for the determination of how well a house maintained a steady temperature. A small standard deviation indicates that the temperature remained fairly constant and did not differ much. A larger standard deviation indicates large temperature swings and poor temperature control. According to Table 4.1, the Illinois House had the best temperature control during the scored competition hours and was a close second over the whole week. Other houses exhibiting good temperature control were those of Ontario/BC, Ohio State and Germany.

Temperatures were measured by the competition organizers are graphed in Table 4.2 and visualized in Figure 4.10. The graph visualizes where peaks and valleys occurred for various teams. As would be expected, most of the peaks occurred during the public tour hours but carried over into the competition measurement periods. Many of the temperature peaks can also be visually correlated to outdoor temperature which is shown along with humidity in Table 4.2. With this graph in mind, it should be no surprise that many teams had severe temperature peaks on the first Friday afternoon when the outdoor temperature peaked around 31°C (87°F).

4.9   SOLAR DECATHLON HOUSE ANALYSIS (2009): EVALUATION OF PASSIVE SOLAR DESIGN RULES OF THUMB BASED ON COMPARISONS

The three houses that most consistently followed the passive solar design rules of thumb according to our reference (Lentz, 2010) were Team Alberta, University of Illinois and Iowa State University. After recalculating all 20 Solar Decathlon houses he assigned a letter grade to all teams for how well each one met each of the rules of thumb. Following his evaluation these three houses as well as three of the winning teams were diagrammed for this chapter.

4.9.1   *Team Alberta: Sol Adobe House*

The Sol Abode House (Figs. 4.12–4.16) had a fairly large amount of solar glazing, the 11th most with a 21.9% solar glazing ratio. This was compensated for with large amounts of thermal mass integrated into the wall finishes of most of the interior. The use of Rundle stone on sunlit and

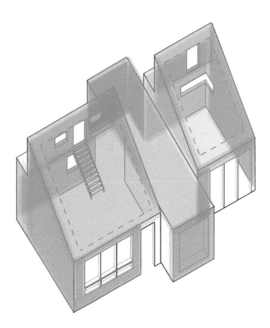

Figure 4.12.   Team Alberta, Solar Decathlon (2009): Sol Abode House – outer wall insulation.

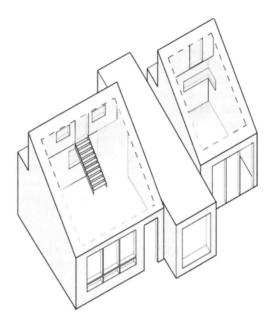

Figure 4.13.   Team Alberta; Solar Decathlon (2009): Sol Abode House – northern clerestory for daylighting.

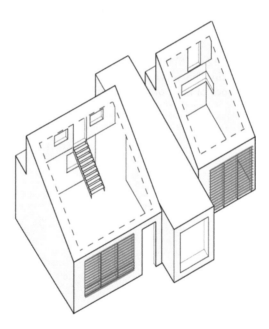

Figure 4.14.   Team Alberta; Solar Decathlon (2009): Sol Abode House – sun shading devices.

shaded walls gave the thermal mass a good orientation to compensate for the large amount of solar glazing. This resulted in only 2.81 m^2 (30.2 ft^2) of unaccounted for solar glazing, the 5th lowest amount. The solar glazing also had a solar heat gain coefficient of 0.39 which is close to the 0.4–0.55 target range. Although for the climate this house will normally inhabit, the *SHGC* should be closer to 0.55 or above (Table 4.3).

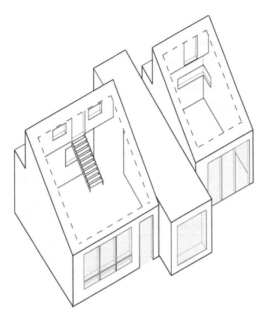

Figure 4.15.   Team Alberta; Solar Decathlon (2009): Sol Abode House – southern fenestration.

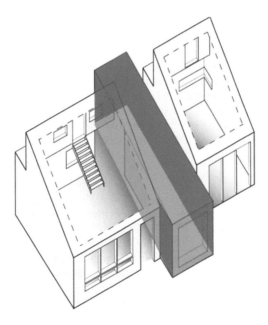

Figure 4.16.   Team Alberta; Solar Decathlon (2009): Sol Abode House – location of thermal mass.

Team Alberta's Sol Abode House was the best ranked in daylighting. The effective aperture of the house was 0.185, only differing from the target value of 0.18 by 0.005. This is indicative of a good balance of windows to wall area. Proper daylighting design will let this house save energy on lights while not losing too much thermal energy through excessive window areas. In terms of oriented window-to-floor percentages, this house had 9.90% north facing, 2.30% east facing, and 3.30% west facing glass. While the north and west percentages were above the targets of

Table 4.3.   Team Alberta: SolAbode House analysis summary.

Passive solar design strategies	Rule of thumb			Ranking
Direct gain	21.9% solar glazing ratio			11th highest
Thermal mass	Large amounts of thermal mass	$2.81 \, m^2$ ($30.2 \, ft^2$) unaccounted for solar glazing		5th lowest
Indirect gain	None			
Shading strategy	Motorized blinds that sit on the outside of the window and are deployed by a building control system			
Heat retention/insulation	Whole envelope R-value	14.6 ($\sim 2.57 \, m^2 K/W$)		6th highest value of all houses
	Wall R-Value	46.1 ($\sim 8.12 \, m^2 K/W$)		
	Roof R-Value	42.7 ($\sim 7.70 \, m^2 K/W$)		
Windows				
Daylighting	0.185	North facing	9.90%	Best ranked
		East facing	2.30%	
		West facing glass	3.30%	

4 and 2% respectively, they were both significantly less than the majority of the other houses. Along with good amounts of windows, the Alberta House also had excellent shading to prevent unwanted solar heat gain. The shading system consists of motorized blinds that sit on the outside of the window and are deployed by a building control system. These blinds meet all five criteria mentioned above for effective shading of windows.

The envelope of this house was very good with respect to energy conservation as well. Although the whole envelope R-value was only 14.6 ($\sim 2.57 \, m^2 K/W$), it was still the 6th highest value of all the houses. This was due largely to good wall and roof R-values that met the rule of thumb criteria. The walls were R-46.1 ($\sim 8.12 \, m^2 K/W$) and the roof was R-42.7 ($7.70 \, m^2 K/W$), exceeding the respective R-22 ($3.87 \, m^2 K/W$) and R-40 ($7.04 \, m^2 K/W$) rule of thumb minimum values.

4.9.2   *University of Illinois: Gable Home*

University of Illinois's Gable Home (Figs. 4.17–4.21) was designed with the goal of being Passiv Haus certified. This design approach is evident in the house's impressive envelope. With a whole envelope R-value of 40.7 ($\sim 7.17 \, m^2 K/W$), this house was by far the best insulated. The high envelope R-value is due largely to R-63.2 ($\sim 11.13 \, m^2 K/W$) walls and an R-70.2 ($\sim 12.36 \, m^2 K/W$) roof which both exceed rule of thumb design criteria. High R-value windows also helped super-insulate this house, drastically reducing the need for heating and cooling systems (Table 4.4).

The direct gain windows were too large by the rule of thumb criteria with a ratio of 22.7% and no additional thermal mass. This was beneficial during the competition week due to low outdoor temperatures. The team also used windows with a 0.52 solar heat gain coefficient which falls in the recommended range. It could cause some problems in the summer months, especially with such a highly insulated envelope. Aside from the direct gain windows, Illinois' window design was the best of all teams. They were the only team to design the north, east and west windows to the rule of thumb criteria. The house only has a 0.67% window to floor ratio for both the east and west walls and keeps the north wall windows below 4%. All of these measures helped reduce heat loss during the winter. The Illinois House also employs shading strategies which meet all five of the design criteria mentioned above.

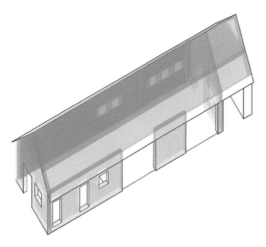

Figure 4.17.   University of Illinois; Solar Decathlon (2009): Gable Home – outer wall insulation.

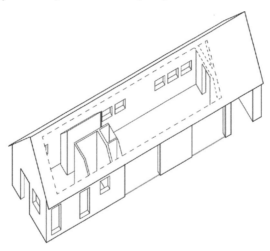

Figure 4.18.   University of Illinois; Solar Decathlon (2009): Gable Home – northern daylight windows.

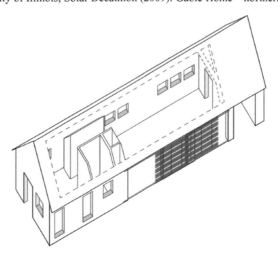

Figure 4.19.   University of Illinois; Solar Decathlon (2009): Gable Home – sun shading devices.

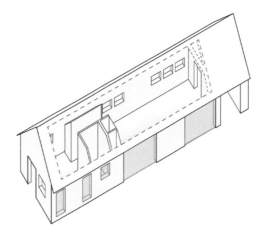

Figure 4.20.   University of Illinois; Solar Decathlon (2009): Gable Home – southern fenestration.

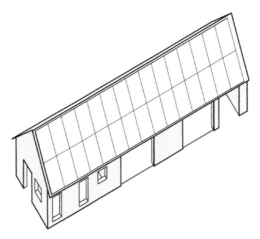

Figure 4.21.   University of Illinois; Solar Decathlon (2009): Gable Home – position of thermal mass.

Table 4.4.   University of Illinois: Gable Home analysis summary.

Passive solar design strategies	Rule of thumb			Ranking
Direct gain	22.7% solar glazing ratio			
Thermal mass	Little	$7.3\,m^2$ ($79\,ft^2$) unaccounted for		
Indirect gain	None			
Shading strategy	All five criteria met			
Heat retention/insulation	Whole envelope R-value	40.7 ($\sim 7.17\,^2$K/W)		Best
	Wall R-Value	63.2 ($\sim 11.13\,m^2$ K/W)		
	Roof R-value	70.2 ($\sim 12.36\,m^2$ K/W)		
Windows	Low U-value	0.52 solar heat gain coefficient		0.67% window to floor ratio
Daylighting	0.11	North facing	3.99%	
		East facing	0.67%	
		West facing glass	0.67%	

This house had an effective aperture of 0.11 which allowed for some daylighting but not the maximum allowed by the rule of thumb. Again, with minimal window area, the house can retain heat well but at the expense of reduced daylighting capabilities. With such reduced non-solar glazing, the majority of lighting is gained through the solar glazing on the south facade. The massive amount of insulation most likely saves a greater amount of energy than is saved by replacing electrical lighting with daylighting, especially if energy efficient fixtures are used.

4.9.3   *Team Ontario BC: North House*

This joint team from Ontario and British Columbia (BC) Canada, most likely the coldest home location of any 2009 team based its strategy for the North House (Figs. 4.22–4.25) on maximizing the area of solar gaining glass. With 44% solar glazing ratio it incorporated one of the highest 'window-to-wall' ratios of all teams. Not all of this glazing was accounted for by thermal mass,

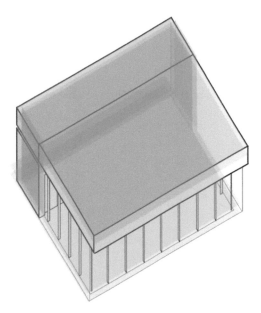

Figure 4.22.   Team Ontario BC; Solar Decathlon (2009): North House – outer wall insulation.

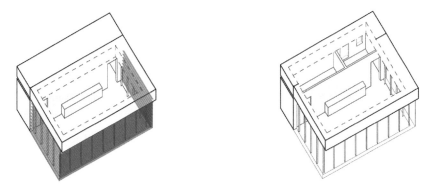

Figure 4.23.   Team Ontario BC; Solar Decathlon (2009): North House – sun shading devices (left) and northern daylight (right).

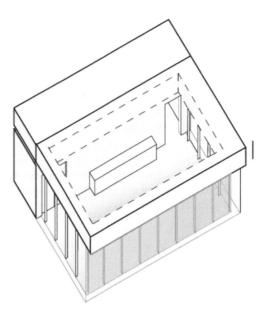

Figure 4.24.   Team Ontario BC; Solar Decathlon (2009): North House – southern fenestration.

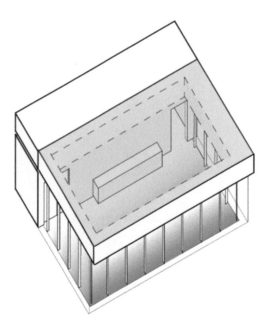

Figure 4.25.   Team Ontario BC; Solar Decathlon (2009): North House – location of thermal mass.

although they also provided amongst the highest amount of thermal mass. Their effective aperture (*EA*) value was therefore also amongst the highest and their insulation value somewhere in the middle of all possible values. The low *R*-values of their windows were compensated by extreme insulation values in the roof and remaining opaque walls, but remained below 20 ($\sim$3.52 m^{2}K/W) in total. The Solar Heat Gain factor was selected within the recommendations, but especially east and west windows were oversized. East and west windows will not produce much gain in

Table 4.5. Team Ontario BC: North House analysis summary.

Passive solar design strategies	Rule of thumb		Ranking
Direct gain	44% solar glazing ratio		Highest value
Thermal mass	Lots	Lots not accounted for	
Indirect gain	None		
Shading strategy	Movable external blinds		
Heat retention/insulation	Whole envelope R-value	13.6 ($\sim$2.40 m^{2}K/W)	
	Wall R-value	44.5 ($\sim$7.84^{2}K/W)	
	Roof R-value	60.9 ($\sim$10.73 m^{2}K/W)	
Windows			
Daylighting	1.357	North facing	6.07%
		East facing	16.48%
		West facing glass	21.97%

winter due to the solar orientation especially in the northern latitudes of Canada but also might not create terrible overheating due to milder climates in these locations. Still shading is deployed on all three glazed envelope surfaces. The house incorporated amongst the highest R-value for windows. In spite of its non-alliance with common sizing requirements for passive solar design, Team Ontario BC were still able to do well in the competition contests due to a large PV array and well-tuned active systems. The shading devices were mostly deployed and the negative impact of the low R-value could not be felt as temperatures were still mild during the competition, much in opposition to the winter climate the house might now encounter at its final location (Table 4.5).

4.9.4 Team Germany

The solar house designed and built by Team Germany won the 2009 competition (Figs. 4.26–4.30). It also performed well on most passive solar strategies. The house incorporated large amounts of

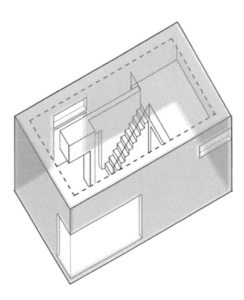

Figure 4.26. Team Germany; Solar Decathlon (2009): Solar Home – outer wall insulation.

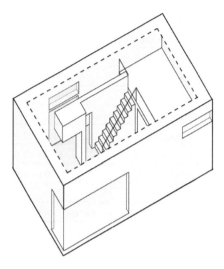

Figure 4.27.   Team Germany; Solar Decathlon (2009): Solar Home – northern clerestory windows for daylighting.

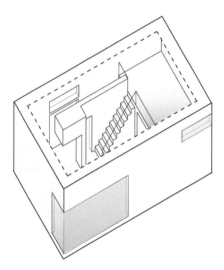

Figure 4.28.   Team Germany; Solar Decathlon (2009): Solar Home – southern fenestration.

south facing solar windows with 24% *WWR*, but all solar glazing was accounted for by thermal mass in the form of novel PCM. This fact highlights the possibility to increase solar glazing with the increase of thermal mass and innovative material solutions. Due to the large amount of solar glazing, also the effective aperture (*EA*) for daylighting was slightly increased, which was compensated for by extremely high *U*-values for windows and doors, the second highest after Illinois. Team Germany (Table 4.6) also designed the house towards the passive house standard, which was reflected in the high *R*-values and the tight envelope. East, west and north window areas were larger than recommended, but all were at least partially shaded. Team Germany was able to keep temperatures within the comfort zone at a high rate with one of the lowest deviation values. The one negative aspect of the German house might have been the dark surfaces of the shading devices, which would be detrimental in climates with very hot summers due to increase in cooling loads especially on the west and east side of the home.

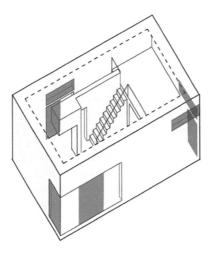

Figure 4.29.   Team Germany; Solar Decathlon (2009): Solar Home sun-shading devices.

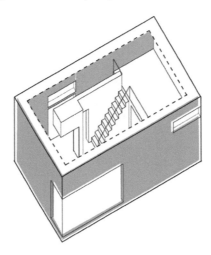

Figure 4.30.   Team Germany; Solar Decathlon (2009): Solar Home – location of thermal mass.

Table 4.6.   Team Germany analysis summary.

Passive solar design strategies	Rule of thumb		Ranking
Direct gain	24% solar glazing ratio		
Thermal mass	Novel PCM	All accounted for	
Indirect gain	None		
Shading strategy	All windows were at least partially shaded		
Heat retention/insulation	Whole envelope *R*-value	27.7 ($\sim$4.88 m^{2}K/W)	
	Wall *R*-value	66.3 ($\sim$11.68 m^{2}K/W)	
	Roof *R*-value	85.1 ($\sim$14.99 m^{2}K/W)	
Windows	Extremely high *U*-values		
Daylighting	0.367	North facing	21.88%
		East facing	14.83%
		West facing glass	10.35%

4.9.5   *Iowa State University: Interlock House*

Iowa State University's Interlock House (Figs. 4.31–4.35) had a reasonable amount of solar glazing with a 9% ratio. This falls within the 7 to 12% range making this house one of only two to stay below 12% and the only house to actually fall in the desired range for solar glazing. The house incorporated a moderate amount of thermal mass with poured concrete counters in the kitchen, bathroom and bedroom. The horizontal orientation of the mass meant that it functioned as unlit floor which does not account for very much solar glazing. Only $0.97\,\text{m}^2$ ($10.5\,\text{ft}^2$) of solar glazing was left unaccounted for, the third lowest amount of all houses (Table 4.7).

This house had the third best overall envelope R-value with 20.4 ($\sim3.60\,\text{m}^2\text{K/W}$). Although the team only designed the R-33.2 ($\sim5.85\,\text{m}^2\text{K/W}$) walls to be above the rule of thumb value, they came close with the R-38.9 ($\sim6.85\,\text{m}^2\text{K/W}$) roof. What really created a highly insulated envelope were the house's windows and doors. The windows had the highest R-value of any at the competition with an R-11.1 ($\sim1.95\,\text{m}^2\text{K/W}$) insulating value due to the krypton gas fill, selective

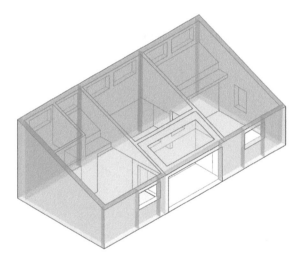

Figure 4.31.   Iowa State University; Solar Decathlon (2009): Interlock House – outer wall insulation.

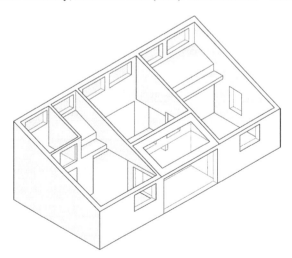

Figure 4.32.   Iowa State University; Solar Decathlon (2009): Interlock House – northern clerestory windows for daylighting.

films, and triple panes. The house also boasted some impressive insulating doors. The prototype doors take advantage of high *R*-value vacuum insulation panels. With the combination of efficient doors and windows, the design team was able to drastically reduce the impact of the traditional weak spots in the building's envelope.

Daylighting was a little short in this house. With an effective aperture of 0.119, there is only slightly more emphasis on daylighting in this house than in the Illinois House. As with the Illinois House, the increased wall area and reduced window area helps retain heat but creates a need for additional electric lighting. Window percentage on the east side is below the rule of thumb threshold at 2.02% while there are no windows on the west side. Only the north windows were too large with a 13.83% ratio. The house also had appropriately designed shading with a combination of indoor curtains and adjustable outdoor aluminum louvers.

This house was one of only two to incorporate a sunspace for indirect solar heating. The sunspace is roughly 10% of the volume of the living space and has a solar glazing ratio of 114.7%. Additionally, the sunspace acts as an airlock for entering the house which minimizes the amount of air infiltration.

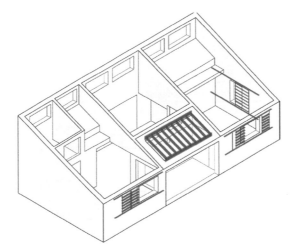

Figure 4.33.   Iowa State University; Solar Decathlon (2009): Interlock House – sun shading devices.

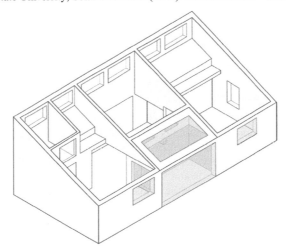

Figure 4.34.   Iowa State University; Solar Decathlon (2009): Interlock House – southern fenestration.

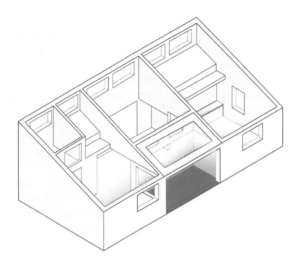

Figure 4.35.   Iowa State University; Solar Decathlon (2009): Interlock House – location of thermal mass.

Table 4.7.   Iowa State University: Interlock House analysis summary.

Passive solar design strategies	Rule of thumb			Ranking
Direct gain	9% solar glazing ratio			
Thermal mass		$0.97\,\mathrm{m}^2$ (10.5 ft^2) of solar glazing was left unaccounted for		
Indirect gain	Sun space with solar glazing ratio of 114.7%	10% of the volume of the living space		Only one of two teams
Shading strategy	All five criteria are met			
Heat retention/insulation	Whole envelope R-value		20.4 ($\sim 3.60\,\mathrm{m}^2\mathrm{K/W}$)	Third best
	Wall R-value		33.2 ($\sim 5.85\,\mathrm{m}^2\mathrm{K/W}$)	
	Roof R-value		38.9 ($\sim 6.85\,\mathrm{m}^2\mathrm{K/W}$)	
Windows		11.1 ($\sim 1.95\,\mathrm{m}^2\mathrm{K/W}$)		
Daylighting	0.119	North facing	13.83%	
		East facing	2.02%	
		West facing glass	none	

4.9.6   *Team Minnesota: Icon House*

Team Minnesota had designed its Icon House (Figs. 4.36–4.37) on the upper limit of solar glazing to floor area ratio with 28% and only partially accounted for this excessive amount with additional thermal mass. The *EA* ratio was close to 0.18 due to highly insulated windows with low visible light transmission. Due to its northern climate the team was very aware of insulation and provided one of the highest R-values for walls and roof and some of the best performing windows, but still only around 15 ($\sim 2.64\,\mathrm{m}^2\mathrm{K/W}$) for the whole envelope (Table 4.8). While reducing its west windows, the house had windows in excess of the recommendation on the north and east. Over the course of the year these windows would lead to more losses than gains. But during the competition week, the house was able to keep temperature fairly homogenous and ranked in the top ten of in the comfort zone competition. Team Minnesota was one of the few teams, which had developed a fixed overhang on the east and west to mitigate the gains for those orientations.

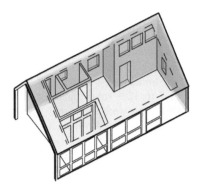

Figure 4.36.   Team Minnesota; Solar Decathlon (2009): Icon House – outer wall insulation.

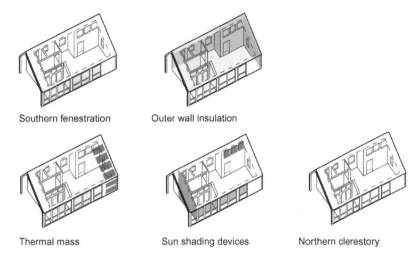

Southern fenestration   Outer wall insulation

Thermal mass   Sun shading devices   Northern clerestory

Figure 4.37.   Team Minnesota; Solar Decathlon (2009): Icon House – passive solar design strategies: southern fenestration, outer wall insulation, thermal mass, sun shading devices, northern clerestory
windows.

Table 4.8.   Team Minnesota: Icon House analysis summary.

Passive solar design strategies	Rule of thumb		Ranking
Direct gain	28% solar glazing ratio		
Thermal mass	Only partially accounted for		
Indirect gain	None		
Shading strategy	Fixed overhang on the east and west to mitigate the gains for those orientations.		
Heat retention/insulation	Whole envelope R-value	15.6 ($\sim$2.64 m^{2}K/W)	
	Wall R-value	52.2 ($\sim$9.19 m^{2}K/W)	
	Roof R-value	70.1 ($\sim$12.35 m^{2}K/W)	
Windows			
Daylighting	0.164	North facing	9.57%
		East facing	9.32%
		West facing glass	1.41%

4.9.7   *Summary of house evaluations*

While many teams did not design houses to meet the passive solar rules of thumb, there was some visible evidence that following these rules increased the performance of the houses. Several of the rules of thumb recommended staying below certain limits. Many of the graphical comparisons showed trends that contradicted the limits set by the rules of thumb. Most notably of these rules would be the direct-gain limit of 7 to 12% solar glazing and the maximum limit of 0.18 for effective aperture. Other rules of thumb were supported by graphed data. The most strongly supported relationship was between increased R-values and improved temperature control and average temperature.

Unfortunately, the nature of the Solar Decathlon does not put an emphasis on passive solar design in the objective contests. The comfort zone competition is more a measure of how well the HVAC system works in a house than how well the house is designed. Additionally, the energy balance competition draws the emphasis away from passive solar design and placed it on purchasing excessively large photovoltaic arrays. This essentially facilitates the status quo and removes the incentive to make the buildings themselves more efficient. These issues of the 2009 competition had been recognized and acknowledged in the 2011 competition. While the architecture and engineering subjective contests allow for credit to be given to passive solar features, there remains no opportunity to effectively measure the performance impact of these design features within the context of the competition. Annual performance evaluations would be necessary to fully evaluate their potential.

As a means of predicting future performance in a Solar Decathlon competition, the design rules of thumb were minimally effective. But still the passive rankings are a good way to get a rough idea of where a house would place in the comfort zone competition but rarely does it exactly predict the actual performance in the competition. Only two houses have the same rule of thumb placement and comfort zone rank. Still the competition is only measuring performance during 8 days in a competition scenario and not over the course of a year. This performance evaluation is the task of the next two sections, which highlights the enormous potential of passive solar architecture with improved envelope technologies. Based on these comparisons, rules of thumb for passive solar house design are still good performance indicators for rough estimations during the design phase especially for cold climates. If further accuracy is desired a computer model should be utilized.

4.10   SIMULATING SOLAR RADIATION IN THE INTERLOCK HOUSE IN WINTER

The Interlock House was designed as a single-story house with 67 m^2 (approx. 720 ft^2) open floor plan with a sunspace facing south which is composed of removable glass walls and skylight as shown in Figure 4.38. The sunspace plays different roles in different seasons. It can be closed to collect solar radiation in winter and opened to natural ventilation in summer. Even with all glass walls closed, there are 3 operable windows connecting the air in the sun space with air in the main space (living room, kitchen and bedroom). An NSF (US National Science Foundation) funded research project studies airflow with solar radiation in the Interlock House in more detail (Iowa EPSCoR, 2013).

In ANSYS FLUENT 14.0 the winter condition was simulated with the following settings: all exterior windows are closed; in the sunspace the two windows above and the one connecting to the bedroom at lower level are open. The U-value is 1 W/m^{2}K for windows and 0.12 W/m^{2}K for walls, roof and floor. Outside the air temperature is 270 K (−3°C) with 748 W/m^2 direct and 232 W/m^2 diffuse solar radiation. The *SHGC* is 0.41 for south and 0.62 for north (He *et al.*, 2012).

In Figures 4.38 and 4.39, the roof; south, east and west walls are removed for better presentation. While the floor and second glass door are heated to 320 K (47°C), the frame wall above the sunspace reaches up to 325 K (52°C) because of the extra solar radiation penetrating through the

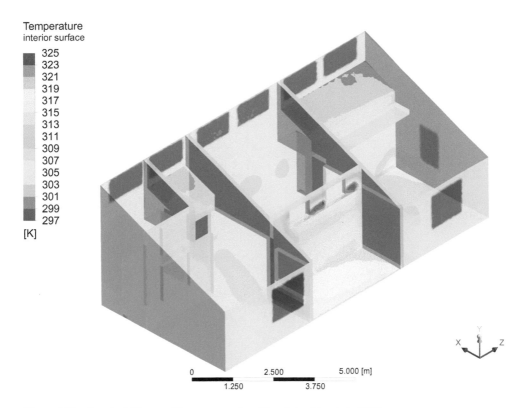

Temperature
interior surface

325
323
321
319
317
315
313
311
309
307
305
303
301
299
297
[K]

Figure 4.38.   Interlock House with sun space closed in winter (He *et al.*, 2012).

skylight. East kitchen and west bedroom are also heated by the solar radiation passing through the south facing windows on each side. However, the high point on the west side is 10 K lower than the east side, which proves that low velocity airflow at corners helps to accumulate heat. There is also an area at the bottom of the north wall with higher temperature than on ambient surface. The heat derives from the solar radiation through the two small operable windows above the sunspace.

As shown in Figure 4.38, solar radiation heats the floor in the sunspace between the two glass doors. The floor behind the second glass door is also heated but not as much as the floor in the sunspace. Most air can be warmed up to 307–310 K (34–37°C) in the sunspace and 305–308 K (32–35°C) in the living room according to these simulations.

Organizing convective air loops through strategically placed windows is an important design strategy to distribute the heat evenly and minimize temperature difference in a closed complex space. In the case of the Interlock House the inlet and outlet are on opposite sides and heights. In the sunspace hot air rises along the second glass door and flows into the main space at a speed of 0.2 m/s as shown in Figure 4.39. Due to the mass equation, as long as air is flowing out air should also be flowing in. In this case the lower opening, which is the operable window connecting sunspace to the bedroom, provides the inlet air to the sunspace at a speed of up to 0.4 m/s. Due to the air loop which connects living room, bedroom and sunspace (see Fig. 4.39), the temperature in these space ranges between 307–312 K (34–39°C). Although the main air stream does not pass through the east kitchen, the air movement in the kitchen is also driven by the solar radiation and forms loops in a relatively isolated zone. Only the southeast corner has 1 degree cooler temperature than other areas. This simulation highlights, that heat is distributed evenly in the open floor plan.

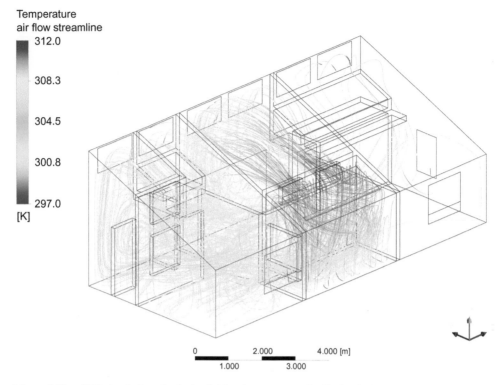

Figure 4.39.   CFD simulation of velocity field and temperature distribution in the Interlock House in winter (He *et al.*, 2012).

With a very sealed envelope, which limits air infiltration, the Interlock House could actually become too hot on a sunny day in winter, so that the users might need to open windows to let in some cold air. Alternatively this simulation indicates that more thermal mass could benefit the energy balance between day and night.

4.10.1   *Interlock Gouse: measured data in winter*

Measured data (Fig. 4.40) supports the effectiveness of passive heating by solar radiation in winter. The figure shows an exemplary situation as recorded on November 23, 2012: the outside temperature was below 0°C (32°F) during most of the day. The solar radiation was at a medium level, median figure around 400 W/m^2. The solar radiation during sunny days in winter could be as high as 900 W/m^2 as recorded by the weather station on the roof. Only with this medium level of solar radiation, the house was heated up and kept around 20°C (72°F) during the day, and the sunspace temperature reached about 30°C. It is certain that good insulation retained energy from the warm days before, which delayed the cooling down of the interior but it still proves that the heat gain from solar radiation can balance the heat loss and warm the house in winter with good insulation and space design (Fig. 4.41).

4.10.2   *Outlook for future simulations and measurements*

The CFD simulations and visualizations of the Interlock House conducted with ANSYS FLU-ENT14.0 reveal the physics behind indoor air movement and highlight the power of solar radiation. The simulation also highlights the importance of spatial composition for heat transfer through

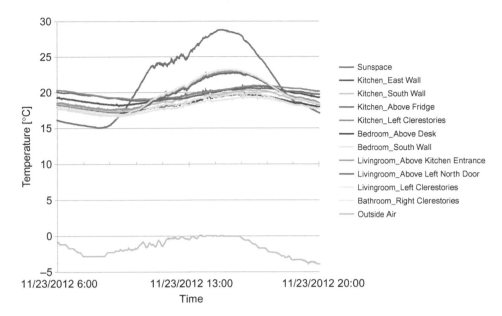

Figure 4.40.   Comparison of temperature recorded in November 2012 by indoor and outdoor sensors.

Figure 4.41.   Solar Sun space of the Interlock House (Iowa State University) (Photo credit: Department of Energy; Jim Tetro).

natural air movements already found important for natural cooling (Stoakes *et al.*, 2011a; 2011b). It might even in some instances create overheating, thus the possibility to modulate the heat is important. Direct solar radiation transmitted through glass always gives a high temperature spot on a certain indoor surface and this surface then serves as the "fireplace" for the whole house. Together with radiation absorbed by water vapor in the air conduction from this hot spot increases the interior temperature in winter. In the process of conduction hot air rises around the hot spot and develops air loops which spread the heat to other zones. To keep the whole space warm, both insulation and air loop matter greatly and great attention should be paid to the design of the air loop, essential to spread heat from one spot to the whole room/house. Once the loop is blocked or air leakage releases the heat to the outside, hot air cannot reach the relatively cold indoor zones and mechanical heating might be needed at a certain point. Refined quantification of passive solar design elements in combination with new innovative materials like phase change materials (PCM) and innovative glass coatings and assemblies has a strong future in energy efficient design and will help to keep active systems small and costs low.

4.11   SOLAR THERMAL STORAGE WALL IN COLD AND CLOUDY CLIMATES

A passive solar design strategy, which was not used by any of the successful 2009 Solar Decathlon teams, was the solar thermal storage wall or Trombe wall.

The un-built entry to the 2011 Living Aleutian Home competition by Jorge Rodriguez and Yue Zhang uses such a device and aims to create a replicable model for the village of Atka in Alaska. The project uses a series of spatial energy strategies that fulfill the standards of the Living Building Challenge (LBC) 2.0 (Living Building Challenge, 2013).

In order to incorporate the Native Aleutian living style, spatial strategies are based on communal rather than individual needs. The traditional indoor spaces of the region, called Barabaras (Nabokov and Easton, 1989), which encourage interaction among family members through one large communal space, served as precedent for the spatial organization. The project thus proposes one large indoor living space which incorporates spaces for different daily routines, such as cooking and spending time together only through separated levels or balconies not separate rooms. The project thus incorporates the minimum program requirement for a four-person family, including three bedrooms, bathroom, arctic entrance, kitchen, dining room, living room, storage room, and laundry space. The dimension of the project is 122.5 m^2 (1316 ft^2), distributed on two floors.

Spatial and energy strategies are intricately linked in this project. The most significant space is an indoor facing south garden, which receives grey water for treatment to achieve a net zero water balance. The space also captures and generates passive solar energy using passive solar collection strategies.

Due to Atka's challenging weather conditions, such as diffused solar light throughout most of the year and the strong, constant wind from the southeast the project proposes a series of interrelated strategies to achieve the net zero energy goal of the challenge. Atka is surrounded by hills and the site receives little direct solar gain from December through February, therefore the house is placed far enough from the southern property line to avoid any solar infringement on neighbors as well as to increase optimal solar exposure.

In order to prevent wind infiltration, increase insulation and reduce wind exposure, the project is semi buried. In addition, this reduces the foundation and depth of the bedrock. To propose a replicable project through the village, the project is placed on the flattest area possible, a condition that can be found in most neighboring sites.

Solar passive strategies of this Atka home (Figs. 4.42–4.45) include:

- A design, square in plan that:
  a) Maximizes the ratio of interior floor space to exterior wall space and
  b) Uniformly distributes the heat throughout the house.
- A south facing sunspace and Trombe-wall that captures and stores solar thermal energy. (The Trombe-wall is made from local, dark stone). In addition, the Trombe wall takes advantages of the basaltic geological condition of the island to build a massive thermal mass. It is built by black local basaltic stones with no concrete mortar and contained by a lumber structure and a metallic net.

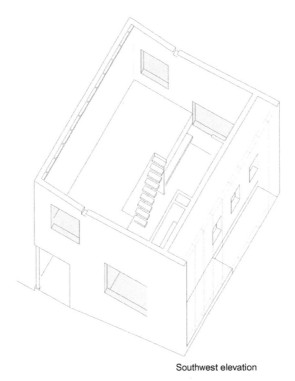

Southwest elevation

Figure 4.42.   Atka Home (Jorge Rodriguez) – daylighting windows.

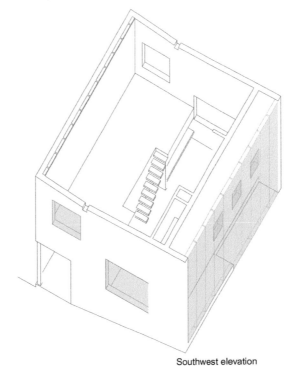

Southwest elevation

Figure 4.43.   Atka Home (Jorge Rodriguez) – southern fenestration.

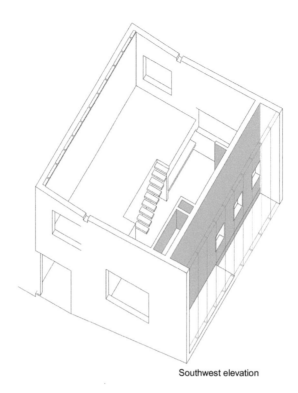

Southwest elevation

Figure 4.44.   Atka Home (Jorge Rodriguez) – location of thermal mass.

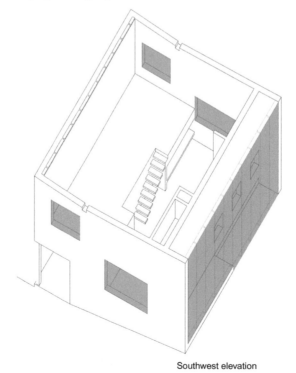

Southwest elevation

Figure 4.45.   Atka Home (Jorge Rodriguez) – sun shading devices.

- The stored energy is released into the living space by radiation and convection and is controlled by insulating panels and operable vents.
- An insulated convection loop of connected chambers circulates stored heat around the outside of the living areas.
- Significant insulation in the walls, ceilings, and floors reduces heat loss to a point where heat input above solar is required for only the coldest months.

4.12 OVERALL CONCLUSION

Passive solar design, like any architectural design is based on geometric proportions and spatial relationships. This chapter intended to highlight the important geometric and proportional relationships required to design passive solar home projects. Thus space and the relationship of opening size to usable area were introduced as powerful design tools. Student competition design projects for the 2009 Solar Decathlon houses and the 2011 Aleutian Home Competition served as case studies to highlight the design challenges in achieving the most appropriate balance between orientation, maximizing solar heat gain, reducing heat transfer losses in winter, adequate daylighting all year round and the prevention of excessive heat gain in summer. Although quantified in measured numbers, these relationships are inherently architectural. The possibilities for diversity are manifold as the design outcome of these seven projects show. With these proportional relationships in mind architecture can harvest solar energy and reduce fossil fuel consumption as well as provide comfortable spaces for living in many challenging climate zones.

ACKNOWLEDGEMENTS

The Solar Decathlon 2009 analysis data is based on Tim Lentz' thesis conducted in 2009 and 2010 (Lentz, 2010). The CFD simulation work done by Shan He was first supported by an ISU College of Design research investment grant and is now supported with funding by the federal US NSF EPSCoR program (Experimental program to stimulate competitive research). The work has first been published with the Association of Collegiate Schools of Architecture (ACSA) in their 2012 Barcelona International conference proceedings (He *et al.*, 2012) and adjusted for inclusion in this chapter. The continuing research on the Interlock House is currently funded by the Experimental Program to Stimulate Competitive Research (EPSCoR) awarded by the National Science Foundation under grant number EPS-1101284. Isabelle Leysens, undergraduate research assistant to the Center for Building Energy Research (CBER) and senior in architecture did a fantastic job diagramming the six 2009 solar decathlon homes as well as the hemicycle house by Frank Lloyd Wright and Jorge Rodriguez has graciously provided his Atka House design drawings and concept description.

REFERENCES

Alberti, L.B.: *On the art of building in ten books (De Re Aedificatoria, 1452)*. Translated by Joseph Rykwert with Neil Leach and Robert Tavernor. MIT Press, Cambridge, MA, 1988.
Ander, G.D.: *Daylighting performance and design*. Van Nostrand Reinhold, New York, NY, 1995.
Balcomb, J.D.: *Passive solar buildings*. MIT Press, Cambridge, MA. 1992.
Chandel, S.S. & Aggarwal, R.K.: Performance evaluation of a passive solar building in western Himalayas. *Renew. Energy* 33 (2008), pp. 2166–2173.
Chiras, D.D.: *The solar house*. Chelsea Green Publishing Company, White River Junction, VT, 2002.
Collins, J. & Nash, Al: Preserving yesterday's view of tomorrow the Chicago world's fair houses. *CRM* 25:5 (2002), pp. 27–31.
He, S., Passe, U. & Wang, Z.J.: Integrating thermo-fluid computational models into understanding sustainable building design. *Proceedings of ACSA International Conference*, 2012.

Iowa EPSCoR: 2013, http://iowaepscor.org/buildingscience (accessed 2/9/2013).

Lentz, T.R.: *Analysis of the passive design and solar collection techniques of the houses in the 2009 U.S. Department of Energy's Solar Decathlon competition*. MSc Thesis, Iowa State University, Ames, IA, 2010.

Living Building Challenge. 2010, https://ilbi.org/lbc (accessed 2/9/2014).

Mazria, E.: A design and sizing procedure for direct gain, thermal storage wall, attached greenhouse and roof pond systems. *Proceedings of the Second National Passive Solar Conference*, Philadelphia, 1978. US Energy Research and Development Administration, Washington DC, 1978.

Mazria, E.: *The passive solar energy book*. Rodale Press, Emmaus, PA, 1979.

Nabokov, P. & Easton, R.: *Native American architecture*. Oxford University Press, New York, NY, 1989.

NREL: About the Solar Decathlon. National Renewable Energy Laboratory, United States Department of Energy, 2010, http://www.solardecathlon.gov/about.cfm (accessed 13/4/2014).

Passiv Haus Institut Germany: 2012, http://www.passiv.de/en/02_informations/01_whatisapassivehouse/01_whatisapassivehouse.htm (accessed 30/7/2014).

Pliny the Younger: *The letters of the younger book two*. Penguin Classics, New York, NY, 1969.

Project Green Home: last updated 2014, http://www.projectgreenhome.org/ (accessed 2/09/2014).

Smith, T.G.: *Vitruvius on architecture*. Monacelli Press, New York, NY, 2003.

Solar Decathlon: last updated 2014, http://www.solardecathlon.gov/past/2009/technical_resources.html (accessed 2/9/2014).

SPSG (Stiftung Preußische Schlösser und Gärten Berlin-Brandenburg). Last updated 2014, http://www.spsg.de/index.php?id=163 (accessed 2/09/2014).

Stoakes, P., Passe, U. & Battaglia, F.: Predicting natural ventilation flows in whole buildings. Part 1: The Viipuri Library. *Proceedings of Building Simulation*, 2011a.

Stoakes, P., Passe, U. & Battaglia, F.: Predicting natural ventilation flows in whole buildings. Part 1: The Esherick House. *Proceedings of Building Simulation*, 2011b.

Storrer, W.A.: *The Frank Lloyd Wright companion*. The University of Chicago Press, Chicago, IL, 1993.

Williams, R.J.: *Passive solar heating*. Ann Arbor Science Publishers, Ann Arbor, MI, 1983.

CHAPTER 5

The human factor in sustainable architecture

Ardeshir Mahdavi

5.1 INTRODUCTION AND OVERVIEW

Buildings and their associated infrastructures constitute a major class of interventions by people in the environment resulting in resource depletion, environmental emissions, and waste generation. To date, such building-related interventions, together with others involving industry and transportation, have comprised the planet's capacity to support the sustenance of future generations. This circumstance has motivated numerous pleas emphasizing the importance of sustainability in architecture. However, as in many other environmentally relevant domains, the theory (how we think the built environment should be) and practice (how the built environment actually evolves) have not converged. Hence, if we are to take the notion of sustainable architecture seriously, we need to approach it critically in the context of the complex and consequential relationships involving people, buildings, and environment. Why do people construct buildings? How can we measure their effectiveness in meeting people's requirements? How can we assess their ecological implications in the context of other human-triggered interventions in the environment?

The primary objective of this chapter is to provide a broad and critical framework to address these and other cognate questions. Toward this end, the chapter is structured as follows. Section 5.2 entails a brief introduction to "Human Ecology" as a fruitful conceptual framework for the discussion of interrelationships between people and their surrounding environment, which is – more often than not – a built environment. Section 5.3 is dedicated to a number of essential background or boundary conditions necessary for a meaningful discussion of sustainability in the building sector. These include, amongst other topics, the central importance of population growth as well as social factors related to standard of living and "life style". Such factors are relevant to the services expected from buildings and the resulting demand for resources. Section 5.4 addresses the indoor environmental conditions necessary to meet the requirements of building users. Thereby, we shall focus on thermal comfort conditions in buildings. In Section 5.5, we deal with people's passive and active influences on buildings' indoor climate and environmental performance. Section 5.6 summarizes the chapter's conclusions.

5.2 HUMAN ECOLOGY

Ecology may be defined as the scientific discipline that deals with the relationships between organisms and their surrounding world. Accordingly, human ecology may be simply defined as the ecology of the *Homo sapiens*. There are many traditions and associated approaches to human ecology. For the purpose of the present discussion, we follow the approach of the "Vienna School of Human Ecology" (Knötig, 1992a; 1992b; Mahdavi, 1996a). Our goal is not the exhaustive treatment of this specific school of thought, but a brief reference to a couple of its useful conceptual instruments.

Construction and operation of buildings and related artifacts may be viewed as an integral part of the totality of largely regulatory operations initiated by human beings as they interact with their surrounding world. Human ecology offers a useful way of thinking about these interactions via a number of somewhat abstract yet versatile concepts. Thereby, a central pair of concepts involves:

- The human beings' *ecological potency*.
- The surrounding world's *ecological valency* (Knötig, 1992a; Mahdavi, 1996b).

Stated in simple terms, *ecology potency* refers to a dynamic human repertoire of capabilities and means of dealing (i.e., coping and interacting) with the surrounding world. On the other hand, *ecological valency* denotes the totality of that surrounding world's characteristics (resources, possibilities, opportunities, challenges, risks, hazards, pitfalls) as it relates to, confronts, or accommodates the human ecological potency repertoire. Note that this latter concept (*ecological valency*) was essentially dealt with in (Uexküll, 1920) and is also akin to the Gibson's concept of *affordance* (Gibson, 1977; 1979).

Given this conceptual framework, we can describe the main consideration in human ecology as the complex and dynamic relationships between the ecological potency of human beings and the ecological valency of their surrounding world. We can thus broadly characterize the entire building construction and operation endeavor in human ecological terms: buildings are constructed and maintained with the (implicit or explicit) intention to favorably influence the relationship between people's ecological potency and the ecological valency of their surrounding world. Such an intention expresses itself, for example, in the "shelter function" of the vernacular architecture (Mahdavi, 1989; 1996c). In today's building delivery process, this intention is often expressed explicitly and formally, for example when desirable indoor environmental conditions are specifically defined and are expected to be maintained in the course of building operation. Provision of desirable occupancy conditions, or in other words, maintaining a high degree of "habitability" may be thus seen as the main objective of building activity. The challenge is to secure habitability with a minimum of resource depletion and environmental impact.

Evaluating the degree of the habitability of the built environment is another problem that can benefit from human ecology's concepts. Specifically, we have a second pair of concepts in mind, which concerns distinct aspects that can be attributed to the relationships between people and their surroundings. Thereby a high-level distinction is made between the *material-energetic* and *information-related* aspects of these relationships (Knötig, 1992a; Mahdavi, 1992; 1996a). These two aspects can be assigned to every entity, state, and process. The *material-energetic* aspect refers to the assumption that nothing exists unless some amount of matter or energy is involved. The information-related aspect refers to the assumption that matter and energy have a certain distribution in space and time, which can be represented in terms of a structure. An information content can be correlated with this structure.

The relevance of this pair of concepts to the present discussion becomes clear if we ask ourselves how the habitability of the built environment is to be measured. An important component of this judgment lies in people's subjective experiences and opinions. We can argue that such subjective evaluation processes of the built environment involve both the material-energetic and the information-related aspects of the relationships between inhabitants and the built environment. A common approach to "operationalize" such evaluation processes in planning and operating involves the use of "psycho-physical" scales. The idea is that exposure to various levels of physical (material-energetic) stimuli translate – in a more or less predictable way – into corresponding subjective experiences. For example, exposure to increasing levels of sound intensity are said to translate into an experience of increased loudness and associated stress (annoyance). But psychophysical scales have been shown to be debatable. It would be highly problematic to postulate a deterministic relationship between measurable environmental factors and occupants' evaluation of environmental conditions (Mahdavi, 1996a; 1996b; 2011a).

In a nutshell, it appears that human evaluation processes are generally easier to describe and predict in exposure situations dominated by the material-energetic aspect of the environmental relationships. In extreme cases of high-intensity exposure, the necessity for protective regulations is self-evident due to the obvious health hazards for the involved individuals (e.g., irreversible physical damage to the organism). It is, thus, not surprising that most efforts toward predicting the outcome of human evaluation processes have focused on the identification of a measurable material-energetic scale (such as sound pressure level) to which subjective judgments (such as the degree of annoyance) are expected to correlate. However, the impact of internal information processing on the degree of expressed dissatisfaction associated with various energetic levels of exposure has been demonstrated in many experimental psycho-acoustic experiments (see Section 5.3 for a further discussion of this point).

5.3 BOUNDARY CONDITIONS

5.3.1 *Motivation*

We will address the more immediate implications of the human factor for the built environment (i.e., human requirements and behavior) in Sections 5.4 and 5.5 of this chapter. But it is important that we approach the sustainability discourse in architecture from a broad and critical standpoint. Thereby, we need to consider the relevant boundary conditions and pertinent criteria for sustainable built environments. Moreover, the expectations from – and potential of – sustainable architecture should be seen in the context of the environmental implications of other areas of social development and human activity (Mahdavi, 2012). Relevant are in this context a number of critical questions such as the following:

- How important is the antecedent treatment of phenomena and processes such as population growth, lifestyle development, as well as agricultural and industrial production for the effectiveness of sustainable building efforts?
- How do the relative resource needs and environmental loads associated with building activity compare to other domains of human activity and production such as industry and transportation?
- To which extent can contextual factors such as urban planning decisions and mobility solutions affect and constrain the energy and environmental performance of individual building objects?
- Do we sufficiently consider and account for the impact of user behavior (including the rebound effect) on the energy and environmental performance of buildings?

These questions are evidently far from trivial. However, we do need to deal with them, if we are to properly treat the subject of the human factor in the context of sustainable architecture discourse.

5.3.2 *Population and life style*

Many treatments of the global sustainability debate often start by a mention of the world population and the implications of its ongoing increase. United Nation's data (medium growth scenario) project for 2050 alone for India and China a combined population number of three billions. Likewise, the population of Africa – slightly over 200 million around 1950 – is projected to approach two billion by 2050. Population increase appears to be taken as a fact of fate. The topic of population growth containment appears thorny and difficult politically. It is hence alluded to in discussions of sustainable architecture, if at all, only briefly as a given boundary condition to be forgotten as soon as the debate has moved to individual instances of some recent building projects and their reputed sustainability features. Even if the population growth is going to be regarded as an inevitable and unalterable process, at least the implications for resources, environment, as well as ecological and social systems should be frankly discussed, rather than evaded. In human ecological parlance, the ecological valency of an ecosystem can sustainably support only a finite number of people of a given ecological potency. Transgressing that limit invariably results in an ecological degradation of the environment.

The ecological strain resulting from population increase is aggravated by a parallel process involving the improvement of living standards – at least for some populations – around the world. For instance, the global primary energy consumption of China is projected to increase – from the value in 2008 – about 230% to reach roughly 200 exajoules by the year 2035 (EIA, 2012). Moreover, whereas in the last twenty years per capita energy use has been stagnating (albeit at a very high level) in countries such as United States and Germany, both per capita energy use and gross national income (GNI) have been increasing in China, India, and Brazil (Databank, 2014). In fact, the gross bational product (GNP) of China and India is projected to increase within a period of 40 years (2009–2049) roughly by a factor of 6 and 9 respectively. Lest these assertions are misunderstood: a rise in people's standard of living is necessary, crucial, and desirable socially and ethically. However, even though not a necessity, per capita improvement in living standard is typically mirrored in per capita increase in resource depletion and environmental impact. For example, within a short period of about 25 years (1976 to 2001) per capita meat consumption in China increased by a factor of 5 to reach a value of roughly 50 kilograms of meat per person per year. The environmental ramifications of such developments can be easily exemplified, if one is reminded of the high animal feed conversion, i.e., kilogram animal feed to kilogram animal product, which is over 4 for chicken and eggs, about 11 for pork and almost 32 for beef (FAO, 2012). Human ecologically speaking, rise in indicators such as GNP and GNI can be interpreted as a population's increased ecological potency. A precarious implication of this rise is the population's higher capacity to rapidly exploit natural resources and intervene in the working of natural systems, resulting in all too familiar negative consequences.

The relevance of this conclusion for the present discussion cannot be emphasized enough. The combined ecological and environmental effects of concurrent population growth and the rise of the so-called standard of living can easily undo incremental advances in efficiency improvement effects via sustainable architecture.

5.3.3   *Buildings, industry, mobility*

The relative share of buildings in energy use and environmental impact, as compared to other areas of human activity and production such as industry, agriculture, and transportation is undoubtedly significant. In the European Union, the relative energy demand for the domains transportation, industry, and buildings (residential and commercial sector) was in the year 2000, 31%, 28%, and 41%, respectively (Janssen, 2004). Hence, to compare buildings' share in resource depletion with other areas is not meant to trivialize it. Rather, such a comparison can serve two purposes: at a global level, the investment costs and efforts for efficiency improvement potential in each domain can be compared to those in the other domains. Thus, public funds and financial incentives such as tax exemptions could be primarily deployed in the areas where maximum effect in energy efficiency improvement and environmental impact reduction can be achieved with minimum investments. At a more detailed level (particularly in cases where behavioral decisions of individuals may matter), building-related energy use and environmental impact issues could be assessed in a broader context, including also individual energy use profiles and life style considerations.

This latter point can be further illustrated if we consider the relative energy allocation to various purposes and activities of middle-class individuals in a European country (Mahdavi, 2010). It seems activities such as driving cars, travelling in airplanes, and using electronic gadgets are not all insignificant as compared to energy requirements for heating of buildings. As mentioned earlier, the so-called developing countries increasingly adapt both production and consumption practices of the so-called developed countries. This invariably leads to increased energy use, resource depletion, and negative environmental impact. A significant case in point in this regard pertains to the automobile industry. It is expected that the global demand for automobiles will increase from currently around 70 million to around 110 million in the year 2020. Thereby, according to some estimates, the highest rate of increase will occur in growing markets such as

China. An even more senseless case of energy and resource usage pertains to the requirements for and implications of the world-wide production and deployment of weapons. Expenditures for weapons dramatically increased in the years 2001 to 2011 (over 80% in USA and Russia: close to 190% in China, around 60% in Saudi Arabia and India).

The decisive role of the human factor in the sustainability discourse is not limited to the fact of population increase. Nor is it a sole function of increased affluence in some populations around the world. A highly important – but insufficiently understood – variable concerns the behavior of individuals and populations. Human behavior denotes here life styles, consumption habits, and interaction patterns regarding energy-consuming technical systems (vehicles, appliances, buildings' technical systems, etc.).

Broadly speaking, people's behavioral tendencies may in certain cases favor solutions and products that are disadvantageous from the energy and environment point of view. For instance, in various analyses of human mobility (see, for example, Knoflacher, 1996), it has been argued that people, confronted with multiple options to embark on a trip, display a tendency to favor the reduction of their personal physical exertion. In other words, people may prefer mobility options such as driving a car that could take more time such as searching for a parking place and require more energy use such as fuel for the vehicle, rather than an option such as walking and biking that would require from them a higher level of physical exertion. On the other hand, the choice of a certain mobility medium also depends on the availability of options. If a regional and urban setting provides an extensive infrastructure tailored for individual transport, it should not come as a surprise if the cars would emerge as the dominant mobility medium. In contrast, increased provision of well-designed and convenient pedestrian routes and bike paths has shown to result in a correspondingly higher rate of trips made on foot or on bike. This complex interdependency of behavioral preferences and infrastructural conditions has arguably a decisive influence on the emergence, evolution, survival, and demise of technical solutions and their respective environmental implications. Behavioral patterns with implications for sustainable development can be influenced, to a certain degree, not only by consciousness raising measures and information campaigns, but also – and perhaps more effectively – via proper design and planning measures and strategies as well as proper economic incentives. To disregard such interdependency based on the clichéd "social engineering" criticism would be shortsighted, if not dissembling.

5.3.4 *Human behavior and the built environment*

In respect to the narrower context of the built environment, only recently more attention is paid to the ramifications of people's presence and actions in buildings. Occupants operate buildings' control devices such as windows, shades, luminaires, radiators, and fans to bring about desirable indoor environmental conditions. These control actions have obviously a significant impact on buildings' performance (Mahdavi, 2011a; 2011b). A better understanding of the patterns of human presence and control-oriented behavior can conceivably facilitate and guide both technical responses such as occupancy-sensitive environmental control systems and information campaigns toward improving buildings' energy efficiency and environmental performance (see Section 5.5 for a detailed treatment of these issues).

A further example for the relevance of human behavior in the sustainable architecture discourse concerns the so-called rebound effect (Sorrell, 2007). This effect refers to the paradoxical circumstance that, under certain conditions, energy efficiency measures may end up increasing the energy use (see Section 5.4). Instances of architecturally relevant rebound effect underline the importance of the human factor in energy efficiency potential assessments and projections. For the expected efficiency effects of technical measures to materialize, behavioral and economical boundary conditions must be taken into consideration. This also explains why some energy and environment experts suggest that the energy prices must be adjusted in tandem with energy efficiency improvements.

5.3.5   *Concluding reflections*

We should derive some conclusions from the previous discussion before turning to a detailed treatment of human requirements and human behavior in the next sections of this contribution. The above discussion of the boundary conditions underscores the need to broaden the scope of the conventional sustainable architecture discourse. They suggest that there may be an extensive set of essential higher-level measures and actions involving human populations and individuals that could be undertaken parallel with, if not prior to measures focused on individual buildings:

- Population increase needs to be addressed in earnest. Whether the focus is on a region, a country, a continent, or the entire planet, the respective ecological valency or the available carrying capacity must be considered: to which extent and for how long could this finite capacity support the product of a given population and its life style? It should be obvious that both cannot be maximized.
- In policy, in education, and all manners of social discourse, the distinction between the standard (or quality) of living and purely economic measures such as GNP must be taken seriously. When a society's entire economical and political system is exclusively focused on production, property, power, and profit, it should not come as a surprise if genuine sustainability concerns can be forgone in favor of superficial and ancillary agenda. Just shifting industrial production away from detrimental and non-essential products such as those serving militarization, arms race, and war would go a long way to contributing to global sustainability objectives.
- The existing mobility paradigm needs to be reexamined. Individual energy and emission intensive modes of motorized transportation need to be radically reduced. It is essential to move away from the practice of designing and organizing cities around cars, instead of around pedestrians. The environmental damage due to energy use and emissions associated with mobility need to be included in the relevant economic calculations. Likewise, unjustified tax exemptions for fuel usage in aviation should be removed. As a general principle, the environmental cost of energy and resource use must be considered in related pricing schemes. Moreover, the economic value of energy and resources should be continuously revisited to incorporate the effect of technological advances in high-efficiency energy and resource deployment.
- Individuals and communities, especially in affluent societies, need to be confronted with the environmental consequences of their life style choices in view of mobility, residency, diet, and recreational activities. Efforts are needed to achieve consensus on required measures and policies that would facilitate beneficial and adequate life style changes.
- In the narrower realm of architecture, overarching sustainability considerations play an important role as well. Are sustainable development concerns considered in the applicable urban planning prerequisites and land use regulations? Is the intended construction activity (e.g., erection of specific buildings) motivated by a genuine necessity? Could such necessity be accounted for through retrofit and adaptation of existing buildings instead of new construction? Only when such questions have been thoroughly and critically addressed, would it make sense to frame the architectural sustainability discussion at the level of individual new building projects.

In the past, such recommendations for action have not been sufficiently taken heed of. Explaining the numerous social, economical, and political factors responsible for this circumstance would go beyond the scope of the present contribution. But this does not bear on their validity and urgency.

5.4   BUILDING PERFORMANCE AND HUMAN REQUIREMENTS

5.4.1   *Introductory remarks*

In building science, the habitability of indoor environments is often treated in distinct domains involving, for instance, thermal, visual, and acoustical aspects of building performance. This is prudent as a matter of expediency, as long as we keep in mind that an aggregate judgment of indoor environment quality is a dynamic process and the complex result of material-energetic and information-related processes pertaining to all such domains. Approaching the topic from a human ecological point of view, we can ask ourselves in this context two interrelated questions, one relevant to people's ecological potency and the other relevant to the ecological valency of their surrounding outside world:

- What features of human organism are relevant to how people cope with and evaluate (thermal, visual, acoustical, etc.) conditions around them?
- What aspects of people's surrounding indoor environment are relevant – and should be primarily targeted in the building design and operation processes – to accommodate people's needs and expectation?

To answer these questions, an understanding of the physics, physiology, and psychology of the processes involved would be required. For the purpose of the present treatment, we focus on the thermal environment, given its importance for buildings' performance in view of sustainability and habitability.

5.4.2   The thermal environment

5.4.2.1   *Heat balance and human body*

Under normal conditions, the human thermoregulation system can maintain a fairly constant core body temperature of roughly 37°C. Toward this end, the human organism needs a long term energy balance in the course of thermal exchange with the environment. Specifically, energy losses to the environment need to be compensated via internal heat production. In thermal comfort literature (see, for example, Fanger, 1972), thermal exchange between human body and its surroundings is commonly described as follows:

$$M - W = Q_{conv} + Q_{cond} + Q_{rad} + E_{skin} + C_{resp} + E_{resp} + S \qquad (5.1)$$

In this formulation, M is metabolic rate and W is mechanical work. Hence, $M-W$ represents the body's net heat production via human metabolism. To specify metabolic rate, the met unit is applied (heat production per unit skin area), whereby 1 met $= 58$ W/m^2. The metabolic rate and the associated rate of heat transfer is primarily a function of the activity level and the temperature of the surroundings (see Fig. 5.1). Further terms in the above equation represent sensitive heat losses for the skin via convection (Q_{conv}) and radiation (Q_{rad}), heat conduction via contact (Q_{cond}), and latent heat transfer via sweat evaporation and moisture diffusion (E_{skin}). Heat transfer via the respiratory system is captured in terms of sensitive (C_{resp}) and latent (E_{resp}) components. Lastly, S denotes the body's heat storage capacity.

Fluctuations in the body's thermal balance with its surroundings occur frequently and are generally harmless as long as they are not long-term or severe. However, if due to a sustained imbalance the body's core temperature cannot be maintained, not only thermal discomfort, but also critical physiological consequences may arise. Multiple mechanisms are involved in human body thermoregulation. For instance, blood flow through skin tissue can be modulated by constricting and dilating blood vessels. Sweating can increase evaporative heat loss from the skin. Heat production can be increased through shivering, which signifies increased metabolism in muscles. Given the importance of long term thermal balance between the human body and its surrounding, human thermal sensation has been interpreted as a kind of warning system. Departure from the thermal equilibrium state would be experienced as thermal discomfort, motivating people to seek conditions or apply measures toward regaining thermal equilibrium.

Thermoreceptors distributed over the skin provide the brain with information on skin temperature. The thermoregulation processes in human body have been suggested to correlate with skin and core temperatures. Changes in skin temperature, depending on their temporal features (i.e., sudden versus slow) are result in thermal sensations. Due to the dynamic feature of thermoreceptor signals, people are more likely to feel cold by falling skin temperatures as by low yet

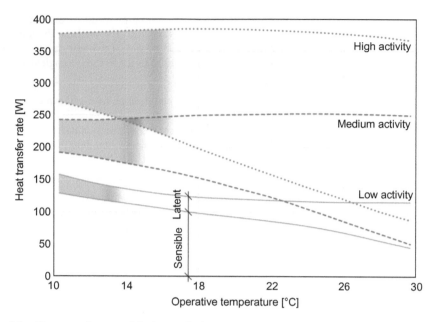

Figure 5.1.   Heat transfer rate of the human body as a function of activity (low, medium, high) and the ambient operative temperature (adapted from Rietschel and Raiß, 1968).

constant skin temperatures. In general, thermoreceptors of the human skin give rise to the conscious impressions of environments as being warm or cold. Thereby, cold and warm sensations may be perceived as positive or negative depending on the body core temperature. If the core is overheated, a cold sensation is likely to be perceived as pleasant. However, if the core is cold, a warm sensation is more likely to be experienced as pleasant.

5.4.3   *Thermal comfort models and standards*

Insights gained based on the study of the human body's thermoregulation and heat balance have informed the attempts to identify (both analytical and statistical) relationships between physiological factors and the experience of thermal comfort. For example, indices such as the *PMV* (predicted mean vote) and *PPD* (predicted percentage of dissatisfied) postulate that people would be likely to be in the state of thermal comfort if their mean skin temperature and sweat secretion are within a certain range (Fanger, 1972). *PMV* is a numerical index that expresses the statistical mean of responses (thermal comfort evaluations) by a large group of people according to numerical values (+3, +2, +1, 0, −1, −2, −3) they assign to thermal sensations (hot, warm, slightly warm, neutral, slightly cool, cool, cold). The correlating *PPD* index is a predictive measure of the percentage of thermally dissatisfied people in a specific thermal environment. Based on studies conducted in climate chambers, optimal thermal comfort conditions were suggested to correlate with personal factors (metabolic rate, clothing) and environmental conditions (air and radiant temperatures, humidity, air flow speed). The results of such studies are frequently structured in terms of equations and associated rules, tables, and graphic means to be used by designers and engineers. Using such tools, for people with a certain level of clothing and activity, the optimal indoor climate conditions in architectural spaces could be inferred. Various standards such as ISO 7730 (ISO 2005) and ASHRAE 55 (ASHRAE, 2004) specify building categories based on *PMV* ranges occurring in them (e.g., category A with *PMV* between −0.2 and +0.2; category B with *PMV* between −0.5 and +0.5; category C with *PMV* between −0.7 and +0.7). As an example, Figure 5.2 shows winter and summer thermal comfort zones based on data in ASHRAE Standard 55 and ISO 7730.

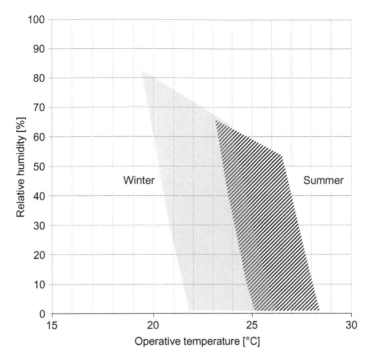

Figure 5.2.   Comfort zones (±0.5 *PMV*) for summer (*clo* = 0.5) and winter (*clo* = 1.0) as a function of indoor operative temperature and relative humidity assuming activity levels between 1 and 1.3 met and air flow speeds of less than 0.2 m/s (based on ASHRAE Standard 55 and ISO 7730).

The *PMV*-based approach is meant to statistically apply to large groups of people. Moreover it primarily pertains to the human body as a whole. Further studies and associated formulas have addressed the local thermal discomfort due to draught risk, radiative asymmetry, cold or warm floor surfaces, and vertical temperature gradients (Fanger *et al.*, 1985; 1988; Olesen, 2002; 2008; Olesen *et al.*, 1979).

The above approach to description and prediction of thermal comfort in buildings has both merits and limitations. Amongst the virtues of the approach is the systematic way in which in a number of personal (level of activity, thermal resistance of the clothing) and environmental factors (air temperature and humidity, air flow speed, radiant temperature) serve as the input information to predict thermal comfort level of people in a room. However, the approach has also been faulted with regard to intrinsic limitations and predictive performance. On the one hand, precise definition of the model's input variables (e.g., the exact determination of people's activity or the thermal resistance of their clothing) is not trivial and may constrain the applicability of the model to practical situations. On the other hand, laboratory-based models may not properly capture conditions in the field, where inhabitants are typically accustomed and adapted to their living and working environments. Moreover, the steady-state assumptions underlying classical thermal comfort models do not apply to circumstances in real – particularly free-running – buildings. Field studies have documented, particularly in free-running buildings, considerable deviations of *PMV*-based thermal comfort predictions from actual comfort votes by the occupants. To illustrate this point, Figure 5.3 includes two set of regression lines taken from a study by (de Dear and Brager, 2002). Each set contrasts the tendency of *PMV*-based predictions and occupancy expressions of indoor comfort temperatures as a function of mean outdoor air temperature. The first set, which pertains to buildings with HVAC (heating, ventilation, air-conditioning) systems, displays good agreement, whereas the second set, which is based on data from naturally ventilated buildings, shows a major discrepancy.

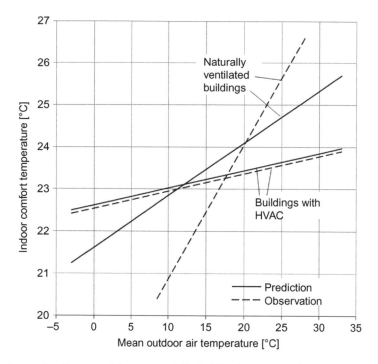

Figure 5.3.   Regression lines pertaining to actual (dashed lines) versus predicted (continuous lines) comfort indoor temperatures as a function of mean outdoor temperatures for naturally ventilated buildings and buildings with HVAC (based on information in de Dear and Brager, 2002).

In this context, human-ecologically relevant psychological factors such as people's expectations as well as their behavioral adaptations have been suggested to play a key explanatory role. People, when thermally uncomfortable, tend to react in such a way as to reduce thermal constraint and thus restore thermal comfort (Auliciems, 1983; Humphreys and Nicol, 1998). Such reflections have led to the conception of an adaptive approach to thermal comfort definition and prediction (Nicol *et al.*, 2012). Numerous actions may be initiated – consciously or unconsciously – by buildings' occupants as response to cold and heat sensations. For example, people can modify body heat generation (e.g., by reducing or increasing the activity level), control body heat loss (e.g., by changing posture or changing the clothing level), regulate the thermal conditions in the environment (e.g., by changing the thermostat settings, or operating windows, fans, and shades), and change location in the space or building. Field surveys and field studies appear not only to point to the limitations of simple heat balance based comfort models, but also reveal certain adaptive relationships. Specifically, neutral (or comfort) temperature expressions collected in various surveys display a significant relationship to the prevailing operative temperatures. In fact, in the operative temperature range of 20 to 25°C, operative temperature values and comfort temperatures are almost equal. This has been interpreted to imply that people can generally match their comfort temperature to the conditions in their environments (Nicol *et al.*, 2012). Moreover, field survey results – particularly from buildings in free-running mode – also point to a strong correlation between neutral temperatures and prevailing outdoor temperatures (Humphreys *et al.*, 2010).

Considerations pertaining to the aforementioned adaptive processes have found their way into comfort guidelines. Thereby, acceptable indoor temperatures in free-running buildings are defined as a function of the outdoor air temperature. Thereby, various methods have been proposed to derive running averages for outdoor temperature. For example, EN ISO 15251 (EN ISO, 2007)

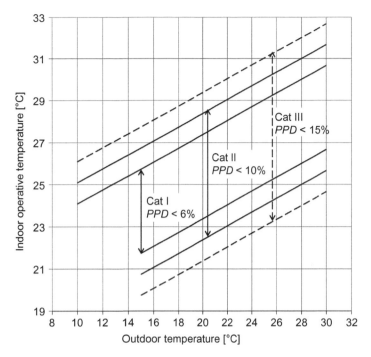

Figure 5.4.   Design indoor (operative) temperatures as a function of – exponentially averaged – running outdoor temperature (Category I, *PPD* < 6%; Category II, *PPD* < 10%; Category III, *PPD* < 15%) according to EN ISO 15251.

includes three categories of thermal comfort in naturally ventilated buildings (see Fig. 5.4) as a function of the running average of – exponentially weighted – daily outdoor temperature.

Views on applicable thermal comfort zones have a major consequence for sustainability considerations in buildings. If the assumptions regarding indoor environmental conditions that would result in acceptable thermal conditions in rooms are too rigid, free-running buildings would not be a viable option. Generally speaking, only buildings with HVAC systems can provide narrowly controlled indoor environmental conditions independent of the outdoor conditions. However, both examples from traditional (vernacular) architecture (Mahdavi, 1996c) and more recent low-energy buildings have shown, that in many climatic zones and over considerable periods of time, properly designed buildings can offer adequate indoor conditions without the need for extensive mechanical systems for heating, cooling, and ventilation. Surveys and field study results underlying the adaptive approach to thermal comfort postulate a more flexible framework for the definition of desirable indoor environmental conditions in buildings. They help thus counter comfort-based (and more recently, productivity-based) arguments for deployment of large-scale energy-intensive mechanical environmental control systems in buildings.

5.4.4   *A concluding note on the information-related aspect of environmental relationships*

While the efforts to ground the adaptive approach to thermal comfort on a solid scientific basis (Nicol *et al.*, 2012) are not particularly convincing, insights from associated studies are highly valuable. They suggest that multi-level complexities involved in human processes of environmental sensation, perception, and evaluation cannot be captured with simple heat balance based models of thermal comfort. Future advances may more successfully explain the combined physiological, psychological, and cognitive underpinnings of thermal comfort evaluation processes.

Meanwhile, however, building design and operation professionals must be sensitive to the considerable variance in thermal responses of people to similar thermal conditions. Human ecology provides a conceptual high-level (and qualitative) perspective toward the understanding of such evaluative variances in terms of the importance of not only the matter-energetic aspect but also the information-related aspect of the environmental perception phenomena.

As mentioned earlier, one should be utterly careful with postulating general deterministic relationships between measurable environmental factors (such as temperature and humidity in a room) and occupants' evaluation of environmental conditions (i.e., the thermal comfort vote). A number of studies have aptly demonstrated the important effects of information-related aspects of environmental exposure situations on perception and evaluation processes (Mahdavi, 2011a). For example, in one experiment, two demographically similar groups of participants provided significantly different assessments of the same acoustical event (recorded white noise). Participants in the first group, who were told the recording was of a waterfall, judged it much more favorably than the second group, who was told the recording was of a factory. People's attitude toward the alleged source of an acoustical event clearly influenced their evaluation of the exposure, despite the absence of any objective difference in the nature of the event (Mahdavi, 2011a). In another experiment (Schönpflug, 1981), participants were exposed to white noise of different intensity while performing certain tasks (time estimations). The participants who received positive feedback about their performance ranked the same acoustical exposure as less annoying than those who received negative feedback concerning their performance. But the feedback messages were manipulated and did not reflect the true performance. Hence, their effect on participants' subjective evaluation of the noise exposure situation cannot be explained in terms of an acoustically induced impairment. The explanation lies rather in the nature of the information processing that was triggered by the combined effect of acoustical exposure and negative feedback. The degree of annoyance due to noise was apparently higher, once it was identified as the reason for one's (alleged) failure.

Such experiments imply that subjective evaluations are not at all fully determined by energetic descriptors of the so-called environmental exposure. Rather, such evaluations emerge through the complex workings of the information processing in human minds.

5.5   BUILDING PERFORMANCE AND HUMAN IMPACT

5.5.1   *The relevance of people's behavior*

To highlight the critical role of the people factor for building performance, one needs only to reflect on a few simple questions (Mahdavi, 2011b) that building designers are typically expected to pose and answer:

- How much energy will be needed to heat, cool, ventilate, and illuminate buildings?
- What kinds of indoor climate conditions concerning thermal comfort and indoor air quality are to be maintained in buildings?
- What level of daylight availability can be expected in the indoor environment under dynamically changing outdoor illuminance levels?
- Will the acoustical environment in indoor spaces provide the necessary conditions for communication and task performance?
- Can occupants be safely evacuated from buildings in case of an emergency such as outbreak of fire?

Obviously, none of these questions can be reliably answered without considering the role of the people living and working in buildings. People affect the performance of buildings, due to their presence and their actions. The energy and thermal performance of buildings is not only affected by the people's presence as a source of sensible and latent heat, but also due to their actions, including use of water, operation of appliances, and manipulation of building control devices for heating, cooling, ventilation, and lighting. User-based operation of luminaires and shading devices in a room affect the resultant solar gains, light levels, and visual comfort conditions. The presence of people in a room and the associated sound absorption influences the sound field and thus the acoustical performance of the room. The safety performance of a building cannot be evaluated without considering the behavior of people under emergency.

People's relevance to performance-based design is, however, not restricted to assumptions regarding user presence and actions in buildings. Frequently, information on user-based requirements in view of thermal, visual, and acoustical comfort must be explicitly reflected in building operation specifications. For example, control settings for air temperatures and illuminance levels in architectural spaces must be defined in accordance with structured knowledge on people's needs and requirements. Moreover, targeted values of performance variables are expected to encapsulate information that is relevant to people's requirements and expectations in view of health, comfort, and satisfaction. Appropriate selection and interpretation of building performance indicators requires thus that the relationship between occupied spaces and their occupants are considered and understood. Given this background, we need to systematically situate buildings' users and occupants in the context of performance requirements. Accordingly, we must deal with the mechanisms and corresponding models of how people's presence and interactions with buildings' environmental systems influence the values of relevant performance indicators.

5.5.2   *Kinds of effects*

Broadly speaking, a useful distinction can be made between "passive" and "active" effects of users and occupants on buildings' performance. Passive effects of people on the hygro-thermal conditions in buildings denote those effects caused by the mere presence of people in the building. For instance, hygro-thermal conditions and indoor air quality in architectural spaces are influenced by such passive people effects: depending on their activity, people release not only various quantities of sensible and latent heat (see Fig. 5.1), but also water vapor, carbon dioxide, and other execrations and odorous substances. Likewise, in the building and room acoustics domain, the presence of people in a space has an effect on the sound field via introduction of additional sound absorption. To capture the passive effects of people's presence in buildings in the design process, we typically rely on existing data such as occupancy load schedules derived from measurement results of people's metabolic rates. This is, as such, a straight-forward process, barring two possible complexities. Firstly, different levels of resolution are conceivable regarding temporal and spatial distribution of such passive effects. Secondly, the passive people effects such as heat emission (as illustrated in Fig. 5.1) may depend on the context, e.g., thermal conditions in occupants' rooms. This interdependence would require the concurrent consideration of the human agent and their immediate environment.

In most buildings, occupants operate control devices such as windows, shades, luminaires, radiators, and fans to bring about desirable indoor environmental conditions. We refer to these control actions as people's active effects. They have a significant impact on buildings' hygro-thermal and visual performance. To predict and evaluate buildings' performance we need good knowledge of such control-oriented user behavior. General information about building type (residential, commercial) and environmental systems (free-running, air-conditioned) as well as organizational and administrative information (e.g., working hours) can only provide rough directions regarding such active effects. More representative people presence and action models require, however, extensive observational data based on empirical studies of occupancy and control-oriented user behavior in a large number of buildings. Thereby, possible relationships between control actions and environmental conditions inside and outside buildings could provide the underlying basis for derivation of user behavior models.

5.5.3   *Empirical observations and models*

There is a substantial and growing body of observational studies to capture the patterns of occupants' presence in buildings and their interactions with buildings' environmental control systems

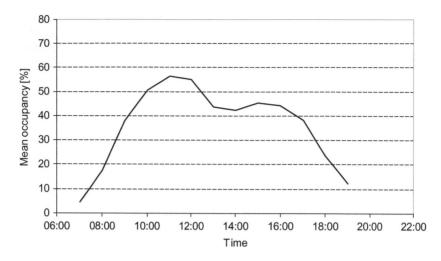

Figure 5.5.   Mean reference work day occupancy based on data from five office buildings in Austria.

such as windows, blinds, and luminaires. Frequently, such studies attempt to establish a link between user control actions or the state of user-controlled devices and measurable indoor or outdoor environmental parameter (see, for example, Boyce, 1980; Herkel *et al.*, 2005; Hunt, 1979; Inoue *et al.*, 1988; Lindelöf and Morel, 2006; Love, 1998; Mahdavi, 2011b; Nicol, 2001; Rea, 1984; Reinhart, 2004). While highly useful, these studies are often variously limited, due – amongst other things – to the small number of buildings and rooms involved, the duration and consistency of data collection, the accuracy of the measurements, the robustness of the analyses, and the clarity of the documentations. Some of these limitations and their implications were addressed in the course of a recent case study, involving a number of office buildings in Austria. Thereby, we systematically collected an extensive set of observational data regarding building occupants' presence and control action patterns pertaining to lighting and shading systems while considering the indoor and outdoor environmental conditions under which those actions occurred (Mahdavi, 2011b; Mahdavi *et al.*, 2008a; 2008b). Some of the lessons learned from this study are summarized below.

It is important to understand that the pattern of people's presence in buildings cannot be simply inferred from building type and function – e.g., residential versus commercial. Nor can it be based solely upon organizational information from building and facility managers. In our study of five office buildings in Austria, the mean occupancy patterns (see Fig. 5.5) was unlike common assumptions in the professional community or presumptions of the organizations involved. Moreover, the five buildings we studied displayed very different occupancy patterns (see Fig. 5.6). Even if all offices in a building belong to the same organization, there could be drastic differences between their occupancy patterns. To illustrate this point, Figure 5.7 shows monitored occupancy patterns in seven offices in one of our case study buildings. The considerable statistical variance of occupants' presence in their offices is further exemplified in Figure 5.8, which shows mean presence level and respective standard deviations.

A building's usage and the functions it supports can repeatedly and considerably change over time, yet again implying variable and hardly predictable occupancy patterns. Moreover, offices can be, in the course of time, assigned to different individuals or user groups with inherently different occupancy tendencies. Ultimately, the same individual occupant might, over time, display varying patterns of presence, given professional or personal circumstances. Such factors lead to the considerable uncertainty in assumptions pertaining to occupancy levels to be expected in buildings.

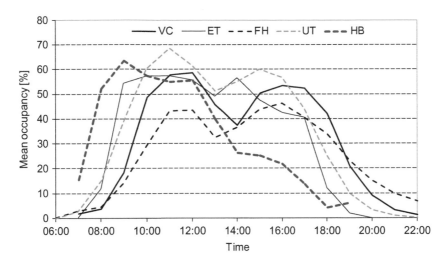

Figure 5.6.   Mean reference work day occupancy patterns for five office buildings in Austria.

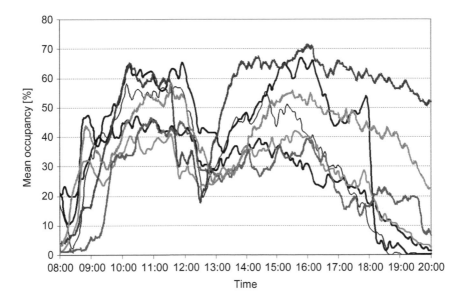

Figure 5.7.   Reference day occupancy levels in seven offices in an office building (FH).

Our specific case study did result in a number of empirically-based statistically significant relationships between the frequency or probability of user control actions involving, for example, lights and blinds operations, or state models of user operated devices on the one hand and some independent variables pertaining to occupancy, indoor environment, or outdoor conditions on the other hand. Figures 5.9 to 5.11 exemplify (office buildings, work days, Austria) instances of these kinds of relationships, namely the light switch on probability as a function of task illuminance level immediately prior to the onset of occupancy (Fig. 5.9), light switch off probability as a function of the duration of absence from the workstation (Fig. 5.10), and shades deployment level as a function of facade orientation and the incident global vertical irradiance on the facade (Fig. 5.11).

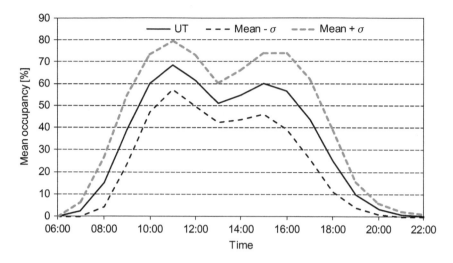

Figure 5.8.   Mean and standard deviation (σ) of occupancy for a reference day in an office building (UT).

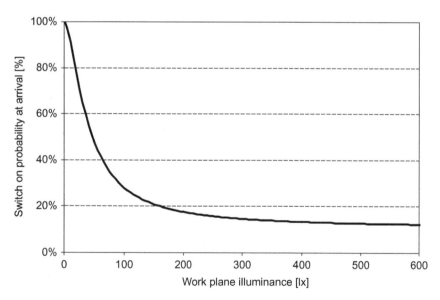

Figure 5.9.   Manual light switch on probability as a function of task illuminance level immediately prior to the onset of occupancy.

Such empirically-based models are sometimes referred to, erroneously, as deterministic. In fact, they are simply observation-based statistically aggregated relationships. They might provide clues and indications concerning the environmental triggers of behavioral tendencies. But they certainly do not represent causal models of human control actions in buildings. Moreover, as with all statistically derived relationships, these kinds of models are limited in at least two regards:

First, they cannot be divorced from the population from which they are derived and simply applied to other contexts, at least not without losing much of their statistical credence. Behavioral tendencies and their dependencies on hypothesized independent variables are influenced by a large number of diverse factors, such as the climate, cultural issues, building type and functions, organizational specifics, building systems peculiarities, space orientation, and interior design

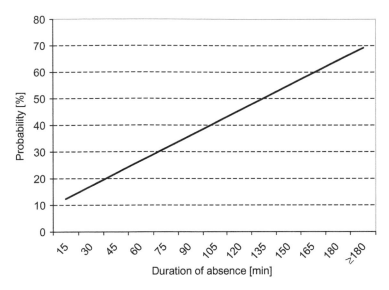

Figure 5.10.   Manual light switch off probability as a function of the duration of absence from the workstation.

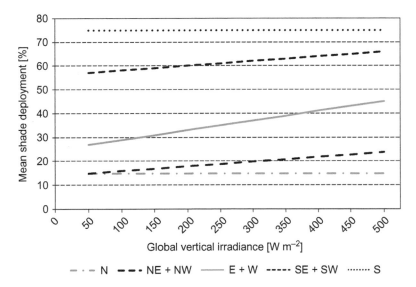

Figure 5.11.   Shades deployment level (in %) as a function of facade orientation (S: south, NE: north-east, NW: north-west, E: east, W: west, SE: south-east, SW: south-west, S: south) and global vertical irradiance on the facade (model derived based on the aforementioned observations in office buildings in Austria).

features influence. Second, aggregate models do not explicitly reflect the inherently probabilistic nature of most control-oriented control actions. Nor do they capture the dynamism of actual processes and events in buildings, as stochastic models can – at least in principle (Fritsch *et al.*, 1990; Nicol, 2001).

The latter models have been used to generate time series of both occupancy intervals and user control actions that "look" similar to actual (real) processes and event sequences. Thus, if grounded in quantitatively sufficient and qualitatively adequate empirical data, stochastic occupancy and control action models, while realistic in their random fluctuations, could represent, in toto, the general occupancy-triggered processes in a building. Such models can be implemented in simulation applications in terms of autonomous agents with built-in methods to generate behavioral patterns that appear realistic (Bourgeois, 2005; Chang and Mahdavi, 2002; Liao *et al.*, 2012). However, the promise of stochastic occupancy needs to be qualified against both reliability and applicability concerns.

An argument can be made for the utility of simple (code-base or descriptive) occupancy-related simulation input assumptions in the design development phase, where calculations can be used to obtain numeric values for a number of aggregate performance indicators such as buildings' annual heating and cooling loads. Such aggregate indicators can support at least two purposes: (i) benchmark a specific building design proposal against applicable codes, standards, and guidelines, or (ii) comparatively assess the likely performance of multiple design alternatives. Thereby, concise inferences are expected concerning the quality of the proposed building "hardware" vis-à-vis design variables pertaining to the building's envelope, massing, orientation, shape, construction, etc. Naturally, this is done under "standardized" conditions pertaining to external climate, which is typically represented in terms of a standard weather file, and internal occupancy-related processes, which are typically represented in terms of fixed, more or less detailed assumptions regarding internal gains, ventilation rates, etc. Theoretically speaking, the use of a probabilistic presence and user action models would generate more or less different occupancy-related input data for each instance of inquiry, resulting in correspondingly different results. This could represent a problem not only for code-based compliance checking, but also for the performance analyses of design alternatives, when the aim is to compare multiple alternative designs irrespective of variance in contextual boundary conditions (weather) and occupancy. Even so, presumably one can argue that the repeated inquiries with a properly calibrated probabilistic occupancy model can also converge to stable values for aggregate performance indicators.

A different circumstance arises, however, if we consider more elaborate building design analysis scenarios, which require us to consider the implications of uncertainties associated with occupancy processes in buildings. Differences in occupancy patterns over time and location can be quite significant. Such differences can be important especially while trying to gauge the variance of thermal loads or conditions in various zones of a building. Information regarding temporal and zonal load variations is critically important, for example when calculations provide essential data for design and sizing of indoor climate control. Thus, rigid models of user presence and behavior that ignore associated stochastic fluctuations and the resulting uncertainties would be rather problematic, if the detailed configuration of a building's mechanical equipment is the main concern: while dealing with the requirement of providing sufficient heating and cooling capacity to different zones of a building, the variability of required thermal loads must be systematically explored. This cannot be based on spatially and temporally averaged occupancy assumptions. In such instances, application of properly calibrated models with probabilistic features may be critical.

It seems as though different approaches to representation of occupancy-related processes in building design support may be appropriate given different scenarios. If consideration of the implications of variance in model input assumptions is evidently critical to a specific performance inquiry, then probabilistic models of occupancy presence and control actions would be appropriate. On the other hand, when the objective of a performance analysis inquiry is to benchmark design proposals against applicable codes and standards or to parametrically compare design alternatives, uncritical inclusion of random variations of boundary conditions and internal processes in inquiries may lead to misleading results.

5.5.4   *A note on the rebound effect*

As noted earlier (see Section 5.4), studies in different fields have shown that energy efficiency improvement measures may miss their targets, sometimes considerably. Paradoxically, energy efficiency measures may even end up increasing the energy use. This circumstance is often referred to as the "rebound effect". It can be defined as the non-materialized fraction of the projected energy saving due to an energy efficiency improvement measure. Related observations have been also reported in the building domain, whereby projected energy performance of new buildings and the energy saving potential of thermal retrofit measures on existing buildings were found to be overly optimistic. The rebound effect is complex and can have multiple roots. Thereby, one of the contributing factors may be attributed to behaviorally relevant circumstances. For example, thermal retrofit measures on a building can principally reduce the heating energy required to maintain certain thermal conditions in that building. However, the reduction potential may not be exploited in actuality, if occupants modify their behavior in a more energy-intensive direction. For example, they may change the temperature settings for heating, or they may ventilate spaces more frequently, or they may turn on the heating in more rooms. Such behavioral phenomena may explain the results of a number of recent studies, which documented a lower than expected energy efficiency improvement effect following thermal retrofit measures pertaining to existing buildings. The rebound effect may also involve the redirection of energy efficiency gains in one area to increased consumption in another. For example, monetary benefits from building-related energy conservation measures in terms of lower heating costs may be redirected toward increased energy use in higher fuel usage for car driving.

5.5.5   *Conclusion*

The importance of people's passive and active effects on building performance (e.g., indoor environmental conditions, energy use) can be significant. Accordingly, many recent and ongoing research efforts attempt to understand and predict passive and active occupancy effects on building performance. Thereby, physiological and psychological descriptions of occupancy as well as empirically-based observational data provide the knowledge base. Specifically, long-term high-resolution empirical data on people's presence and control-oriented actions in buildings can support the generation of general patterns of user control behavior as a function of indoor and outdoor environmental parameters such as temperature, air flow, air quality, illuminance, and irradiance. These patterns can be expressed either in terms of typologically differentiated aggregate occupancy and control action models or realized in terms of emergent behavior of a society of computational agents with embedded stochastic features. Future developments in this area are expected to facilitate detailed computational models of environmental processes in buildings via comprehensive multiple-coupled representations that dynamically capture the states of occupancy, building, and context.

5.6   CONCLUDING REMARKS

In this contribution, we reviewed the role of people in sustainable architecture from very different vantage points. Thereby, we explored the potential of human ecology as a conceptual framework for our inquiry. In this framework, building activity may be interpreted as an instance of largely regulatory activities by human beings to better match the ecological valency of their immediate surroundings with their ecological potency. In our discourse, we argued that such building-related interventions, together with others involving industry and transportation, have resulted in resource depletion, environmental emissions, and waste generation. Moreover, we suggested that if the notion of sustainable architecture is to be taken seriously, it must critically address not only "narrow" matters pertaining to the technical quality of individual buildings and the requirements and behavioral patterns of their inhabitants, but also the broad ramifications of population growth, life style development, climate change, social policies, investment priorities, and urban planning.

  Given this multi-faceted treatment, it is fitting to conclude a chapter dedicated to the role of people in the sustainable architecture with a critical remark concerning a specific group of people, namely those involved in research, education, development, production, and management activities pertaining to building delivery and operation processes. This group includes, amongst others, scientists, teachers and trainers, developers, architects, engineers, and facility managers. Similar to professionals in other domains, this group of people has a two-fold standing with regard to the sustainability question. In a narrowly conceived professional sense, members of this group should of course make every attempt to target habitability and sustainability of the built environment via knowledge-based design decision making, ecologically informed selection of building materials, components, and systems, as well as in-depth consideration of occupants' functional and indoor-environmental requirements. But viewed more broadly, we need to remind ourselves that the professionals in the domain of built environment do not operate in a vacuum. Rather, they are constrained by a large number of economical, procedural, and political forces and boundary conditions. In many instances, the role of professionals in building projects starts only after a number of crucial decisions – with decisive sustainability ramifications – have been made and preconditions have been fixed. If concrete building projects start at the point where it is already decided which new buildings have to be erected, where they should be located and under which urban setting (density, traffic and energy systems, etc.), then the degrees of freedom for building designers and engineers to contribute meaningfully to sustainability efforts are already severely limited. If responsible building professionals are consulted and involved only late in the consequential chain of environmentally relevant decisions in the building sector, their role would be reduced to "damage control". But as responsible citizens, those involved in the building delivery and operation process need to enter the socially relevant sustainability discussion at the earliest opportunity, where processes are initiated and decisions are made that fundamentally influence the habitability and sustainability of the built environment.

REFERENCES

ASHRAE Standard 55: Thermal environmental conditions for human occupancy. American Society of Heating, Refrigerating and Air-Conditioning Engineers, 2004.

Auliciems, A.: Psychological criteria for global thermal zones of building design. *Int. J. Biometeorol.* 27 (1983), pp. 69–86.

Bourgeois, D.: *Detailed occupancy prediction, occupancy-sensing control and advanced behavioral modeling within whole-building energy simulation.* PhD Thesis. Université Laval, Quebec, Canada, 2005.

Boyce, P.: Observations of the manual switching of lighting. *Lighting Res. Technol.* 12:4 (1980), pp. 195–205.

Chang, S. & Mahdavi, A.: A hybrid system for daylight responsive lighting control. *J. Illum. Eng. Soc.* 31:1 (2002), pp.147–157.

Databank World Bank: Last updated 2014, http://databank.worldbank.org/ddp/home.do?Step=1&id=4. (accessed February 2012).

de Dear, R.J. & Brager, G.S.: Thermal comfort in naturally ventilated buildings: revisions to ASHRAE Standard 55. *Energy Build.* 34:6 (2002), pp. 549–561.

EIA: International energy outlook 2011. http://www.eia.gov/forecasts/ieo/ieo_tables.cfm. 2012.

EN ISO 15251: Indoor environmental input parameters for design and assessment of energy performance of buildings addressing indoor air quality, thermal environment, lighting and acoustics, 2007.

Fanger, P.O.: *Thermal comfort.* McGraw-Hill Book Company, 1972.

Fanger, P.O., Ipsen, B.N., Langkilde, G., Olesen, B.W., Christensen, N.K. & Tanabe, S.: Comfort limits for asymmetric thermal radiation. *Energy Build.* 8 (1985), pp. 225–236.

Fanger, P.O., Melikov, A.K., Hanzawa, H. & Ring, J.: Air turbulence and sensation of draught. *Energy Build.* 12:1 (1988), pp. 21–39.

FAO: "ENERGY-SMART" Food for people and climate issue paper. 2012, http://www.fao.org/docrep/014/i2454e/i2454e00.pdf.

Fritsch, R., Kohler, A., Nygard-Ferguson, M. & Scartezzini, J.L.: A stochastic model of user behaviour regarding ventilation. *Build. Environ.* 25:2 (1990), pp. 173–181.

Gibson, J.: The theory of affordances. In: R. Shaw & J. Bransford (eds): *Perceiving, acting, and knowing.* Lawrence Erlbaum Associates, 1977.

Gibson, J.: *The ecological approach to visual perception.* Lawrence Erlbaum Associates, 1979.

Herkel, S., Knapp, U. & Pfafferott, J.: A preliminary model of user behavior regarding the manual control of windows in office buildings. *Proceedings of The Ninth International IBPSA Conference, Building Simulation*, Montréal, Canada, 2005, pp. 403–410.

Humphreys, M.A. & Nicol, J.F.: Understanding the adaptive approach to thermal comfort. *ASHRAE Transactions* 104:1 (1998), pp. 991–1004.

Humphreys, M.A., Rijal, H.B. & Nicol, J.F.: Examining and developing the adaptive relation between climate and thermal comfort indoors. *Proceedings of Conference on Adapting to Change: New Thinking on Comfort*, Windsor, UK, 2010, http://nceub.org.uk.

Hunt, D.: The use of artificial lighting in relation to daylight levels and occupancy. *Build. Environ.* 14 (1979), pp. 21–33.

Inoue, T., Kawase, T., Ibamoto, T., Takakusa, S. & Matsuo, Y.: The development of an optimal control system for window shading devices based on investigations in office buildings. *ASHRAE Transaction* 94 (1988), pp. 1034–1049.

ISO 7730: Ergonomics of the thermal environment – analytical determination and interpretation of thermal comfort using calculation of the PMV and PPD indices and local thermal comfort criteria, 2005.

Janssen, R.: Towards energy efficient buildings in Europe (final report). EuroACE (the European Alliance of Companies for Energy Efficiency in Buildings), 2004.

Knoflacher, H.: Zur Harmonie von Stadt und Verkehr: Freiheit vom Zwang zum Autofahren. Böhlau Verlag, Vienna, Austria, 1996.

Knötig, H.: Human ecology – the exact science of the interrelationships between *Homo sapiens* and the outside world surrounding this living and thinking being. *The sixth meeting of the Society for Human Ecology "Human Ecology: Crossing Boundaries"*, Snowbird, UT, 1992a.

Knötig, H.: Some essentials of the Vienna School of Human Ecology. *Proceedings of the 1992 Birmingham Symposium; Austrian and Britisch Efforts in Human Ecology*. Archivum Oecologiae Hominis, Vienna, Austria, 1992b.

Liao, C., Lin, Y. & Barooah P.: Agent-based and graphical modelling of building occupancy. *J. Build. Perform. Simul.* 5:1 (2012), pp. 5–25.

Lindelöf, D. & Morel, N.: A field investigation of the intermediate light switching by users. *Energy Build.* 38 (2006), pp. 790–801.

Love, J.A.: Manual switching patterns observed in private offices. *Light. Res. Technol.* 30:1 (1998), pp. 45–50.

Mahdavi, A.: Traditionelle Bauweisen in wissenschaftlicher Sicht. *Bauforum* 132 (1989), pp. 34–40.

Mahdavi, A.: Acoustical aspects of the urban environment. *Aris; Journal of the Carnegie Mellon Department of Architecture* 1 (1992), pp. 42–57.

Mahdavi, A.: Approaches to noise control: a human ecological perspective. *Proceedings of the NOISE-CON 96 (The 1996 National Conference on Noise Control Engineering)*, Bellevue, WA, 1996a, pp. 649–654.

Mahdavi, A.: Human ecological reflections on the architecture of the "well-tempered environment". *Proceedings of the 1996 International Symposium of CIB W67 (Energy and Mass Flows in the Life Cycle of Buildings)*, Vienna, Austria, 1996b, pp. 11–22.

Mahdavi, A.: A human ecological view of "traditional" architecture. *Hum. Ecol. Rev. (HER)* 3:1 (1996c), pp. 108–114.

Mahdavi, A., Mohammadi, A., Kabir, E. & Lambeva, L.: Occupants' operation of lighting and shading systems in office buildings. *J. Build. Perform. Simul.* 1:1 (2008a), 57–65.

Mahdavi, A., Kabir, E., Mohammadi, A. & Pröglhöf, C.: User-based window operation in an office building. In: B. Olesen, P. Strom-Tejsen & P. Wargocki (eds): *Proceedings of Indoor Air 2008 – The 11th International Conference on indoor Air Quality and Climate*, Copenhagen, Denmark, 2008b, paper 177.

Mahdavi, A.: Was kann das Plusenergiehaus – Eine kritische Betrachtung über das "Plusenergiehaus". *Baumagazin* 5 (2010), pp. 4–7.

Mahdavi, A.: The human dimension of building performance simulation. In: V. Soebarto, H. Bennetts, P. Bannister, P.C. Thomas & D. Leach (eds): *Building Simulation 2011 – IBPSA 2011*, Sydney, Australien, 2011a, pp. K16–K33.

Mahdavi, A.: People in building performance simulation. In: J.L.M. Hensen & R. Lamberts (eds): *Building performance simulation for design and operation*. Spon Press, 2011b.

Mahdavi, A.: Sustainable buildings: some inconvenient observations. *Proceedings First International Conference on Architecture and Urban Design*; 1-ICAUD", EPOKA Univ.; Dept. of Arch. (ed), Epoka University Press, 2012.

Nicol, J.F.: Characterising occupant behavior in buildings: towards a stochastic model of occupant use of windows, lights, blinds, heaters, and fans. *Proceedings of The Seventh International IBPSA Conference, Building Simulation*, Rio de Janeiro, Brazil, 2001, pp. 1073–1078.

Nicol, J.F., Humphreys, M. & Roaf, S.: *Adaptive thermal comfort.* Routledge, 2012.

Olesen, B.W.: Radiant floor heating in theory and practice. *ASHRAE J.* 44:7 (2002), pp. 19–24.

Olesen, B.W.: Radiant floor cooling systems. *ASHRAE J.* 50:9 (2008), pp. 16–22.

Olesen, B.W., Schöler, M. & Fanger, P.O.: Discomfort caused by vertical air temperature. In: P.P. Fanger & O. Valbjorn (eds): *Indoor climate.* Danish Building Research Institute, Copenhagen, Denmark, 1979, pp. 561–579.

Rea, M.S.: Window blind occlusion: a pilot study. *Build. Environ.* 19:2 (1984). pp. 133–137.

Reinhart, C.: LIGHTSWITCH-2002: a model for manual control of electric lighting and blinds. *Solar Energy* 77 (2004), pp. 15–28.

Rietschel, H. & Raiß, W.: *Lehrbuch der Heiz- und Lüftungstechnik.* 15th edition, Springer, Berlin/Göttingen/Heidelberg, Germany, 1968.

Schönpflug, W.: Acht Gründe für die Lästigkeit von Schallen und die Lautheitsregel. From Akustik zwischen Physik und Psychologie. In: A. Schick (ed): *Akustik zwischen Physik und Psychologie.* Klett-Cotta, Stuttgart, Germany, 1981.

Sorrel, S.: *The rebound effect: an assessment of the evidence for economy-wide energy savings from improved energy efficiency.* UK Energy Research Centre, Sussex Energy Group for Technology and Policy Assessment, 2007.

Uexküll, J.: *Kompositionslehre der Natur.* Edited by Thure von Uexküll, Frankfurt am Main, 1920.

CHAPTER 6

Sustainable construction materials

Kenneth Ip & Andrew Miller

6.1 INTRODUCTION

The construction industry is a major contributor to the world economy contributing approximately 10% of the world GDP and employing 7% of the global workforce (Economy Watch, 2010). It is a major consumer of natural resources including energy, water and materials and therefore has major responsibilities to utilize these resources in a sustainable manner.

Nearly half of the materials consumed throughout the world are used in construction (CIB, 2008) and in the UK construction material consumption is equivalent to seven metric tons per person per year (Lazarus, 2004). The demands of the construction industry vary with the economy but overall the demand is rising with increasing requirements as the global population is growing and developing areas of the world improve their quality of life. The processing of raw materials and manufacture of building products consumes energy with inevitable emissions of carbon dioxide and implications for climate change.

In order to achieve sustainability it is essential that we optimize the use of the world's resources conserving the use of finite materials such as minerals and consuming renewable resources in a controlled manner. Sustainability will not only require management of the material resources but also the harvesting and manufacturing processes that have their own environmental impacts and consume both energy and water creating further pollution.

If sustainability of construction materials is to be achieved we will need to understand the environmental implications of the materials we use. It will require construction systems that incorporate less waste, development of innovative building materials and systems that enable re-cycling and re-use of components that have not reached the end of their useful life when the building is demolished. It will require a full lifecycle approach and an optimization of the potential of all our natural resources.

The aim of this chapter is to develop an understanding of the key issues affecting the sustainable provision of construction materials to meet increasing demands. It identifies the demands on world resources and the drivers for controlling their use. The chapter concludes with two examples of natural building materials illustrating their potential for use in building construction.

6.2 MATERIALS IN CONSTRUCTION

Construction materials are selected primarily for a purpose that may be structural, functional or aesthetic. This chapter will focus on materials for the construction of buildings where the chosen material may fulfill more than one purpose. Structural elements may be an aesthetic feature of the internal or external design and finishes to the envelope may provide thermal or acoustic insulation as well texture and color.

The selection of materials is often driven by cost and may consider lifetime costs including maintenance and replacement costs within the lifetime of the building as well as the initial cost of the material and its installation. The concept of lifecycle costing is well established and can be used to evaluate options of different materials of durability appropriate to the usage and life expectancy of a particular building to achieve the best value for money. However the issues are often clouded when different routes to building procurement are used or when initial and operational costs are devolved to different budget holders.

In order to optimize the sustainability of material resources the same whole life environmental impact approach needs to be taken. The impacts of winning the raw materials, including mining and harvesting; processing, transporting and erecting; maintenance and replacement; disposal and the depletion of finite resources all need to be evaluated. Quantification of each impact is complex and the integration of the different impacts such as damage to local habitats, consumption of energy and water and the pollution their consumption causes is extremely difficult. Nevertheless the building designer is left with the task of optimizing both financial and environmental impacts when producing a sustainable building.

Many environmental assessment methodologies have been developed that facilitate the evaluation of the overall environmental impacts of different building materials and construction systems. These methodologies are only as good as the quality of data on which they are based and on the priorities of the design team. Both priorities and the environmental impacts vary around the world with needs for development and availability of materials. The impacts are constantly changing with improved technology and increased efficiency of production and changes in methods of energy generation and the greenhouse gas emissions associated with them.

Building designers are often driven down the route to sustainability through regulation and client demand for energy efficient buildings achieving high standards of comfort with minimum energy consumption. The client may be seeking minimum energy bills but reducing the consumption of fossil fuels will have the advantage of reducing greenhouse gas emissions.

6.2.1 *Traditional construction methods*

Early buildings were constructed in order to provide shelter from the surrounding climate. They utilized local resources, initially caves, and then natural materials such as stone, timber and animal skins gradually refining the processes and improving the standards of comfort inside. Modern buildings use a wide range of natural and processed materials that are sourced from around the world. They also include sophisticated engineering for heating and cooling and consume more natural resources in the form of fossil fuels.

The level of comfort achieved in modern buildings is far higher than the original buildings and there are no doubt implications for health and quality of life for the occupants. However, the environmental impacts of the early buildings were far lower than the modern ones and there is a lot to be learnt from understanding the traditional materials and evaluating their use in modern construction.

6.3   DEMANDS FOR MATERIAL RESOURCES

The construction industry in the UK is responsible for the consumption of over 420 million metric tons of material resources, equivalent to 7 metric tons per member of the population, each year. In addition 8 million metric tons of oil is used each year for the manufacture and transportation of building materials and the waste generated (Smith *et al.*, 2002).

The world reserves of some building materials such as aggregates, clay, gypsum and stone are considered to be plentiful although others, particularly metals such as copper, iron, tin and zinc have reserves limited to decades. The data for Table 6.1 have been abstracted from a more comprehensive table of non-renewable resources used in the manufacture of building materials published by Berge (2009). It identifies the estimated 'reserves' and 'reserves base' in years for different resources where 'reserves' identifies the sources of supply that are currently economic to obtain whereas 'reserves base' includes reserves that are not currently economically viable.

The reserves that are defined as 'large' and 'very large' are often considered as infinite yet there are regional variations across the world and therefore implications for the resulting environmental

Table 6.1.   World reserves of common building materials (Berge, 2009).

Raw material	Reserve [years]	Reserve base [years]	Annual growth in consumption [%] (1999–2006)
Aggregate (sand and gravel)	Very large	Very large	
Clay	Very large	Very large	
Copper	31	61	3
Gypsum	Large	Large	
Iron	95	219	10
Silica	Large	Large	
Stone	Very large	Very large	
Tin	22	40	4
Zinc	22	46	4.5

impacts. As available reserves become depleted so the long term management and development planning for further resources that have been geologically mapped but not yet mined becomes increasingly important. There can be social and economic impacts of ensuring vital resources do not become locked beneath other developments.

In the UK the demand for sand in construction is approximately 50 million metric tons per year, 10% of which is silica sand for glass manufacture and the remaining 90% for other construction products such as mortars and concrete (Smith *et al.*, 2002). The silica sand required for glass making is of a higher quality than ordinary construction sand having a 95% silica (SiO_2) content and less impurities like clay and iron oxides.

Silica sand is mined in various locations across the UK, primarily in England but with increasing demand for Scottish supplies. In 2007 however the permitted reserve, the mass of silica sand with valid planning permission for extraction, was just over 31 million metric tons, equivalent to less than ten years supply (BGS, 2009).

The demand for copper has risen because of its suitability for electrical wiring and for plumbing. Currently around 10 million metric tons per year are produced worldwide with approximately 25% from Chile. Several tons of waste is generated from the production of one ton of copper although the spoils can be landscaped and replanted. The production of copper is high in embodied energy with added implications for transportation. However, the relatively high cost of copper has meant that in many parts of the world much is reclaimed and recycled but still with relatively high embodied energy. Trends in the plumbing industry may however decrease the demands on copper as advances in plastic piping facilitate its use on both hot and cold water systems (Smith *et al.*, 2002).

6.3.1   *One planet living*

Expectations of the luxuries of modern life and advances in technology have led to an ever increasing demand on the earth's resources, so much so that on a worldwide scale we are consuming natural resources at a rate that cannot be sustained. In Western Europe we are consuming resources at a rate equivalent to more than three planets and in other developed areas the rate is even greater.

There is only one planet earth and we must therefore look to consuming resources in a manner that is equitable to everyone and in a manner that is sustainable for future generations. The concept of One Planet Living therefore highlights the impacts of human occupation on the environment and sets out principles for living in a sustainable manner.

The concept of One Planet Living was developed by Bioregional and the World Wildlife Fund (WWF) (Bioregional, 2013) and provides the principles through which we can work towards conserving the earth's resources and living sustainably. The original demonstration project,

Beddington Zero Energy Development (BEDZED) included 82 homes together with work units and community space was designed by architect Bill Dunster and constructed in Beddington, south London in UK with completion in 2002 (Lazarus, 2004).

In order to achieve sustainability there is a need for an holistic approach evaluating whole life cycle impacts of all aspects of our lives including food, shelter, work and leisure, optimizing our use of resources and harmful impacts on the planet. Evaluation must consider the implications of waste and the opportunities for re-use and recycling of materials that have not reached the end of their useful life. BEDZED endorses these principles through low energy buildings, utilization of renewable energy, and careful selection of materials for construction and general encouragement of a sustainable life style. The development included live/work units and an electric car club to reduce the environmental impacts of transportation and encouraged the use of locally sourced food through co-operative organization.

The approach to selection of materials includes consideration of winning the raw materials, manufacturing construction components, transportation, construction and eventual disposal and their resulting environmental impacts. In practice difficulty arises in appreciating all of the processes involved, in evaluating the environmental impacts and particularly in evaluating a compound impact that might include destruction of habitat or biodiversity, greenhouse gas emissions and depletion of finite resources.

True sustainability includes social and economic impacts as well as environmental so our new building developments must be affordable, comfortable and provide for our health and social well-being. The focus of this chapter is however on the environmental impacts where construction materials have significant impact on the thermal performance of the building but also upstream in winning the raw materials and manufacturing the building materials.

6.4   LIFE CYCLE OF BUILDINGS

The demand for new buildings has inevitable consequences for the supply of construction materials but a life cycle approach to the selection of these materials will enable their potential to be optimized, the depletion of finite materials to be carefully monitored and the use of renewable materials to be properly managed. Throughout their life cycle the impact the materials have on the environment will depend upon the processes undergone and the nature of the materials themselves.

The stages in the life cycle of building materials are illustrated in Figure 6.1. Between each of the stages and often within the individual stages there is a need for transportation which may be worldwide depending on the source of the material and the processing required. Transportation implications can be significant especially if the materials are bulky and heavy. They should also be considered carefully when considering the use of recycled materials as some materials are transported to and from distant countries because of the economics of re-cycling.

6.4.1   Stages in the life cycle

6.4.1.1   Raw materials
Construction materials can be taken from the earth, either by mining or excavating from the surface, or harvested from renewable resources such as plants, trees and animals. The raw materials are unevenly distributed throughout the world, thus the demand for some common building materials such as copper often have environmental impacts in regions far away from where the building is to be built. As the reserves of the materials deplete so the difficulties of winning the materials increase as will the cost and the impacts on the environment. Advances in technology enable deeper and safer mining yet still create greater spoils and changes to the local landscape and habitat. Raw materials also need to be transported from their point of origin to the construction site, often with intermediate journeys through different stages in their refinement and manufacture to building materials.

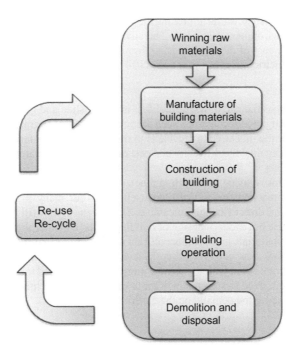

Figure 6.1.   Life cycle of building materials.

The impacts of taking raw materials can be seen to be social and economic as well as environmental. Nevertheless a common way of evaluating the environmental impact of materials is the consideration of embodied energy in terms of the energy consumed and the greenhouse gases emitted throughout the life cycle of the material. The embodied energy can be conveniently expressed in kg of carbon dioxide or equivalent greenhouse gas emitted, but the figures need to be evaluated in the context of the specific materials under consideration as well the accuracy and completeness of the data base used to determine the embodied energy.

6.4.1.2   *Waste in construction*

The construction industry is notorious for its wastage of materials with millions of tons of materials being sent to landfill every year. The wastage during construction is caused through over ordering, damage caused in handling, faulty goods and poor workmanship as well as off-cuts due to standard delivery sizes. The wasted materials have already caused the upstream impacts on the environment and now have further impacts in disposal. They have cost money in purchase with further costs associated with handling and disposal.

UK building regulations require any construction project, other than the very small ones, to have a site waste management plan ensuring that the significance of waste is appreciated and appropriately managed. The benefits can be clearly demonstrated in terms of savings of finite resources and reductions in both environmental impacts and costs.

It is estimated (BRE, 2008) that 13% of waste from construction sites is new unused materials and 60% of materials in waste skips is packaging of one kind or another. Changing attitudes within the industry has meant that some suppliers will take back unwanted, undamaged goods for resale and packaging for re-use. Careful design can lead to less wastage from off-cuts, good workmanship can reduce wastage from faulty or sub-standard work, and good site management can reduce damage in handling and storage.

These measures will help towards improving the efficiency of use of building material resources and reducing their environmental impacts. The advantages can be immediately experienced in financial terms as the cost of disposal is rising through increases in the landfill taxes which are set to rise by £8 per metric ton every year until 2014 when it will reach £80 per metric ton.

6.4.1.3 *End of life*
When a building reaches the end of its useful life it is demolished and generally the materials are disposed to landfill. High value materials such as metals are often separated and sold for scrap to be recycled. However the incentive to recycle is in the value of the material on the scrap market which must be traded against the difficulty and cost of retrieving the material from the total mass of demolished materials.

There is potential for re-using or recycling much of the demolished building as the materials themselves have not reached the end of their useful life, but many materials are difficult to separate and there is no direct financial incentive. The advent of landfill tax has provided some encouragement to reuse but modern design of buildings must consider the process of demolition and disposal and use materials and components that are easier to separate and re-use or recycle.

Traditional brickwork was constructed with lime mortar which made it relatively easier to clean and re-use the bricks than with the cement mortar generally used today. Fixing mechanisms and adhesives therefore need to selected in order to optimize the potential of the materials being demolished, thus reducing the demand on the earth's finite resources.

6.4.1.4 *Recycling/reuse*
The hierarchy of waste (Environment Agency, 2011) dictates that we – reduce, re-use, re-cycle – the materials we use. Reduction of the amount of material used in a building is a feature of the design providing for the needs of the building owner while optimizing space and therefore materials.

There is an opportunity for re-use of some components either unwanted at the construction stage or in good condition at refurbishment or demolition stages. However there needs to be an infrastructure in place for the collection, storage and redistribution of these materials. Similarly re-cycled materials need to be efficiently collected and re-manufactured.

Demolition materials are commonly used as hard core as temporary road approaches at landfill sites. However, the potential of the materials can be improved to provide good quality aggregates for fresh concrete using some filtering and modern concrete mixes. The introduction of magnesium based cement has such an impact. Magnesium cement can be used to create concrete with a far smaller proportion of Portland cement than common concretes and tolerates a greater proportion of re-cycled aggregates. It therefore has the double advantage of increasing the potential of demolished masonry material and also reducing the embodied energy of the new concrete as high amounts of energy are consumed in the manufacture of Portland cement.

The environmental impacts of individual building materials need to be aggregated together when evaluating the environmental performance of a building. Optimization of the use of an individual material must therefore be seen in the context of that of a whole building, its operation throughout its useful life and eventual disposal.

A new building can enhance the townscape and provide much needed amenities for the community, commerce or industry. However, throughout its life cycle the building can be viewed as consuming resources and as producing emissions and waste that can have negative impacts on the environment.

6.5 LIFE CYCLE ASSESSMENT

The use of a life cycle assessment (LCA) methodology can facilitate comparisons between materials. It can be used to evaluate different environmental impacts and burdens within the longer life cycle of the buildings themselves. However, many assumptions are made in establishing LCA

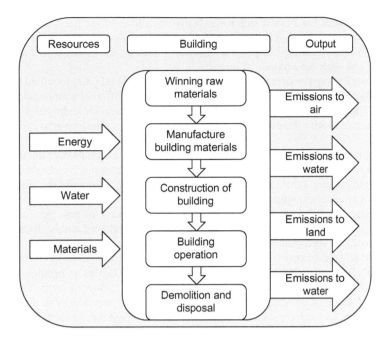

Figure 6.2.   Life Life cycle environmental impact of buildings.

results and comparisons between results in different databases can be misleading without clear definition of the assumptions made. The environmental impacts are illustrated in Figure 6.2.

Life cycle assessment methodology is generic and is described as having four phases:

- Goal and scope definition
- Inventory analysis
- Impact assessment
- Interpretation

Standard procedures are defined in BS EN ISO 14040:2006 Environmental Management Life Cycle Assessment – principles and framework.

Life cycle analysis of a building enables the material to be assessed in the context of the whole life of the building including the need for maintenance and replacement and consequently a second set of environmental impacts. It may be particularly important when selecting materials for a building which will itself have a useful life far longer than the individual component.

A softwood window frames may need replacement after 15 years whereas hardwood may not need replacement until after 25 years or a PVCu frame may last even longer and require less maintenance. This then needs to be considered in the context of the useful life of the building and the number of replacements in a lifetime of 60 or 100 years.

6.5.1   *Environmental impact assessment methods*

The imperative of living within the environmental resource capital of the planet and not creating pollution that will ultimately destroy life as we know it should be incentive enough for us to use our materials in a sustainable manner. However, the consequences of profligate use are neither tangible nor immediately recognizable and without such evidence there is a natural tendency towards increasing demands without real concern for the future. There is also the additional fear of the need for behavior change and additional costs associated with sustainable buildings.

The Kyoto protocol has been a step towards the realization of the need to act, but even this is not universally recognized. There are however many initiatives throughout the world driving the industry towards creating buildings with lower impact on the environment through energy consumption and other environmental impacts.

In Europe the Energy Performance of Buildings Directive (EPBD) has provided the framework upon which legal requirements of the building regulations in different member states have been based. In the UK successive editions of the building regulations are demanding higher standards of thermal performance with the eventual aim of all new buildings requiring zero carbon emissions in operation.

The question of the embodied energy of the original building materials does not directly feature in the analysis although some environmental assessment methods employed to evaluate the performance of the building fabric as well as the operation of the building use the embodied energy of materials as an element within their evaluation.

Although environmental assessment is not a legal requirement for new buildings it can be a requirement of clients wishing to establish their environmental credentials. It may also be an expectation by Local Authorities in acquiring planning permission.

Environmental assessments such as the 'building research establishment environmental assessment method' (BREEAM, 2013) and the 'code for sustainable homes' (Communities and Local Government, 2006) bring together analysis of different factors impacting on the overall environmental performance of a building. Different assessments relate specifically to different building types although the methodologies are broadly similar. The 'code for sustainable homes' is used here as an example.

6.5.2   Embodied energy

The embodied energy of a material refers to the amount of energy consumed in providing that material. There are however many interpretations of the term embodied energy and in order to make useful comparisons between different materials it is necessary to have a clear understanding of the definition being employed.

A common definition of embodied energy is the energy consumed up to the end of the manufacturing process (cradle to gate). However, it may include delivery to the construction site (cradle to site) or even the construction processes into the completed building. The latter two definitions are clearly site specific and relate only to an individual building.

The definition is further complicated by the consideration of the scope of the analysis which can include the gross energy requirement (GER) or more simply the process energy requirement (PER). The inclusion of energy consumed for transportation of materials between processes is a further element requiring clarification before figures can be compared.

6.5.2.1   Gross energy requirement
The gross energy requirement is a measure of all the energy inputs to a specific material. It can be considered as the total energy consumption for which that material has been responsible. Thus it includes not only all the energy consumed in winning the raw materials, transporting them and their manufacture into building materials and components, but also the manufacture and maintenance of the plant used in winning and processing those materials, the transportation of the labor force for all of these activities and the repair of the environmental damage caused by these processes.

6.5.2.2   Process energy requirement
The process energy requirement is the energy consumed in the processes directly undergone by the building material or product. It will include the energy consumed by the plant and machinery throughout the processing, but not that of the second and higher generation consumptions of the plant that made the plant and the repair of damage caused by the processes.

There are however different approaches to the evaluation of these figures and care is required when using published data in order to make comparisons. Common issues where clarity of assumptions and methodology is required are:

- Apportioning energy where more than one product undergoes the same process or one is a by-product of another.
- Inclusion of transportation and how return journeys of the delivery vehicle are considered.
- Apportioning transportation energy for a part load on a ship or other mode of transportation.

The most comprehensive document currently available to assist in identifying assumptions made in published figures is the ICE published by BSRIA based on the research of Hammond and Jones (2011) at Bath University in the UK.

6.5.2.3 *Embodied carbon*

Although embodied energy and embodied carbon are directly related the impact of any material on resource depletion and on greenhouse gas emissions may be very different. It will depend upon the primary fuel consumed and the means of generation of electricity. Consumption of renewable energy may be considered to have zero emissions provided the embodied energy of the collectors and generators are neglected. Similarly nuclear energy will also have zero carbon emissions. The embodied carbon is therefore dependent upon the fuel mix in the location where the processing takes place.

Some materials can even be considered as having negative embodied carbon when carbon sequestered during their growth has been accounted for. Trees and short term crops used for building materials sequester atmospheric carbon dioxide during their growing period, the weight of which may be greater than the emissions during manufacture.

6.5.3 *Code for sustainable homes*

The 'code for sustainable homes' (Communities and Local Government, 2006) has become a popular method to assist designers in achieving a good standard of environmental performance and a useful means of demonstrating the performance that has been achieved. Homes are rated according to a star system with one star being allocated at a basic level and six stars to the best current practice. Whilst not a legal requirement many social housing providers are demanding level three or four stars as the norm.

Assessment of the building is made under nine categories: energy and carbon dioxide emissions, water, materials, surface water run-off, waste, pollution, health and wellbeing, management and ecology. Credits are awarded for performance in each category and then a weighted average is calculated to provide a points score from which the star rating is awarded.

The weighting factors allocated to the individual categories are given in Table 6.2. It is interesting that the category materials is ranked only fifth in the hierarchy of environmental impacts and contributes only 7.2% of the total points allocated. However, it should be appreciated that materials selection has greater impact in the earlier stages of the building life cycle.

The credits allocated to materials are awarded for the environmental impacts of the materials themselves and for the responsible sourcing of both the structure and the finishings. Credits for responsible sourcing are awarded for procuring materials through an accredited scheme such as that certified by the Forest Stewardship Council (FSC).

Credits for the environmental impacts of materials are based upon 'environmental profiles' established by the Building Research Establishment (Howard *et al.*, 1999) and published in the Green Guide to Specification (Anderson *et al.*, 2009) which produces a rating from A+ for the materials with the least environmental impact down to E for those with the greatest.

The 'environmental profiles' system involves consideration of 13 separate impacts which are then weighted into a single rating. Details of the individual issues and the weightings employed are presented in Table 6.3.

Table 6.2.   'Code for sustainable homes' weighting factors for
environmental impact categories.

Categories of environmental impact	Weighting factor [%]
Energy/carbon dioxide	36.4
Water	9.0
Materials	7.2
Surface water	2.2
Waste	6.4
Pollution	2.8
Health and wellbeing	14.0
Management	10.0
Ecology	12.0

Table 6.3.   Environmental impact categories, issues measured and weightings for 'environmental profiles' (Anderson *et al.*, 2009).

Environmental impact category	Environmental issue measured	Weighting [%]
Climate change	Global warming or greenhouse gas emission	21.6
Water extraction	Mains, surface and groundwater consumption	11.7
Mineral resource extraction	Metal ore, mineral and aggregate consumption	9.8
Stratospheric ozone depletion	Emission of gases that destroy the ozone layer	9.1
Human toxicity	Pollutants that are toxic to humans	8.6
Ecotoxicity to fresh water	Pollutants that are toxic to freshwater ecosystems	8.6
Nuclear waste (higher level)	High and intermediate level radioactive waste from nuclear energy industry	8.2
Ecotoxicity to land	Pollutants that are toxic to terrestrial ecosystems	8.0
Waste disposal	Material sent to landfill or incineration	7.7
Fossil fuel depletion	Depletion of oil, coal or gas reserves	3.3
Eutrophication	Water pollutants that promote algal blooms	3.0
Photochemical ozone creation	Air pollutants that react with sunlight and NOx to produce low level ozone	0.2
Acidification	Emissions that cause acid rain	0.05

6.6   NATURAL BUILDING MATERIALS

The term natural building material is commonly used to refer to materials that can be easily obtained from nature. The inference is that they are renewable and sourced locally and therefore have minimal environmental impact. In the early stages of building construction, when man was moving from cave dwelling to constructing permanent shelter, modern materials like steel and concrete were not available and natural materials were more commonly used. There is therefore much to be learnt from traditional construction methods and advances in technology have enhanced their potential application in modern buildings.

Timber is a natural building material although, depending on the species, it may have a growing cycle of hundreds of years. In common with all plants it grows through a process of photosynthesis using light energy to convert carbon dioxide from the air to carbon and oxygen, fixing the carbon as the structure of the wood and releasing oxygen back to the atmosphere. Trees can be seen to have the advantages of being a renewable source of building materials which lock in carbon from the atmosphere, thus reducing pollution. They can be considered as the lungs of the earth, breathing in carbon dioxide and breathing out oxygen. The carbon is locked in the wood for as long as it remains as wood, but at the end of its useful life it may be burnt to release heat energy or decompose naturally, but in each case the carbon is re-released to the atmosphere.

There are many other natural materials that have shorter growing times, some even down to a few weeks, which can be used as construction materials for structure as well as specialist insulations, finishes and furnishings. The natural life cycles of these materials can mean less impact on the environment and less embodied energy. In some cases, such as the ability of natural fibers to buffer moisture and heat, they demonstrate enhanced properties over manufactured products.

One of the advantages of natural materials is considered to be their low embodied energy because of the minimal processing required. Embodied energy however should also include transportation energy and it is therefore important to consider local materials and local manufacturing plant in order to reduce transportation energy. In the case of large projects it may be appropriate to consider setting up a 'flying factory' to manufacture building products local to the construction site.

The use of natural materials in the modern construction industry is relatively new and thus the infrastructure for growing, manufacturing and distributing is still developing and the products can be more expensive than the alternatives. An holistic approach needs to be applied to the development of the natural materials industry as their cultivation may mean displacement of other plants such as food crops. They may also provide an opportunity for the utilization of agricultural land that is currently underused.

A thorough evaluation of the properties of natural materials and their potential for use in the construction industry was undertaken by the Construction Industry Research and Information Association (CIRIA) and published as a handbook (Cripps *et al.*, 2004). The handbook includes plant fibers for structure and insulation as well as pigments for dyes and paints.

Traditional use of local plants as building materials has developed based on experience since the first construction of shelter. However modern construction requires clear definition of material properties and performance. There is now scientific interest in analyzing these materials and academic publications provide useful data on a range of materials including corn cob, date palm wood and bamboo amongst many others (Agoudjil *et al.*, 2011; Pinto *et al.*, 2012; Van der Lugt *et al.*, 2006).

In the next section case studies of two natural building materials, hemp and timber, will be considered. They evaluate the life cycle environmental impacts of using the materials in the UK. In other parts of the world their local availability and application need to be considered in terms of requirements as well implications of manufacture and transportation.

6.6.1 *Hemp*

Industrial hemp (*Cannabis sativa*) is a fast growing annual plant that has been used for centuries because of the quality of its fibers. The plant, together with its close relatives has been used in paper, textiles, food, medicines and construction for centuries. The variety of uses for different parts of the plant and opportunity of using leftovers as biomass has heightened appreciation of the life cycle efficiency of this plant.

Tom Woolley has been a key player in the testing and promotion of using hemp as an environmentally benign material. His book (Bevan and Woolley, 2008) has identified a wide range of benefits of using hemp/lime composites including: exceptional high levels of air tightness achieved through the monolithic wall construction; improved air quality due to the hygroscopic properties of the wall; and lower energy consumption attributed to the heat transfer process involving sensible and latent thermal capacities of hemp-lime walls. Evrard and De Herde (2009) also showed in their experimental measurements relatively higher comfort levels were achieved for the same energy input when compared to other types of wall construction.

In recent years there has been increased use of hemp in construction with the dissemination of research studies on its construction and performance, such as studies commissioned by the National Non-Food Crops Centre (Hawthorn Environmental Services Ltd., 2010) and the establishment of a UK based growing supply chain for production and distribution of hemp-lime construction products. Furthermore, its application in high-profile building projects such as the Renewable House at BRE Innovation Park (Pritchett, 2009) has established the status of hemp-lime construction as a main stream green technology in the UK.

Interest in hemp has generated a number of research studies related to their life cycle environmental impact. Norton *et al*. (2009) undertook a full LCA of natural fiber in insulation materials made of hemp and recycled cotton and established the embodied energy and global warming potential (GWP). González-García *et al*. (2010) in their LCA comparative study on the environmental impact of hemp and flax as raw materials for non-wood pulp mills showed both plants had relatively lower input and impact than compared with the same materials obtained from other agricultural crops. They also identified that the production and use of fertilizers for hemp cultivation were responsible for as much as 80% of the greenhouse gas emissions.

6.6.1.1   *Life cycle of hemp/lime wall*

Hemp is an extremely robust annual crop that is planted in late spring when the soil is warming up. Seed drilling takes place from April up to the end May after plugging and harrowing of the land. It has a remarkable rate of growth, often reaching heights in excess of 4 m by August when it is cut with specially developed equipment. After 2–3 weeks, the dry hemp straw is baled and stored on farm until required for processing, where the fiber is separated from the woody core. Yields of hemp vary depending on site conditions and agricultural processes.

Most hemp grown in the UK utilizes seed imported from France and therefore requires transportation, usually be road and ship. Further transportation energy is consumed by agricultural machinery in plugging, fertilizing and harvesting, and for transportation from farm to processing plant and eventually to construction site.

Once sown the plants are fertilized using between 60 and 200 kg/ha (ADAS 2005, MAFF, 2000) ammonium nitrate, triple superphosphate and potassium chloride depending on the required final crop.

In the UK, most hemp-lime walls are constructed using a blend of specially prepared hemp hurd and lime based binder. The hemp and lime binder are mixed on site by an electric mixer. The hemp-lime mixture can be sprayed using a modified dry spray concrete system or alternatively, poured into the timber shuttering and manually tamped. The wastage is negligible as spillages can be re-used during construction process.

Softwood timber used for the structural support can be sourced from a conifer forest where it is sawn and kiln dried to around 10% of moisture with a density 460 kg/m³. For a small scale construction sawn timber is delivered by lorry and the shutter and timber posts are fixed manually. Water, hemp hurd and the binder are mixed in an electric mixer. The hemp-lime mixture is poured into the timber shuttering and successively tamped to make sure it is properly packed in. The shutter is removed manually after the wall is set and dry.

Life cycle analysis of a hemp lime wall (Ip and Miller, 2012) has determined the total GHG emissions are 46.63 kg CO_2e per square meter. Taking into consideration the carbon sequestrated by hemp hurd of 45.82 kg CO_2e, lime binder of 28.55 kg of CO_2e and timber of 8.34 kg of CO_2e, the net greenhouse gas emission is −36.08 kg of CO_2e. The negative value indicates it produces a net reduction in greenhouse gases over its useful life.

6.6.2   *Timber*

It is estimated that there are over 100,000 species of tree flora in the world which can be divided into softwoods and hardwoods (Miller, 2004). Softwood is fast growing and produced from coniferous trees generally in the northern hemisphere in boreal or temperate forests. Hardwood is produced from broadleaved trees either from temperate forests in the north or tropical forests around the equator and in the southern hemisphere.

The majority of timber used in the construction industry in the UK is from Temperate and boreal forests with small quantities of tropical hardwood, usually selected for its structural or aesthetic properties. The environmental significance of importing tropical hardwoods has been raised as a key issue with respect to deforestation and its impact on climate change. The deforestation has been due not only to the demand for timber but also for fuel and clearing for agriculture. It is therefore essential that these sources of timber are appropriately managed and sustainably harvested.

All timber used in construction should be certificated both for sustainable forest management and for chain of custody to ensure that timber from certified forests is not mixed with timber from other sources in transportation and stages in manufacture. Systems such as that offered by the Forest Stewardship Council (FSC) provides certification of both forest management and chain of custody and is accepted by many environmental assessment methods such as the 'building research establishment environmental assessment method' (BREEAM).

The environmental credentials of timber as a construction material are very good. It sequesters carbon during its years of growth, it often requires little in the way of energy for manufacture and at the end of its useful life it has inherent energy which can be released by combustion.

The photosynthesis process by which plants and trees grow absorbs carbon dioxide and water from the atmosphere and emits oxygen back to the atmosphere. In general terms it is estimated that a tree is close to 50% carbon and in growing one metric ton of timber a tree will sequester 500 kg of carbon or 1800 kg of carbon dioxide. During its growth period a tree therefore has a positive impact on the environment through the reduction in greenhouse gases.

The life-cycle of timber from sustainable sources can be traced through every stage from cradle to factory gate and beyond to construction, maintenance and to disposal when appropriate. However the cradle to factory gate analysis is most common. It includes:

- Seed gathering and propagation.
- Seedling planting and forest management including fertilizing, protection from animals and thinning.
- Harvesting the mature trees including forestry activities.
- Drying and seasoning felled timber.
- Processing slab wood and rough sawn timber.
- Secondary processing timber joinery.
- Transport within and between each of these stages.

Through appropriate forest management and production efficient use can be made of the material discarded from the felled trees. It can be used for composite timber products and garden mulches and the supply of timber for the construction industry can be fully sustainable when harvested timber is replaced through re-planting.

In the UK a high proportion of construction timber is imported, yet some regions including Sussex in the south east of England are heavily wooded. It is often quoted that the climate in England is not good for the growth of construction timber as the trees grow too quickly to achieve the required structural properties. Whilst rapid growth means faster absorption of CO_2 the resulting timber cannot be used directly for construction. However processed timber products such as glulam (glue laminated timber) provide excellent alternatives to natural timber sections and require less embodied energy than alternative structural materials.

The process of glulam manufacture involves the use of coppiced forest material, generally hardwood such as sweet chestnut which is harvested in cycles of 12–20 years. After harvesting the root system sprouts several new poles from a single stem increasing the rate of production. Traditional coppicing techniques would have been carried out to supply poles for fencing, hop poles and thatching stakes as well as charcoal burning.

Coppiced timber is of insufficient diameter to be used for structural elements in construction. However, initial grading and milling to lengths of uniform cross section (25 mm × 75 mm) are inspected and any knots or other imperfections are cut out leaving sections of good quality timber of varying lengths which are then finger jointed together. The jointed material can then be planed and sawn to required lengths before being glued and laminated to the required dimensions. It is then possible to bend the assembled product to form arches or curved beams.

Traditional glulam processes use phenol-resorcinol-formaldehyde (PRF) adhesives that require the timber to be dried to a moisture content of 15%, however modern adhesives based on polyurethane are more tolerant to moisture and can be used with green timber. These adhesives therefore reduce the production time, reduce the embodied energy and provide the opportunity for utilizing timber that would not normally meet the requirements of quality for construction timber.

The Building Research Establishment (Cornwell *et al.*, 2005) have published a feasibility study for the green gluing of timber where they conclude that it has the potential for improving the use of forest resources in the UK.

6.7   CONCLUSION

This chapter has explored some of the key issues with respect to the sustainable use of materials in construction. It has identified the need to produce quality buildings that are comfortable and healthy to inhabit whilst being energy efficient and utilizing natural resources to their optimum potential. It has highlighted the need to evaluate the use of materials with a whole life cycle approach considering impacts of using the material from cradle to grave, from initial mining or harvesting to eventual disposal including the potential for recycling at interim stages.

Methodologies such as 'environmental profiles' and the 'code for sustainable homes' are available in the UK to facilitate improvement of environmental standards, however it is necessary for each material to be considered in the context of its location and availability as they will affect the resource depletion and the requirement for transportation. A material that can be used sustainably in one part of the world may not be environmentally benign in another.

Local materials generally have minimal embodied transport energy and fast growing renewable local materials can have further advantages of carbon sequestration during their growth period thus reducing their net greenhouse gas emissions. Much scientific research is being undertaken to define the physical properties of local crops such as hemp, straw, date palms and pineapple leaves.

Understanding individual materials and the processes they have undergone must be key to the optimal use of the earth's resources. A re-cycled product may not necessarily be the best environmental option if it has been transported around the world to undergo an energy intensive re-cycling process.

As ever the designer must make a decision that will involve compromise in order to optimize the environmental impacts of the building. The decision will also be driven through economic constraints but the selection of sustainable construction materials will benefit from the availability of good data bases and an appreciation of the complex inter-relationships of the individual environmental impacts of materials.

REFERENCES

ADAS: UK flax and hemp production: the impact of changes in support measures on the competitiveness and future potential of UK fibre production and industrial use. ADAS Centre for Sustainable Crop Management, Cambridge, UK, 2005.

Agoudjil, B., Benchabane, A., Boudenne, A., Ibos, L. & Fois, M.: Renewable materials to reduce building heat loss: characterization of date palm wood. *Energy Build.* 43:2 (2011), pp. 491–497.

Anderson, J., Shiers, D. & Steele, K.: *The green guide to specification*. 4th edition, BRE Press, Watford, UK, 2009.

Berge, B.: *The ecology of building materials*. 2nd edition, Architectural Press, Oxford, UK, 2009.

Bevan, R. & Woolley, T.: *Hemp lime construction – a guide to building with hemp lime composites*. HIS BRE Press, 2008.

BGS (British Geological Society): Mineral planning fact sheet – silica sand. BGS, London, UK, 2009.

Bioregional and WWF: 2013, www.oneplanetliving.org (accessed 2/1/2013).

BRE: *Sustainable construction – simple ways to make it happen.* IHS BRE Press, 2008.

BREEAM, Building Research Establishment Environmental Assessment Method. www.breeam.org (accessed 1/03/2013).

BSI. ISO 14040: 2006 Environmental management – life cycle assessment – principles and framework. BSI, UK, 2006.

CIB Report 318: Construction Materials Stewardship – the status quo in selected countries. J.B. Storey (ed): CIB W115, 2008.

Communities and Local Government. Code for sustainable homes: a step change in sustainable home building practice. Communities and Local Government London, 2006, www.communities.com (accessed 1/03/2013).

Cornwell, M., Thorpe, W. & Cooper, G.: *Green gluing of timber: a feasibility study.* IP 10/05 IHS BRE Press, 2005.

Cripps, A., Handyside, R., Dewar, L. & Fovargue, J.: *Crops in construction handbook.* CIRIA, London, UK, 2004.

Economy Watch: Construction industry trends. 29 June 2010, http://www.economywatch.com/world-industries/construction/trends.html (accessed 17/12/2012).

Environment Agency: Guidance on applying the waste hierarchy to hazardous waste. 30 November 2011, http://www.environment-agency.gov.uk/business/regulation/129223.aspx (accessed 1/3/2013).

Evrard, A. & De Herde, A.: Hygrothermal performance of lime-hemp wall assemblies. *J. Build. Phys.* 34 (2009), pp. 5–25.

González-García, S., Hospido, A., Feijoo, G. & Moreira, M.T.: Life cycle assessment of raw materials for non-wood pulp mills: hemp and flax. *Resour. Conserv. Recy.* 54 (2010), pp. 923–930.

Hammond, G. & Jones, C.: *Embodied carbon: the inventory of carbon and energy (ICE).* Building Services Research and Information Association (BSRIA), Bracknell, Berkshire, UK, 2011.

Hawthorn Environmental Services Ltd.: An investigation of the potential to scale up the use of renewable construction materials in the UK. NNFCC 09-009, National Non-Food Crops Centre, 2010, http://www.nnfcc.co.uk/tools/investigation-of-the-potential-to-scale-up-the-use-of-renewable-construction-materials-in-uk-nnfcc-09-009 (accessed 10/09/2011).

Howard, N., Edwards, S. & Anderson, J.: *BRE methodology for environmental profiles of construction materials, components and buildings.* BRE Press, Watford, UK, 1999.

Ip, K. & Miller, A.: Life cycle greenhouse gas emissions of hemp-lime wall constructions in the UK. *Resour. Conserv. Recy.* 69 (2012), pp. 1–9.

Lazarus. N.: *BEDZED construction materials report. Toolkit for carbon neutral development* – Part 1. Bioregional, 2004.

MAFF: Project NF0307 – Final project report: *Hemp for Europe manufacturing and production systems.* Ministry of Agriculture, Fisheries and Food, London, UK, 2000.

Miller, P.: *Procuring legal and sustainable timber – a construction industry guide.* The Chartered Institute of Building, 2004.

Norton, A., Murphy, R., Hill, C. & Newman, G.: The life cycle assessment of natural fibre insulation materials. *Proceedings of the 11th International Conference on Non-conventional Materials and Technologies (NOCMAT)*, 6–9 September. Bath, UK, 2009.

Pinto, J., Cruz, D., Paiva, A., Periera, S., Tavares, P., Fernanderes, L. & Varum, H.: Characterization of corn cob as a possible raw building material. *Constr. Build. Mater.* 34 (2012), pp. 28–33.

Pritchett, I.: Hemp and lime composites in sustainable construction. *The Journal of the Building Limes Forum* 16 (2009), pp. 34–37.

Smith, R.A., Kersey, J.R. & Griffiths, P.J.: The construction industry mass balance: resource use, wastes and emissions. Viridis Report VR4, Viridis, 2002.

Van der Lugt, P., Van den Dobbelsteen, A.A.J.F. & Janssen, J.J.A.: An environmental, economic and practical assessment of bamboo as a building material for supporting structures. *Constr. Build. Mater.* 20:9 (2006), pp. 648–656.

CHAPTER 7

Water use and conservation

Alfredo Fernández-González

7.1 INTRODUCTION

The increase in human population and the rapid expansion of urban infrastructure have negatively impacted the hydrologic cycle. As societies become more affluent and their lifestyles continue to evolve, humans around the globe are now striving for additional comfort which demands an abundance of available resources, one of which is water. The extraction of natural resources and the mass production of industrial products result in large amounts of chemical discharge that affect the quality of surface and groundwater. Furthermore, the expansion of urban infrastructure is swallowing agricultural land, wiping out forests, and confining water streams. Impermeable, human-made surfaces also redirect natural flows of both surface and groundwater. As a result, urban areas are dealing with increased storm water runoff that is often times poorly managed, thus accentuating water pollution, urban heat island, the out-migration of local species, and an overall impairment of local ecosystems. The increase in impermeable surfaces is also a trigger for accelerated erosion and sedimentation. Impermeable materials used for construction and the paving of roads reduce groundwater infiltration and therefore have an impact on plant life. These relationships, illustrated in Figure 7.1, directly influence the natural hydrologic cycle.

As a way to reduce the negative impacts of urbanization, the concept of integrated site and building water management (ISBWM) was developed by mimicking the basis of the hydrologic cycle. ISBWM uses a series of onsite strategies that complement prudent decisions of water usage in daily human activities.

The systems and strategies employed by the ISBWM approach can be applied to an individual site or to a series of sites in a neighborhood or building tract to take maximum advantage of all viable water harvesting and treatment sources to minimize the dependence on municipal, centralized water supply systems. The harvested water resources may be used for a variety of purposes, such as irrigation of landscaped areas through digitally controlled, high-efficiency systems, cooling tower make-up, and toilet flushing. ISBWM strategies, along with other advanced principles of onsite water treatment, can help reduce the water footprint in urban areas and restore the natural balance of local ecosystems.

7.2 WATER DISTRIBUTION, SHORTAGES, AND OTHER CHALLENGES

The shortage of fresh water resources and their unequal distribution throughout the planet are important issues to consider for future urban developments. Even though almost 3/4 of the earth's surface is covered with water, 97.5% of it is salty ocean water. As illustrated by Figure 7.2, of the small percentage of fresh water left, only a fraction of it (~0.007%) is accessible for direct use as the rest of it is part of the icecaps of Antarctica and Greenland. This small amount of fresh water becomes available on a sustainable basis by being constantly renewed through the hydrologic cycle. Fresh water resources comprise the water beds of surface-water and the shallow underground sources that are viable to be tapped for human use.

The availability and access to fresh water have become increasingly important national and international issues, as the demand for fresh water in many regions of the world already exceeds

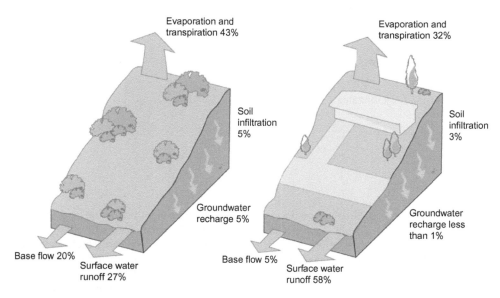

Figure 7.1.   Pre-development water circulation on site (left) and post-development influence of built environment on the water circulation on site (adapted from: Level Organization, New Zealand, 2007).

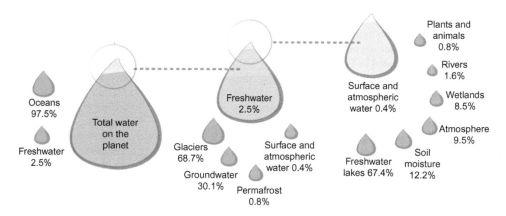

Figure 7.2.   Availability of water in the planet (adapted from: UN World Water Assessment Programme, 2003).

fresh water supplies. According to the United Nations, more than 50% of the countries in the world will face water stress or shortages by 2025 (UN World Water Assessment Programme, 2003).

Water stress and water shortages are not unique to developing nations. Figure 7.3 illustrates the results of a national survey of state water managers conducted by the US Government Accountability Office (GAO) in 2003. This survey revealed that at least 36 states are bound to experience local, regional, or state-wide water shortages under average climate conditions (GAO, 2003).

Early symptoms of a potentially serious water crisis in the US were evidenced by the severe drought experienced in 2007 throughout the southeast, and by the continued droughts in the west that are seriously threatening the Colorado River system, which is the source of fresh water to approximately 30 million people in seven states. According to the Scripps Institution of Oceanography, the Colorado River system is currently operating at a deficit and there is a

Figure 7.3.   States expecting water shortages over the next decade (adapted from: US Government Accountability Office, 2003).

50% chance that Lake Mead, the source of 90% of the water supplied to Las Vegas, could run completely dry by 2021 if the climate changes as it has been predicted and future water demands are not reduced (Barnett and Pierce, 2008).

In addition to anticipated shortages, the number of water pollutants is rising as new products are being introduced to the market. While modern industrial production in the US is monitored by the Department of Health and other relevant agencies, a steady amount of chemical discharge is being released into the sewer on daily basis. Contaminated waterways pose a serious health risk to human and animal populations. Furthermore, the laws regulating the quality of drinking water are greatly outdated. The list of contaminants regulated by the Safe Drinking Water Act has not been updated from the 91 items identified in year 2000. The deficiencies of this list become a major red flag since more than 60,000 chemicals are used in the US according to Environmental Protection Agency estimates.

7.3   THE WATER AND ENERGY NEXUS: A SYMBIOTIC RELATIONSHIP

Water is vastly used throughout the energy sector. Proper management of this resource has to be emphasized in the areas of resource extraction, refining and processing, electric power generation, storage, and transport. Any effort to resolve the water supply crisis has to include the energy and environmental impacts of the proposed solution. Furthermore, the increasing demand for electricity is placing additional pressure on already tight fresh water supply resources. According to the Energy Information Agency, in 2006 approximately 97% of the electricity generated in the United States was produced by thermoelectric or hydroelectric power plants (EIA, 2007). These power plants evaporate large amounts of water as they generate electricity. In 2005, thermoelectric cooling alone represented 41% of the water withdrawn in the USA and 6% of the water consumed nationally (Carter, 2013). The water consumed by the electric generation processes increases in hot-dry regions. For example, in the state of Nevada, thermoelectric power plants produce approximately 18,104 million kWh per year with an evaporation rate of 2.12 liters per kWh produced (Torcellini *et al.*, 2003). While hydroelectric generation produces less pollution and emits less greenhouse gases, its evaporative losses are much higher than those found in thermoelectric generation. In the state of Nevada, hydroelectric power produces approximately 2,510 million kWh per year with a whopping evaporation rate of 277.58 liters per kWh produced

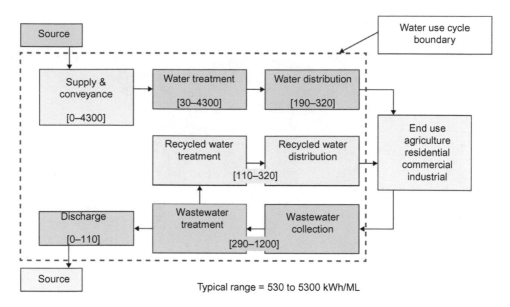

Figure 7.4.   Water use cycle energy intensities in kWh/million liters (adapted from: Krebs, 2007).

(Torcellini *et al.*, 2003). The overall water use for electricity production in Nevada is 27.44 L/kWh (Torcellini *et al.*, 2003).

Similarly, a significant amount of energy is required to convey, store, treat, and distribute municipal water, and to subsequently collect, treat, and discharge wastewater. In the US it is estimated that between 4 and 13% of the national electricity generation is consumed by water-related uses (Copeland, 2013). This variation is primarily related to regional differences and can be significant. For example, in the state of California, as much as 19% of the state's electricity consumption is devoted to pumping, treating, collecting and discharging water and wastewater (Copeland, 2013).

Furthermore, according to the California Energy Commission, water related energy use (also known as "water embodied energy") also accounts for approximately 30% of its natural gas and 333 billion liters of diesel fuel per year (Krebs, 2007). Figure 7.4 illustrates the range of typical water cycle energy intensities in the USA.

While 70% of California's supplied water is used for agricultural purposes, about 70% of the water embodied energy is used to satisfy urban water needs (Krebs, 2007). This important point is clearly illustrated by the figures reported by the Southern Nevada Water Authority (SNWA), a seven-member cooperative agency created to manage regional water resources in southern Nevada, which used in 2012 approximately 1 million megawatt hours to treat and deliver nearly 530 billion liters of water to its mostly urban costumer base (SNWA, 2013).

7.4   SITE AND BUILDING WATER MANAGEMENT ISSUES

In order to properly tackle water management issues, it is important to use a holistic approach that includes an understanding of water qualities in every step of the process. For example, wastewater is a general term that fails to take into consideration the individual properties of the various types of wastewater that could be potentially treated and reclaimed in order to reduce the overall consumption of centralized municipal water and also to reduce the load of water treatment plants. Table 7.1 provides useful definitions of the various wastewater sources typically found in residential and commercial buildings.

Table 7.1.   Definition of wastewater sources in buildings.

Blackwater	Greywater	Stormwater	Foundation drain water	Process water	Swimming pool backwash
Wastewater or sewage containing fecal matter, urine, or food waste	Wastewater from lavatories, showers, and washing machines	Water originating from natural precipitation, surface runoff	Sub-surface water diverted from foundations through a drainage system	Air-conditioning condensate, cooling tower blow down, distillation and RO reject water	Water harvested from swimming pool filters' backwashing

Table 7.2.   Characteristics of wastewater sources available for harvesting.

Water type	Water quality	Cost of recovery	Yield potential
Blackwater	Low	High	95 liters/day/occupant
Greywater	Medium	Medium	Showers, bathtubs and lavatories: 95 liters/day/occupant Clothes washer: 57–87 liters/load
Rain water	High	Low	1 liter/sq. meter of collection surface area (for every 1 mm of rainfall)
Stormwater	Medium	Medium	Volumetric run-off coefficient: sandy soils 0.02; silty soils 0.11; clay soils 0.24
Foundation drain water	Medium	Low	Varies based on water table, geology, climate, soil texture and construction methods.
Air conditioner condensate	High	Low	Dry areas 0.00–0.22 liters-kW of AC/hour Temperate 0.05–0.32 liters-kW of AC/hour Humid areas 0.11–0.54 liters-kW of AC/hour
Blow down water	Low	High	Cooling tower blow down water consumes about 25% of total water used by the system.
Distillation and RO reject water	Medium	Low	Reject water may range from 10–60% of the raw water pumped for treatment
Pool filter backwash	Medium	Low	190–1140 liters/week during operation. Winterization: each fall discharges 9085–13,627 liters.

Wastewater sources are typically not harvested because of local regulations, health concerns, and/or a lack of understanding of the treatment options required to make some of these sources available for onsite reuse. However, sources such as rainwater or air-conditioning condensate are in fact of high quality and have an overall low cost of recovery.

Table 7.2 summarizes the wastewater sources available for onsite harvesting along with some important properties to help determine their potential for reuse. As can be seen, there is a great opportunity for improved efficiency in the entire water treatment process when water nutrient properties and subsequent reclamation opportunities are considered during the design stages of a project. Several successful buildings provide robust case studies that illustrate how non-potable water reuse demands can be matched to the amount of wastewater generated within the site or building, thus eliminating discharge into the sewer. Some elements of a closed loop treatment system have a higher cost of recovery (e.g., blackwater or cooling tower blow down). However, with the expansion of decentralized systems and recent technological advancements, reclamation of these traditionally difficult wastewater sources may find its rightful place within the integrated site and building water management strategies.

The first step towards successful water management is to establish conservation goals to reduce daily consumption. On average, a US residence (consisting of 2.58 occupants) may use over 609,400 liters of water per year. Table 7.3 illustrates typical domestic water use values for the USA.

Both residential and commercial end users need to implement conservation strategies in order to tie in the big picture. Common practices would include low flow fixtures, aerators, drain heat recovery pipes, etc. Figure 7.5 illustrates the comparison of water usage between typical plumbing fixtures and low flow or energy efficient fixtures.

As noted, the building sector consumes great amounts of water on a daily basis. To tackle this issue, there are several certification options and programs available that offer recommendations for achieving optimal water usage in new construction or renovations. Whether they aim to be a model

Table 7.3.   Domestic water use (adapted from Mayer *et al.*, 1999).

Residential water use (liters per capita per day)	
Toilet	70.0
Shower	43.9
Faucets	41.2
Bath	4.5
Washing machine	56.7
Dishwasher	3.7
Other domestic use	6.0
Leaks	35.9
Outdoor	381.5
Unknown exterior	6.4

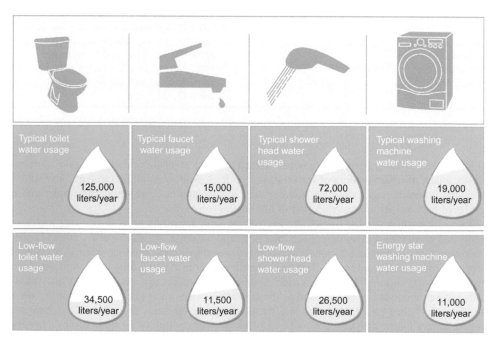

Figure 7.5.   Comparative savings on water usage between traditional and low-flow fixtures (Source: EPA, 2009).

for appropriate water management or to provide incentives for developers and investors through advanced plumbing checklists, these recommendations are an integral part of modern, green building. Among the most common programs and certifications are: Living Building Challenge, LEED, Green Globes, and the WaterSense Program.

Rehabilitation of the economic sphere will speed up the rate of infrastructural development. Lowering the ecological footprint of the urban infrastructure through water conservation, wastewater management, and onsite advanced reclamation techniques will play a critical role in this fast expansion.

7.4.1   *The Living Building Challenge*

The 'Living Building Challenge' is comprised of seven performance areas that need to be met in order for a project to be certified. The seven building performance areas are: site, water, energy, health, materials, equity and beauty. Each area is subdivided into imperatives that give recommendations for advanced management of the relevant resources. With respect to water, the Living Building Challenge (version 3.0) has two imperatives (The International Living Future Institute, 2014):

- *Net positive water* – Project water use and release must work in harmony with the natural water flows of the site and its surroundings. One hundred percent of the project's water needs must be supplied by captured precipitation or other natural closed loop water systems, and/or by re-cycling used project water, and must be purified as needed without the use of chemicals.
- All stormwater and water discharge, including grey and black water, must be treated onsite and managed either through re-use, a closed loop system, or infiltration. Excess stormwater can be released into adjacent sites under certain conditions.

7.4.2   *LEED*

LEED credits for water efficiency involve several levels of water usage reduction and award the number of points accordingly. Mandatory water use reduction requires a management strategy resulting in a 20% decrease of potable and non-potable water on site. Water saving plumbing fixtures and fitting efficiency are also a part of base level water management.

To achieve the next level of water conservation (see Table 7.4 for additional information) a 50% reduction of water usage is required. This level of water conservation includes the limitation of potable water usage for landscape irrigation.

Table 7.4.   LEED credits for water efficiency (adapted from: US Green Building Council, 2014).

Construction category	Credits
New construction (NC)	10 Points
Prereq 1: Water use reduction – 20% reduction required	
Credit 1.1: Water-efficient landscaping, reduce by 50%	2 points
Credit 1.2: Water-efficient landscaping, no potable use or no irrigation	2 points
Credit 2: Innovative wastewater technologies	2 points
Existing building (EB)	10 Points
Prereq: Minimum indoor plumbing fixture and fitting efficiency – required	
Credit 1.1: Water performance measurement, whole building metering	1 point
Credit 1.2: Water performance measurement, sub-metering	1 point
Credit 2: Additional indoor plumbing fixture and fitting efficiency	1–3 points
Credit 3: Water-efficient landscaping	1–3 points
Credit 3.1: 50% reduction	1 point
Credit 4.1: Cooling tower water management, chemical management	1 point
Credit 4.2: Cooling tower water management, non-potable water source use	1 point

The highest level of water conservation (see Table 7.4) does not allow potable water use or any type of irrigation for landscape on site. In its newest version (LEED v4), points are also awarded for cooling tower water management for all types of construction, as significant savings ought to occur from increasing the number of cycles through which water is being re-circulated before it is removed by blowdown. Furthermore, projects connected to municipal steam systems could earn a point if that system either recaptures steam or if it allows condensate to drain after passing through a heat recovery system or system that cools it with reclaimed water (Enck, 2013).

7.4.3   *Green Globes*

Green Globes is a comprehensive green building evaluation program developed by Green Building Initiative (Green Building Initiative, 2014). The certification process involves an environmental assessment protocol – a software tool for online assessment. After achieving a minimum threshold of 35% in the preliminary self-evaluation, new and existing buildings are eligible to seek a Green Globes certification and rating. Upon registration, the process utilizes third-party assessors to provide on-site feedback. The certification consists of several green building objectives, one of which is water usage and conservation. The water section of the certification includes three sub categories – water performance, water conserving features (sub metering and water efficient fixtures) and on-site water treatment. The water treatment recommendations are targeted towards integration of greywater collection (i.e., storage and distribution system to collect, store, treat and redistribute laundry and bathing effluent for toilet flushing, irrigation, janitorial cleaning, cooling and car washing). Moreover, where feasible, the program recommends integration of a biological waste treatment system for the site and building such as peat moss drain fields, constructed wetlands, aerobic treatment systems, solar aquatic waste systems (or living machines), and composting or ecologically-based toilets.

7.4.4   *WaterSense program*

WaterSense, a partnership program developed by the US Environmental Protection Agency (EPA), provides information to the end users regarding water management strategies and water-efficient products. Moreover, WaterSense is a certification label for plumbing fixtures that are on average 20% more efficient than standard fixtures and provide measurable water saving results, meeting the efficiency criteria set up by the EPA. As an example, a WaterSense certified fixture would be a high-efficiency toilet (HET) that uses less than 4.9 liters per flush. The program seeks to help consumers make smart water choices, decrease water use, and reduce strain on water resources and infrastructure (EPA, 2014). The program has been evolving since 2007, when it was first released with specifications for WaterSense certified toilets, and throughout the years has included specifications for landscape irrigation controllers and pre-rinse spray valves and water softeners.

7.5   CENTRALIZED VS. DECENTRALIZED WATER SUPPLY SYSTEMS

Population growth and rapid urban expansion are important challenges for the provision of water in urban areas around the world. As a result, the cost of water procurement and treatment, and therefore potable water delivered to end users is escalating. Moreover, traditional water sources vastly depend on rainfall patterns that are subject to change due to a shifting climate.

The increased pressure to provide sufficient fresh water supplies for energy production, agricultural use, and domestic consumption has sprung a strong interest among water districts and municipalities in sea water desalination. According to the Pacific Institute, a think-tank focused on water issues, the world's desalination capacity in 2005 was approximately 35.9 billion liters per day (Cooley *et al.*, 2006). That amount constitutes less than 2.4% of the total demand for fresh water in the United States, and aside from several serious environmental drawbacks, the desalination process requires more than 3.24 kWh/1000 L of fresh water supplied.

While water desalination or the transportation of water from distant regions are the approaches generally favored by water districts and municipalities that advocate centralized water supply, decentralized (often times privately owned) systems that reclaim and reuse water on-site should be given serious consideration as they have several advantages over traditional water supply and treatment systems. For example, decentralized direct water reuse systems oftentimes recognize the need for different qualities in the water supply to match specific uses. Due to its high level of purity, potable water has high water embedded energy content and it is therefore expensive to produce. When looking at water uses in buildings, potable water should only be used for drinking, cooking, and human cleansing. Other uses such as irrigation, toilet flushing, laundry uses, and make-up water for cooling towers could be satisfied with non-potable water coming from a decentralized direct water reuse system. Less pure, reclaimed non-potable water that is properly handled can be used to significantly reduce a building's potable water consumption, wastewater discharge, and waste loads. In addition, decentralized direct water reuse systems can reduce the water industry's energy use and the need for expensive infrastructure as cities continue to grow.

Decentralized direct water reuse is generally accomplished on a distributed system basis, where small to medium size facilities are built on-site to harvest, treat, store, and distribute the reclaimed water. Moreover, decentralized systems allow for short distribution lines from harvested water to its end users, reducing the systems' costs. In addition to other benefits, when used to irrigate landscape, harvested water replenishes groundwater supplies and helps reduce the flow to storm water drains. These systems can play a big role in reducing summer peak demands as well as delay the expansion of water treatment plants, thus mitigating the ecological footprint of this type of infrastructure.

Today, there are more than 30 decentralized direct water reuse projects operating in the northeast of the United States (Clerico, 2007). These projects cover a wide range of building types and are primarily built in urban areas where an abundant supply of wastewater is available for treatment and reuse.

According to Clerico (2007), some of the most important benefits of this approach are:

- A significant reduction (up to 95%) in potable water consumption when compared to a conventional design.
- A significant reduction (up to 95%) in wastewater discharge and waste loads.
- A positive environmental impact in cities with combined sewer overflow issues.
- Environmentally and economically viable systems that use waste as a resource.
- These systems reduce the need for large capital projects associated with conventional centralized water and wastewater systems.
- Improved energy efficiency relative to conventional centralized water and wastewater treatment systems due to on-site reuse.

7.6 INTEGRATED SITE AND BUILDING WATER MANAGEMENT (ISBWM)

An increasing number of buildings (or multiple-building developments) are harvesting and treating on-site their wastewater for later use in landscape irrigation, toilet flushing, laundry uses, and/or cooling-tower makeup. Integrated site and building water management (ISBWM) is a set of coordinated strategies aimed at reducing the consumption of municipally-supplied potable water, while simultaneously reducing (or even eliminating) the discharge of wastewater to sewer systems. ISBWM therefore relies on a "closed loop" approach to the management of water flow within the site/building.

An effective strategy to reduce excessive consumption and water waste is to match source water to its intended usage. The required volume and quality of water should be considered in

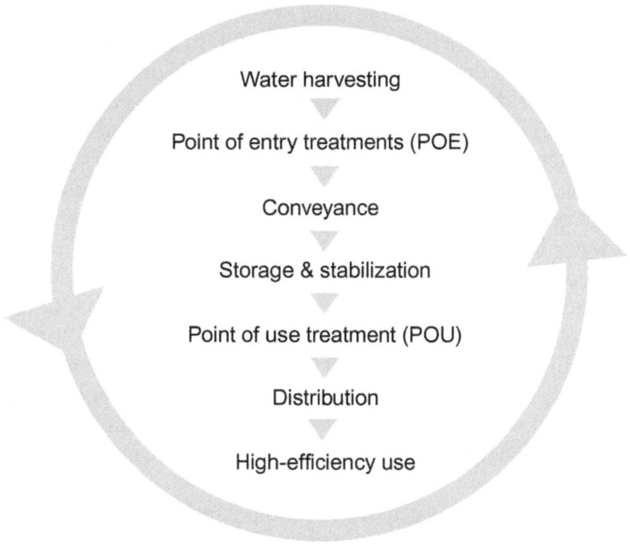

Figure 7.6.   Integrated site and building water management (ISBWM) approach.

terms of requirements of use, seasonal considerations, and scheduling its availability based on demand. In addition to these usage considerations, local regulations and codes should be reviewed to determine any possible regulatory restrictions for either harvesting or usage and the required quality of harvested water.

The closed loop approach cannot be fully achieved on certain locations or building types, as it depends on local codes and the feasibility of certain systems. However, the holistic approach suggested in Figure 7.6 should be pursued during the early stages of design of each new project.

7.7   CASE STUDIES

7.7.1   *Lloyd Crossing Urban Design Plan, Portland, OR*

One of the most ambitious projects with respect to integrated building water management was the 218,508 m^2 Lloyd Crossing Sustainable Urban Design Plan proposed by the city of Portland, OR. The redevelopment proposal for Lloyd Crossing represented an early, bold attempt to achieve the holistic integration of nature, users, infrastructure and its systems. The goal of this proposal was to lower the environmental footprint that this large development would have. The strategies employed were aimed to drastically reduce energy consumption and offset it, in part, with onsite harvesting. The plan also attempted to restore the natural balance with urban greenery and mimic natural watershed characteristics to preserve water. The project called for a 'water neutral' community where the reliance on imported water would be minimized, meaning that the neighborhood would rely on a decentralized water filtration system that would potentially meet all the non-potable use requirements. Neighborhoods in this plan would feature dual supply systems for potable and reclaimed water, rainwater harvesting systems, and a blackwater treatment facility shared by multiple buildings to collect and treat wastewater on site. With these strategies in place, potable water use would be reduced by 62% through the use of an integrated water system that includes stormwater management and treatment, as well as greywater and blackwater treatment and reuse for non-potable purposes.

Furthermore, by combining the systems described in the previous paragraph with strategies for water usage reduction such as efficient fixtures, the projected demand reduction would result

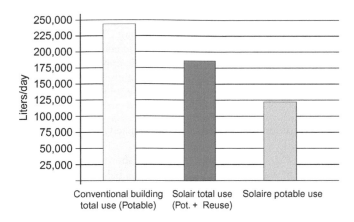

Figure 7.7. Average daily water use of Solaire Building (adapted from: Clerico, 2008).

in approximately 89% of savings in water use. The stormwater strategy was combined with the planning of intersections and streetscapes to maximize the treatment through natural water flow. Water metrics show that in order to achieve the desired performance, the demand for potable water in a neighborhood of this size would have to be reduced from 606 million to approximately 220 million liters per year. The wastewater reductions would have to come down from 545 million liters to 197 million liters per year (Mithun Architects+Designers+Planners, 2004).

7.7.2 *Solaire Apartments, New York, NY*

One of the first examples of the ISBWM approach is the Solaire Apartments in New York City. The Solaire, a 28-story building with 293 dwelling units, received LEED-NC Gold certification due to the many innovative systems included in the project. In terms of water management, the building features a membrane bioreactor process to treat, store, and reuse wastewater for toilet flushing, irrigation (the building also features a pesticide-free green roof), and cooling tower makeup. Moreover, the building effectively reduces nutrient loads to city sewers and the overall demand on both New York City's water and wastewater systems.

While stormwater was harvested in the Solaire building, its contribution to the overall average water consumption is marginal because stormwater was kept separate from the water reuse system as it is strictly dedicated to irrigating the building's green roof (Clerico, 2007). The results published about this building, summarized in Figure 7.7, indicate a significant reduction in the consumption of potable water (Clerico, 2008):

- Reduced potable water requirements by 48%.
- Reduced wastewater discharge by 56%.
- Reduced waste load to the sewer by 41%.
- Stormwater is strictly dedicated to irrigating the building's green roof.

Several other decentralized direct water reuse projects in New York City (e.g., the Helena and Tribeca Green) have improved designs and performance after the lessons learned from the Solaire's experience.

7.7.3 *Bullitt Center Building, Seattle, WA*

The Bullitt Center in Seattle, WA is a six-story, 4645 m^2 facility designed to become the greenest commercial building in the world. The design premise was to improve long-term environmental performance and promote broader implementation of energy efficiency, renewable energy, and other green building technologies.

Strategies for on-site renewable energy production, water conservation and harvesting, etc., were implemented to meet the goals of the Living Building Challenge. As discussed previously, to be certified as a Living Building, a facility is required to completely offset its water usage and energy needs for 12 successive months and meet 20 specific imperatives within seven performance areas or "petals" (The International Living Future Institute, 2014). At the Bullitt Center, the performance criteria for water management were achieved through the following systems (also illustrated in Figure 7.8):

- A rainwater roof collection system with a storage capacity of 211,983 liters before water is filtered and disinfected.
- A 1514 liter cistern for storing greywater from sinks, showers and floor drains.
- Water treatment system with constructed wetlands on a balcony.
- Composting system to address blackwater waste.

The rooftop PV array, with a 230,000 kWh a year production, will also contribute to the rainwater collection by channeling the rainfall to the roof membrane and drains below. The excess harvested water will be released to the ground and atmosphere. Providing that the State Department of Health and Seattle Public Utilities clear the regulatory obstacles, all of the water usage needs will be met through rainwater collection.

7.7.4   Audubon Center at Debs Park, Los Angeles, CA

Another good example of integrated building water management is the Audubon Center at Debs Park, a 467 m^2 environmental education and conservation center with LEED-NC Platinum certification (see Fig. 7.9). The conservation center treats 100% wastewater onsite, eliminating its connection to the municipal sewer. Treated wastewater is used for toilet flushing, achieving an overall reduction of 70% in its potable water use.

The water management system consists of a hybridized anaerobic/aerobic treatment and filtration process and a peracetic acid and ultraviolet light advanced oxidation disinfection process. Two-stage, low-flow toilets allow for different flow options and may use recycled greywater. Furthermore, the landscaping will require no permanent irrigation. Stormwater is kept on-site and diverted to a water quality treatment basin before being released to help recharge groundwater (US Department of Energy, 2014).

The water use data reported for this building is:

- Indoor potable water use: 268,000 L/year.
- Outdoor potable water use: 68,100 L/year.
- Total potable water use: 336,000 L/year.
- Potable water use per unit area: 720 L/m^2.

7.7.5   The Omega Center for Sustainable Living, Rhinebeck, NY

The Omega Center for Sustainable Living (OCSL) is a natural water reclamation facility and is one of the first two projects to receive both a LEED Platinum and Living Building certifications (see Fig. 7.10). The OCSL features an "Eco Machine" that uses nature-influenced technology (see Figs. 7.11 and 7.12). The system is comprised of solid settlement tanks, equalization tanks, anoxic tanks, constructed wetlands, aerated lagoons, recirculating sand filters and dispersal fields. This water treatment system manages the water from the main facility of the Omega Center with a capacity to process up to 196,840 liters of water per day.

The OCSL building is also a net-zero energy facility that only uses harvested renewable energy on site. Since the plant is located on the side of a hill, gravity aids the water's flow, reducing energy demand. Moreover, the system doesn't use any chemicals to process the wastewater. This concept has a goal to serve as a model for responsible water management in the community by preserving natural resources and keeping the environment clean.

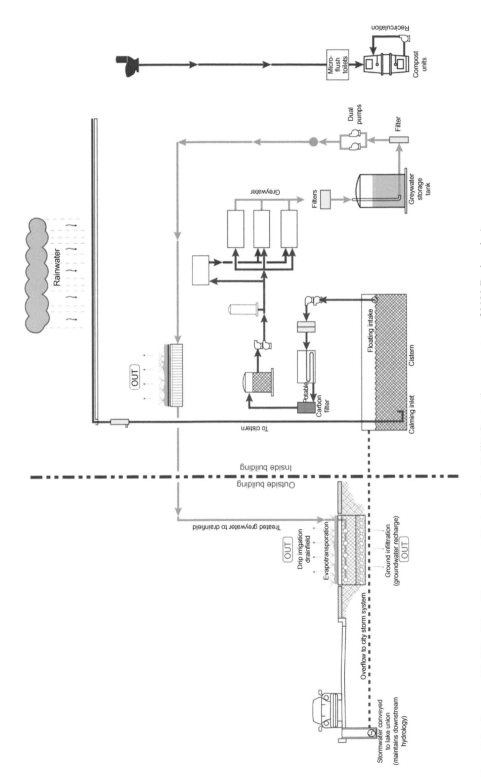

Figure 7.8.   Integrated Building Water Management Strategy for the Bullitt Center (image courtesy of 2020 Engineering).

Figure 7.9.   Aerial view of the Audubon Center at Debs Park, Los Angeles, CA (image courtesy: Alfredo Fernandez-Gonzalez).

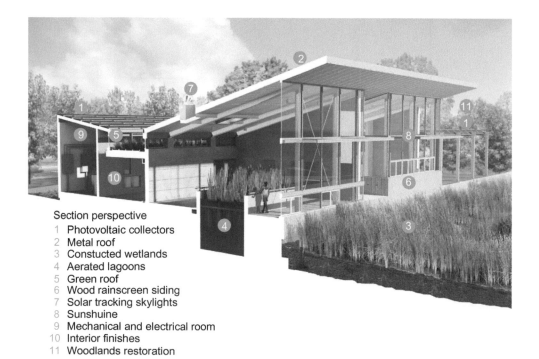

Section perspective
 1 Photovoltaic collectors
 2 Metal roof
 3 Constucted wetlands
 4 Aerated lagoons
 5 Green roof
 6 Wood rainscreen siding
 7 Solar tracking skylights
 8 Sunshuine
 9 Mechanical and electrical room
10 Interior finishes
11 Woodlands restoration

Figure 7.10.   Section-perspective of the Omega Center for Sustainable Living highlighting several of its ISBWM strategies (image courtesy: BNIM Architects Inc.).

Figure 7.11.   Indoor view of "Eco Machine" (image courtesy: Assassi).

Figure 7.12.   View of constructed wetlands (image courtesy: Assassi).

The following are the water management and conservation design features of this project:

- Underground wells are supplying the site with potable water.
- Rainwater is used for cleaning purposes and toilet flushing. Flushed water is processed by the 'Eco Machine'.
- Water conservation features include low flow fixtures and aerators.
- Water for all other purposes is being purified in the 'Eco Machine'.

7.8   THE LIVING OASIS: AN INNOVATIVE APPROACH TO ISBWM

Water reuse is recognized as a safe, technically feasible, and increasingly cost-effective practice. In the United States, about 6435 million liters of water are reclaimed on a daily basis. While most water reuse projects in the USA are centralized facilities designed to provide non-potable water for outdoor irrigation, this chapter makes the case for an integrated site/building water management approach to directly reuse reclaimed water onsite, thereby reducing the consumption of potable water in buildings. Other benefits of this approach are the reduction of hydraulic and to a lesser extent solid discharges to the municipal sewer, ultimately saving municipal energy by reducing the amount of potable water that needs to be pumped and wastewater that needs to be treated at a municipal facility. While the economics of this approach are best suited for large buildings or campuses, decentralized direct water reuse systems can be designed to address the needs of small residential and/or commercial buildings, as illustrated in the case studies of the Audubon Center at Debs Park and the Omega Center for Sustainable Living.

The approach advocated in this chapter, called the *Living Oasis*, was developed in response to the growing concerns that exist with the use of potable water for outdoor irrigation (Fernández-González, 2009). The *Living Oasis* concept consists of seven points listed below and also illustrated in Figures 7.13 and 7.14:

1. Harvest all economically available sources of on-site water supply.
2. Integrate, filter, and treat the harvested water.
3. Provide 24/7 on demand storage.
4. Provide digitally controlled distribution of the reclaimed water.
5. Drip/subsurface irrigation for roofs, walls, and landscaped areas.
6. Minimize the dependence on municipal water supply – reuse and limit waste.
7. Embrace the Architecture 2030 Challenge as a building investment strategy.

7.8.1   *Harvest all economically available sources of on-site water supply*

The geographical location of a project and the size and type of a building determine the kinds of opportunities available for economically feasible on-site water harvesting. The most obvious sources of on-site water harvesting may be the collection of greywater from showers and sinks (on a regular basis) and rainwater (on an intermittent basis, and in highly variable amounts).

However, there are other equally viable and economically attractive sources for on-site water harvesting such as the collection of air-conditioning condensate (particularly in hot-humid climates), cooling-tower blow down (i.e., water intentionally drained off to prevent high concentrations of minerals and other contaminants in the water used for cooling), washer-extractors (i.e., large commercial laundry systems which are often equipped with the option of recovering and reusing onsite water), and the extraction of water from organic waste (this innovative system also resolves one of the greatest pollution problems of the world, as it diverts organic waste from going to landfills by converting it into a high value organic fertilizer through the use of vermicasting).

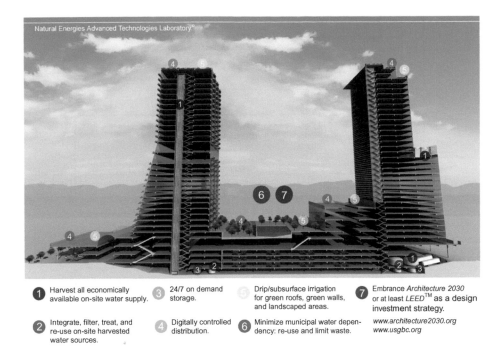

① Harvest all economically available on-site water supply.

② Integrate, filter, treat, and re-use on-site harvested water sources.

③ 24/7 on demand storage.

④ Digitally controlled distribution.

⑤ Drip/subsurface irrigation for green roofs, green walls, and landscaped areas.

⑥ Minimize municipal water dependency: re-use and limit waste.

⑦ Embrace *Architecture 2030* or at least *LEED*™ as a design investment strategy.

www.architecture2030.org
www.usgbc.org

Figure 7.13.   The Living Oasis Concept with seven points of water management strategies.

7.8.2   *Integrate, filter, and treat the harvested water*

Maintaining high-quality harvested water once the various on-site sources listed above are integrated is an important challenge that must be properly understood for the *Living Oasis* concept to be a success. Air-conditioner condensate (which is essentially distilled water) and rainwater are both inherently pure and may be reused without the need of extensive treatment. Greywater, on the other hand, may contain organic matter (such as hair, textile fibers, etc.) and/or highly diluted chemicals (present in cleaning products) which should be filtered out before reuse. Greywater should be used quickly before water decomposition switches from an aerobic to an anaerobic process. Cooling-tower blow down presents an additional challenge as it contains high concentrations of minerals and bacteria, including Legionella. Cooling-tower blow down water must be treated before mixing it with other on-site harvested water sources.

In addition to water quality issues, direct water reuse systems also need to address the varying flow rates from the water sources being considered. A system that integrates rainwater, thereby improving the building's overall stormwater performance, will have a higher upfront cost due to the need for large stormwater storage tanks. In the arid areas of the United States' southwest, the use of stormwater may not be as cost effective as some of the other water sources listed above.

7.8.3   *Provide 24/7 on demand storage*

Once the various harvested water sources have been filtered or treated as necessary, the reclaimed water should be pumped to one or several storage tanks for future use. Ensuring that high-quality non-potable water is always available in sufficient quantities for use in the various indoor and/or outdoor applications where non-potable water is acceptable is key for the success of the *Living Oasis* concept.

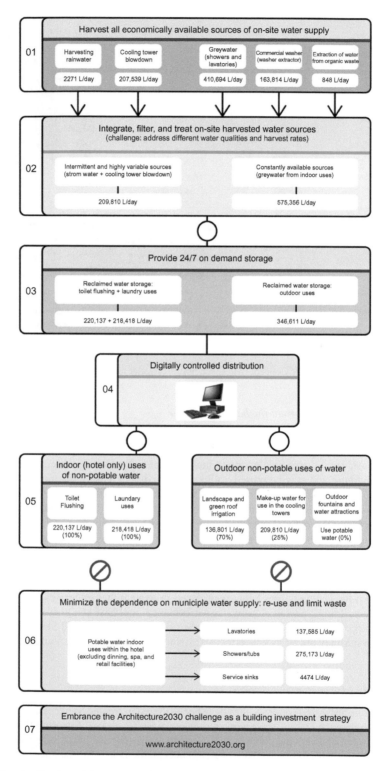

Figure 7.14.   Application of the *Living Oasis* model to a hypothetical hotel in Las Vegas.

7.8.4   *Provide digitally controlled distribution of the reclaimed water*

Whether it is used for toilet flushing, laundry uses, landscape and/or green roof irrigation, or make-up water for cooling towers, a digitally controlled distribution system will ensure that reclaimed water is delivered in the right amount and at an appropriate pressure for each specific application. In the case of irrigation systems, it is important to integrate soil moisture sensors and weather predictions to control the amount of water required by each irrigation zone.

7.8.5   *Drip/subsurface irrigation for roofs, walls, and landscaped areas*

According to the American Water Works Association Research Foundation, residential outdoor water uses in hot-dry places such as Las Vegas or Phoenix constitute 59 to 67% of the total water use. In the United States' southwest, traditional landscape irrigation practices oftentimes waste large amounts of water due to evaporation, piping leaks, and poorly targeted spray nozzles. Practices like xeriscaping have helped reduce the need for irrigation. However, buildings featuring green roofs/walls, sports facilities, recreational parks, or other outdoor amenities requiring vegetated areas still necessitate irrigation to keep the plant material alive. For this reason, the *Living Oasis* concept advocates an efficient use of non-potable water through drip/subsurface irrigation. Drip irrigation reduces evaporation losses, over-watering, and inadequate delivery of water. Subsurface irrigation goes one step further delivering water directly to the root zone, significantly reducing evaporation losses. Drip and/or subsurface irrigation work best when coupled with a digitally controlled distribution system that adjusts water delivery according to soil moisture and weather predictions.

7.8.6   *Minimize the dependence on municipal water supply – reuse and limit waste*

The *Living Oasis* concept reduces the use of municipal water supply by harvesting and continuously reusing economically available on-site water supply sources. According to Clerico (2008), non-residential direct water reuse systems can reduce water consumption by 80%, with residential systems achieving a water consumption reduction of 50%.

More importantly, centralized water supply infrastructure makes cities and buildings extremely vulnerable in the event of catastrophic failure of such infrastructure. An earthquake, a terrorist attack, a power outage, or any other unforeseen situation affecting water supply infrastructure during the summer time in the United States' southwest could kill thousands of people in a matter of hours. By virtue of being decentralized, the *Living Oasis* concept could address this very important issue given the fact that direct water reuse systems currently in operation already use membrane bio-reactor (MBR) technology that in some cases is coupled with reverse osmosis. By adding to these systems ultra violet light and hydrogen peroxide disinfection, the treated water could easily exceed most United States' drinking standards, thus producing "passive survivability" conditions, a term coined by Alex Wilson that refers to the ability of a building to maintain livable conditions when electricity, heating fuel, or municipal water are lost for extended periods of time.

7.8.7   *Embrace the Architecture 2030 Challenge as a building investment strategy*

The Architecture 2030 Challenge points out the urgency to reduce global greenhouse gas emissions in order to avoid catastrophic climate change. Since buildings are the major source of demand for energy, the Architecture 2030 Challenge asks the global architecture and building community to adopt the following targets:

- All new buildings, developments and major renovations shall be designed to meet a fossil fuel, greenhouse gas-emitting, energy consumption performance standard of 60% of the regional (or country) average for that building type.

- At a minimum, an equal amount of existing building area shall be renovated annually to meet a fossil fuel, greenhouse gas-emitting, energy consumption performance standard of 60% of the regional (or country) average for that building type.
- The fossil fuel reduction standard for all new buildings shall be increased to:
  - 70% in 2015
  - 80% in 2020
  - 90% in 2025
- Carbon-neutral in 2030 (using no fossil fuel greenhouse gas-emitting energy to operate).
- These targets may be accomplished by implementing innovative sustainable design strategies, generating on-site renewable power and/or purchasing (20% maximum) renewable energy and/or certified renewable energy credits.

7.9   CONCLUSIONS

The increase in human population and the unremitting migration to urban centers pose an important challenge for the provision of potable water and the adequate sanitation of cities. Furthermore, as cities continue to grow, their footprint negatively affects the hydrologic cycle, worsens runoff problems, and aggravates urban heat island effects.

To address the effects of the rapid urbanization we are witnessing in the XXI century this chapter proposes a decentralized approach for the provision of potable water and the treatment of waste water. The *Living Oasis* concept presented in this chapter is an option that advances direct water reuse, suggesting a clear integrated site and building water management vision that puts sustainability at its core while also providing innumerable business opportunities (as well as challenges) for the private sector as the concept begins to be implemented in different scale projects around the world.

While performance data and a comprehensive economic analysis of the *Living Oasis* concept are not available at this time, the case studies presented in this chapter suggest that direct water reuse in the United States is both feasible and could be highly beneficial particularly in areas susceptible to drought and/or water shortages.

Furthermore, given the great amounts of water consumed by the building sector on a daily basis, certification programs such as the Living Building Challenge, LEED, Green Globes, and WaterSense are challenging the design professions to solve the water and sanitation challenges in urban environments with new approaches that emphasize decentralized solutions.

The information presented in this chapter should be considered during the early stages of the design process, as it is extremely difficult to integrate site and building water management strategies once a building design has been developed without regard of these issues.

ACKNOWLEDGEMENTS

The author wishes to acknowledge and thank Milica Tajsić for her assistance in the preparation of this chapter and for coordinating the illustrations that accompany its text.

REFERENCES

Barnett, T.P. & Pierce, D.W.: When will Lake Mead go dry? Scripps Institution of Oceanography. La Jolla University of California San Diego, CA, 2008.

Carter, N.T.: Energy-water nexus: the energy sector's water use. Congressional Research Service, CRS Report for Congress R43199, Washington, DC, 2013.

Clerico, E.A.: The future of water reuse in New York City (White Paper). 2007, http://www.allianceenviron mentalllc.com/pdfs/AE_NYC_Incentives_White_Paper_final.pdf (accessed 24/5/2014).

Clerico, E.A.: Current status of water reuse (White Paper) 2008. http://www.allianceenvironmentalllc.com/pdfs/AE_Papers_CurrentWaterReuse.pdf (accessed 17/1/2014).

Cooley, H., Gleick, P.H. & Wolff, G.: Desalination, with a grain of salt: a California perspective. Pacific Institute for Studies in Development, Environment, and Security, Pacific Institute, Oakland, CA, 2006.

Copeland, C.: Energy-water nexus: the water sector's energy use. CRS Report for Congress R43200, Congressional Research Service, Washington, DC, 2013.

Enck, C.: New ways to save water in LEED v4. *EDC Magazine* (2013), http://www.edcmag.com/articles/94999-new-ways-to-save-water-in-leed-v4 (accessed 10/1/2014).

Energy Information Administration: Annual Energy Review 2007. US Energy Information Administration, Report No. DOE/EIA-0384, Washington, DC, 2007.

EPA: Water on tap: what you need to know. Office of Water, US Environmental Protection Agency EPA 816-K-09-002, Washington, DC, 2009.

EPA: WaterSense: An EPA Partnership Program, 2014. http://www.epa.gov/watersense/index.html. (accessed 20/2/2014).

Fernandez-Gonzalez, A.: Turning your building into a water-saving "Living Oasis": an innovative approach to integrated building water management. *Living Architect/Monitor*. 11:1 (2009), pp. 26–29.

Green Building Initiative: Green Building Programs: Green Globes® NC Overview 2014, http://www.thegbi.org/green-globes/new-construction.shtml (accessed 24/9/2014).

Krebs, M.: Water-related energy use in California. Assembly Committee on Water, Parks and Wildlife, 2007, http://www.energy.ca.gov/2007publications/CEC-999-2007-008/CEC-999-2007-008.PDF (accessed 10/1/2014).

Level Organization, New Zealand: Change in surface water runoff due to site development. 2007, http://www.level.org.nz/fileadmin/downloads/Water_Use/LevelDiagram100.pdf (accessed 9/3/2014).

Mayer, P.W., DeOreo, W., Opitz, E.M., Kiefer, J.C., Davis, W.Y., Dziegielewski, B. & Nelson, J.O.: Residential end uses of water. American Water Works Association, AWWA Research Foundation, Denver, CO, 1999.

Mithun Architects+Designers+Planners: Lloyd Crossing Sustainable Urban Design Plan and Catalyst Project – Portland, Oregon 2004, http://issuu.com/mithun/docs/lloyd_crossing#signin (accessed 20/2/2014).

SNWA (Southern Nevada Water Authority): 2012 Annual Report 2013, https://www.snwa.com/assets/pdf/about_reports_annual.pdf (accessed 20/5/2014).

The International Living Future Institute: Living Building Challenge[SM] 3.0 – a visionary path to a regenerative future. 2014, http://living-future.org/sites/default/files/reports/FINAL%20LBC%203_0_Web Optimized_low.pdf (accessed 20/5/2014).

Torcellini, P., Long, N. & Judkoff, R.: Consumptive water use for US power production. National Renewable Energy Laboratory, NREL/TP-550-33905, Golden, CO, 2003.

UN World Water Assessment Programme. Water for people, water for life: a joint report by the twenty three UN agencies concerned with freshwater. World Water Assessment Programme, UNESCO Pub., Berghahn Books, New York, NY, 2003.

US Department of Energy: Building Technologies Program – Zero Energy Buildings Database 2014, http://zeb.buildinggreen.com/overview.cfm?projectid=234 (accessed 26/5/2014).

US Government Accountability Office, GAO: Freshwater supply: states views of how federal agencies could help them meet the challenges of expected shortages. US General Accounting Office, Report GAO-03-514, Washington, DC, 2003.

US Green Building Council: LEED Checklists. http://www.usgbc.org/resources/checklists (accessed 24/9/2014).

CHAPTER 8

Energy-efficient HVAC systems and systems integration

Walter Grondzik

8.1 INTRODUCTION

A high-performance building (under whatever name it may be described as-green, net-zero-energy, carbon-neutral) will necessarily involve an energy-efficient climate control system. The most energy efficient climate control options are those based upon passive heating and cooling principles. Unfortunately, passive climate control systems are not effective in all climates and/or in all building types. When a passive heating or cooling approach cannot meet the owner's project requirements (sometimes described as design intent and criteria) an active approach to climate control will likely be considered and implemented.

The owner's project requirements (abbreviated OPR in building commissioning documents) comprise a definitive statement of those project outcomes that will be considered important to defining a successful project. Such pre-defined requirements, which will shape design, construction, and operation decisions, commonly include targets for energy use (such as an EUI, energy utilization intensity, value), acceptable indoor air quality conditions, expected thermal comfort parameters, desired operations and maintenance procedures, and very often a desired green building rating level. Failure to clearly define what a successful project must accomplish places the design team at a distinct disadvantage and often leaves the owner dissatisfied with project outcomes.

Active climate control systems are collectively termed HVAC (heating, ventilating and air-conditioning) systems. In larger buildings requiring active climate control, owner requirements have historically led to inclusion of a system that meets the definition of air-conditioning promoted by the American Society of Heating, Refrigerating and Air-Conditioning Engineers (ASHRAE) – namely a system that can simultaneously control air temperature, relative humidity, air distribution, and air quality. An "AC" system, by definition, includes the "H" and "V" aspects of HVAC.

8.2 HVAC SYSTEM EXPECTATIONS

Looking at this contextual foundation more specifically, an HVAC system will typically be expected or required to provide thermal comfort and acceptable indoor air quality. These are two substantially different outcomes – one involving heat and moisture transfer and the other chemical and physical manipulation of air. As discussed below, designers are starting to question the one-system-provides-all approach that is currently typical of the majority of HVAC system designs – preferring instead to provide a thermal control system and a separate air quality control system. Another common expectation for successful HVAC systems is reasonable-to-excellent energy performance. Reasonable-to-good environmental (green) performance may also be added to the list of system expectations.

8.2.1 *Thermal comfort*

Thermal comfort is a commonly expected outcome for an HVAC system. When occupant comfort is not a concern, this consideration is often replaced by demands for thermal process conditions (such as in some hospital spaces, manufacturing facilities, or product storage areas). Process requirements (which are project specific) will be conveyed to the HVAC designer by the facility owner and will not be further discussed in this chapter. Meeting thermal comfort requirements will normally be in the hands of the design team and warrants further review.

Thermal comfort is a human perception regarding the suitability of thermal conditions for some use or purpose. This perception, which can differ from person to person within the same space, is strongly influenced by the following variables:

- Air dry-bulb temperature.
- Air relative humidity.
- Air speed.
- Radiant temperatures of surrounding surfaces (often expressed as *MRT* – mean radiant temperature).
- The level of thermal insulation afforded by an occupant's clothing ensemble.
- The metabolic activity level of an occupant.

Secondary factors that might affect thermal comfort perceptions include: asymmetry of thermal conditions (a warm window surface in a room with a chilled beam cooling system); transient changes in thermal conditions (unstable conditions that vary quickly over time); individual mood (stress, irritation); other sensory inputs (colors, noise); acclimatization (thermal or cultural); and potentially other issues. These secondary factors can usually be safely ignored during design unless they differ substantially from the norm. The primary factors must all be considered for a successful design outcome. The tendency to think of thermal comfort as simply a dry-bulb air temperature phenomenon is just wrong in the majority of built spaces.

Since thermal comfort is a human perception, the best way to ascertain comfort responses is to ask building occupants if they are comfortable. Procedures for such post-occupancy evaluations have been developed and standardized. During the design phase of a project, however, there are no occupants to ask and no actual building space in which to place them. Fortunately, thermal comfort response trends from existing spaces can be used to inform the design process. In North America, ASHRAE Standard 55 (*Thermal Environmental Conditions for Human Occupancy*; ASHRAE, 2010) is the most commonly used guide for establishing appropriate thermal conditions during design. Standard 55 endorsed values can become the comfort design criteria for a project. A similar comfort standard is used in Europe (ISO 7730: *Ergonomics of the Thermal Environment*; ISO, 2005) and local guidance may be found in other geographical locales. Adaptive comfort models are increasingly being used to inform the design process for passively cooled buildings. Adaptive comfort analyses are not currently used for passive heating systems or energy-efficient active HVAC systems (although the concept most likely has merit).

8.2.2 *Indoor air quality*

Appropriate indoor air quality (IAQ) conditions have joined thermal comfort as an expected HVAC system outcome over the past several decades. Acceptable indoor air quality is typically defined in terms of a health component and a comfort component. The health component addresses control of indoor air contaminants that cannot be detected by occupants yet would have adverse health effects. The comfort component addresses air contaminants (such as odors) that are more of an annoyance than a health hazard, and which can be rationally evaluated by occupants. In North America, ASHRAE Standard 62.1 (*Ventilation for Acceptable Indoor Air Quality*; ASHRAE, 2013a) is the predominant air quality design guidance document. In Europe, ISO 16814 (*Building Environment Design – Indoor Air Quality*; ISO, 2008) plays a similar role in defining IAQ design criteria. As noted in the title, the main focus of Standard 62.1 is ventilation (the use of outdoor air to dilute contaminants), a mitigation strategy that involves the potential for substantial energy impacts. Other IAQ control strategies include filtration (removal of contaminants by straining or adherence) and source control (reduction of contaminant concentrations by spot exhaust or skillful selection of materials and finishes). Ventilation, filtration, and exhaust are all HVAC system concerns and would be addressed during the design of an air conditioning system.

When a single HVAC system is tasked with control of both thermal comfort and indoor air quality, design compromises may need to be made that will affect one or both of these desired outcomes. Such compromises, skillfully made, may be acceptable for many projects. For projects with high-performance outcomes, such compromises may negatively affect system energy efficiency and also system effectiveness.

8.3 TERMINOLOGY

Before looking at HVAC systems in more detail, a discussion of contextual terminology related to system performance is useful.

Energy: energy is described as the ability to do work; much of the effort to develop energy-efficient, green, and/or sustainable projects is focused on reducing the amount of energy required to accomplish a given function (such as illumination or control of air quality); energy can be described as non-renewable (derived from fossil fuels) or renewable (derived from solar radiation, wind, the flow of water); renewable energy sources typically do not substantially contribute to greenhouse gas emissions (and thus to climate change); units are kWh (Btu). Power is the instantaneous rate of energy flow; units are kW (Btu/h).

Efficiency: in its simplest formulation, efficiency is defined as output/input (in consistent units); as efficiency is improved the same output or effect can be provided for less energy input; some HVAC systems have apparent efficiencies greater than 100% (which is technically impossible), to avoid listing an efficiency of say 150% the performance of such systems or equipment is described in terms of coefficient of performance (*COP*) defined as output effect/input energy (in consistent units); efficiency and *COP* are dimensionless.

Effectiveness: effectiveness describes the success of a system in delivering what it was designed (and constructed and operated) to deliver; in the case of an HVAC system this is typically thermal comfort and indoor air quality; effectiveness is qualitative (but should not be dismissed by engineers due to difficulty in reducing it to numbers).

Conservation: conservation is a term that has been saddled with some negative connotations in North America (sometimes conveying a sense of doing without); conservation simply means not wasting resources; it is not the same as deprivation (doing without a resource); conservation can play an important role in the design of high-performance buildings as a means of ensuring that valuable resources are not wasted; the concept of conservation is qualitative, but its results can be quantified (in various units).

Effectiveness and efficiency are often seen as interchangeable concepts by designers. This is not the case. A system can be exceptionally effective (it does its intended job quite well) but highly inefficient. Conversely, a system can be very energy efficient, but dismally ineffective. Effectiveness and efficiency must be balanced during the design of an HVAC system. Effectiveness expectations are defined by the project OPR; efficiency outcomes reside in the skill of the design team. In general, effectiveness should come before efficiency. It makes little sense to install a system that uses energy sparingly but does not deliver the effects anticipated.

Considering conservation allows the element of time to be considered along with effectiveness and efficiency. Conservation can often deliver substantial energy savings by ensuring that systems that are not needed at a given time are not operating and that systems that do not need to run at 100% capacity are appropriately operated at part loads. The concept of conservation also plays a role when considering non-energy resource aspects of an HVAC system, such as water use. In many cases water systems are 100% efficient (output = input), yet the output (or consumption) is greater than necessary. Water flow from a faucet is an example; 100% efficient, yet potentially wasteful.

The key to a high-performance HVAC system is to design the system to ensure effectiveness, maximize efficiency, and control for conservation. From a thermal comfort perspective, move only whatever heat transfer medium is necessary and only when necessary. From an IAQ perspective, consider designing to reduce the need for ventilation through wise selections of materials, furnishing, and finishes. A life-cycle cost analysis will typically be required to arrive at an optimal HVAC system solution because the lowest first-cost solution is usually not the best choice over any reasonable time frame (beyond 2–3 years of project life). Attention must be paid to the construction process and to facility operations. High performance can be value-engineered out of a good design; it can also be misoperated out of a great design. Five years into the life of a building, high-performance will persist only through attention to design, construction, and operation. The design team has a role to play (even if unconventional) in each of these phases.

8.4   COMMISSIONING OF SYSTEMS

The commissioning process (Grondzik, 2009) is highly recommended as a means of ensuring and maintaining quality outcomes through the life of a project. In North America, the defining characteristics of this process are described in ASHRAE Guideline 0 (*The Commissioning Process*; ASHRAE, 2013b) and in code-language Standard 202 (*Commissioning Process for Buildings and Systems*; ASHRAE, 2013c). Similar guidance for the implementation of this process is available through a number of European resources. In its essence, commissioning is a quality assurance process that overlays the building acquisition process to attempt to deliver a project that meets the owner's project requirements fully and upon initial project delivery. Sadly, this type of quality outcome is not as common as might be expected by many owners, and decreases in likelihood with increasing project expectations and complexity.

The commissioning process should begin in pre-design (with the development of a strong OPR document) and continue through the design, the construction, and the occupancy and operations phases of project acquisition. Vitally important characteristics of the commissioning process include documentation, communications, and verification. These are also characteristics of a conventional facility acquisition process, but are undertaken with much more formality and rigor under the eyes of the commissioning process. Verifications occur throughout the commissioning process, with key benchmarks that include verification of the OPR document, of major design decisions, of construction documents, of the installation of equipment and components, of the training of owner's operations personnel, and of actual facility operations for the first year (or so) of use.

With the exception of small systems installed in very small buildings, the vast majority of HVAC systems consist of collections of individual components selected, arranged, and controlled to meet the specific needs of a specific building. Thus, HVAC systems tend to be one-offs (custom installations not quite the same as similar systems in similar buildings). This customization permits the design of high-performance systems; but also allows for the design of low-performance systems. Achieving high performance involves design skills, design integration, and often additional design time and expense.

8.5   HVAC SOURCE EQUIPMENT

The major divisions of HVAC equipment include: source components, distribution components, delivery components, and control components. These, respectively, produce heating and/or cooling effect, move this effect around a building, introduce it into the spaces being conditioned, and regulate the magnitude and timing of such effects (ASHRAE, 2012).

Source components in an HVAC system are intended to provide heat (heating effect), coolth (cooling effect), and/or both. Climate, building function, and design decisions dictate whether heating, cooling, or both are needed for a given building. The interactions between climate and function as mitigated (or amplified) by building layout and enclosure are varied and complicated (ASHRAE, 2013d). The end results of these interactions are the required maximum capacity and load factors for heating equipment and cooling equipment. Sophisticated computer simulations are commonly employed to determine both capacity and projected energy consumption for HVAC systems during the design process. A demand for high performance suggests the wisdom of using such simulations iteratively to optimize building and HVAC system coordination rather than simply using a simulation as a one-pass system sizing tool.

8.5.1   *Heat sources*

There are four main categories of heat source options for an HVAC system. These include:

- On-site combustion.
- Electric resistance.
- Heat transfer.
- Energy capture.

An on-site combustion source burns a fuel on the building site to produce heat. Such fuels may be renewable (such as wood or biomass), but most commonly are non-renewable fossil resources (such as natural gas, propane, oil, or coal). On-site combustion typically involves local pollution concerns, as well as the release of carbon dioxide that will trigger indirect (but important) climate effects. Electric resistance simply taps into the heat production inherent in passing electrical current through a resistor. The electricity is most commonly produced off site at a power production facility, which may use renewable or non-renewable resources depending upon the regional circumstances. Coal, natural gas, fuel oil, nuclear reactions, and hydropower are used to produce electricity. Wind and solar energy are also becoming economically viable sources for electricity production. There is a wide range of equipment of varying configurations and capacities that employs these two means of heat production. Efficiency is a reasonable metric for gauging the performance of such equipment.

Heat transfer as an HVAC system heat source involves identifying a body of usable heat (in or near a building) that can be accessed and used to provide space heating. Such usable heat may be found in air being exhausted from the building, in the air surrounding a building, or in the soil under and/or surrounding a building. Heat exchangers and heat pumps are the most common equipment manifestations of heat transfer as a heat source. There is a reasonable range of equipment types and capacities by which to tap into heat transfer. Efficiency is typically used to describe heat exchanger performance; coefficient of performance is used to describe heat pump performance.

Energy capture is currently the least used heat source option. An energy resource that is not heat, but that can be readily converted to heat, is intercepted on the building site and used to provide space heating. Examples include a solar thermal system, a photovoltaic (PV) system, and a micro-scale wind system. Efficiency is often used to describe the performance of such systems, but this may be of questionable value in making economic decisions as the energy being converted (whether more or less efficiently) is free.

Selection of a heat source (when important to HVAC design due to site climate) will have a potentially large impact on building energy performance. Efficiencies for commonly available heat sources can range from around 85% to around 750% (with the higher values being expressed as *COP*). This is almost a ten-fold range of efficiencies and the impact of this decision on overall building performance must be considered for projects in heating-dominated climates. In addition, choice of a heat source will have an impact on the quantity of carbon emissions attributable to HVAC system operation. There is a similarly wide range of carbon values (as with efficiencies). An additional consideration is whether efficiency and carbon accounting are to be done on a site basis or on a source basis. There can be upwards of a three-fold multiplier for some heat sources if traced back to the source. Site accounting is currently the most common approach (which tends to follow energy efficiency code expectations).

8.5.2   *Coolth sources*

There are also several fundamental options for cooling effect sources (ASHRAE, 2014). As with heat sources, these basic options expand into numerous equipment capacity and configuration selections. The three basic (non-passive) means of producing a cooling effect for a building are:

- Vapor compression refrigeration.
- Absorption refrigeration.
- Evaporative cooling.

These fundamental options present a wide range of thermal efficiencies and carbon emissions. The most thermally efficient (potentially approaching infinity) and almost-carbon-neutral approach is evaporative cooling. This means of producing a cooling effect is, however, substantially constrained by climate conditions (being most effective in hot dry climates); although its applicability can be extended by indirect equipment configurations. Evaporative cooling can also consume a fair amount of water, which may be of concern in an arid climate.

Vapor compression refrigeration is by far the most commonly used means of producing a cooling effect. In smaller configurations this refrigeration cycle is close-coupled and set up as an air-cooled and direct expansion (DX) system. A window air-conditioning unit is such an example. In larger applications the vapor compression cycle is often set up in a water-cooled chiller configuration (using a cooling tower for heat rejection and producing chilled water for use in air handlers or fan-coil units). COP values for vapor compression systems can vary substantially (roughly three-fold) depending upon capacity and equipment details; the most efficient option is a ground-source heat pump configuration.

The absorption refrigeration cycle operates on the basis of chemical reactions and is less efficient (has a lower *COP*) than comparable vapor compression systems (which rely on a compressor, usually electrically driven). The absorption cycle, however, can be driven by solar-generated hot water or waste heat. When using renewable energy resources as the motive force, efficiency is not necessarily a useful selection criterion. Direct-fired absorption chillers that are driven by combustion of natural gas are also available. The carbon emissions from this type of refrigeration system are greatly dependent upon the driver employed.

8.6   HVAC DISTRIBUTION AND DELIVERY OPTIONS

HVAC arrangements can be classified by scale or scope as local, central, or district systems. A local system is typically a small-capacity system, organized as a single all-inclusive package, and intended for use in a single thermal zone. An electric baseboard radiator, a through-the-wall air-conditioner, and a standard split system air-conditioner are examples of local HVAC systems. The four primary functional components of an HVAC system (source, distribution, delivery, and controls) are housed in a single (or perhaps two-part) package that is installed directly in the zone being conditioned. In a central HVAC system, cooling and heating effect are generated in a dedicated location (a mechanical plant room) and these effects are then distributed across multiple zones in a building. A district system distributes heating/cooling effect across multiple buildings (such as on a university campus).

It is difficult to generalize about the energy efficiency potential of local versus central versus district systems. In some applications, where individual control of systems operation provides an opportunity to use systems only when required (such as in a hotel room, dormitory room, or apartment unit) local systems can be energy conserving (even if not terribly efficient). Under different circumstances, the higher efficiency available from the larger equipment typically employed in a central or district system can provide energy benefits. Careful matching of system capabilities with the owner's project requirements is necessary.

A thermal zone – mentioned above and in the following discussion – is a portion of a building (often a room or space) that must be separately controlled if design criteria are to be met. A zone may consist of a portion of a space, a single space, or multiple spaces. The zoning process is absolutely critical to a successful and energy efficient HVAC system and should involve the owner and architect as well as the HVAC system designer. Key zoning considerations include the timing of loads (for example solar heat gains, equipment loads, occupancy loads) and the proposed operational hours of the various spaces in a building.

The basic components typically found in a central HVAC system can be arranged in different ways to develop a range of specific system types. Organizational options involve coupling a heat/coolth source with one of three distribution schemes, any number of deliver devices, and an almost infinite variety of control scenarios. Major distribution system types include all-air, air-water, and all-water configurations.

8.6.1   *All-air systems*

In an all-air system, air that is conditioned in a central location is distributed throughout a building to meet heating, cooling, and air quality needs. This approach is logical, since control of air conditions (as established in an air-handling unit) is the ultimate goal of an HVAC system. In addition, conditioned air can be simply (although skillfully) dumped into the various thermal zones without need for an in-space heat exchanger. The primary drawbacks to an all-air distribution approach are the often extensive volume required to move heating/cooling effect around a building through ductwork – due to the low density and specific heat (thermal capacity) of air – and the sometimes high energy cost associated with moving large volumes of air over long distances.

A number of specific all-air system configurations have been developed over the past half century. These include:

- Single zone constant volume.
- Terminal reheat constant volume.
- Multi-zone constant volume.
- Dual duct constant volume.
- Variable air volume (VAV; with or without reheat).

A full review of each of these systems (characteristics, pluses, and minuses) is beyond the scope of this chapter. Some comments regarding energy efficiency and effectiveness are, however, appropriate. From a thermal control and IAQ perspective, a terminal reheat system is hard to beat; unfortunately it is also hard to design a less energy-efficient system. Many energy codes prohibit the use of a constant volume terminal reheat distribution system. VAV systems have become the default distribution system for many building types. Care must be taken in their design and operation so that the energy efficiency potential of this system type does not override the indoor air quality expectations. Balancing energy-efficient thermal comfort and IAQ is not impossible, but can be a challenge. The other all-air distribution systems noted typically find applications in niche markets. The single zone system is very commonly used in residential applications and large-volume single-zone spaces such as the interior core of an open-plan office building. Hybrid systems combining two of the basic system types are not unusual.

8.6.2   *Air-water systems*

An air-water approach to central system distribution essentially tries to draw upon the best aspects of both air and water as a distribution medium. The benefits of air were discussed above. The primary benefit of water is its high thermal capacity – resulting in smaller distribution volume demands relative to air (small pipes versus large ducts – for similar capacity). The primary drawback to water is that it cannot be used to supply ventilation to a space. Commonly used air-water systems typically deliver the majority of heating/cooling effect to a zone via hot and/or chilled water, while supplying only enough air to the zone to meet indoor air quality ventilation requirements. Water does what it is good at; air does what it is good at. A fan-coil system with central air is a commonly encountered air-water HVAC system. The main architectural concern with an air-water system is the need for a water-to-air heat exchanger (a terminal device) in, above, or adjacent to each zone being served.

8.6.3   *All-water systems*

All-water HVAC systems tend to have limited practical applicability in high-performance buildings. The key (and quite serious) drawback is the inability of an all-water system to provide ventilation air for IAQ mitigation. Although there are passive and/or local workarounds for this shortfall, they tend to be unconventional and often dependent upon weather conditions for success.

8.6.4   *Delivery components*

Delivery components in an HVAC system are installed to introduce heating/cooling effect into the various zones and spaces in a building. In a local system, the delivery components are integral to the system package. In a central system, the delivery component is custom selected to meet the thermal loads experienced by a space while meeting design criteria for air motion, noise, and aesthetics. Typical examples of delivery components include air diffusers and registers, fan-coil units, radiant ceilings, and similar devices. Design of delivery devices can have a substantial impact of energy efficiency through the appropriate distribution of heating or cooling effect (delivering where loads need to be mitigated), control of surface temperatures (primarily on exterior enclosure elements such as windows), and control of air speed.

8.7   HVAC SYSTEM CONTROLS

Controls are the fourth major functional component of an HVAC system. The basic purpose of a control device is to turn a device on/off, to modulate the capacity of a device, and/or to protect a device from unsafe conditions. Multiple interconnected control devices become a control system. The default control scheme in most buildings is a ddc (direct digital control) system. Analog systems (using variations in air pressure or electric voltage) were previously common, but are being replaced by digital controls that involve the transmission of more complex information between sensors, control centers, and controlled devices. Computer programs provide the controlling logic in a ddc system.

   The assembly of sensors, computer logic, and controlled devices is variously called an energy management system (EMS) or a building management system (BMS) – depending upon the scope of control functions (which may include lighting controls, building access control, and fire protection in addition to operation of an HVAC system). The logic designed into a BMS can be simple or amazingly complex depending upon the owner's project requirements. Examples of more complex logic include trend logging (producing useable and easily interpreted records of system performance parameters), system or equipment self-diagnostics (recognizing performance indicators that suggest a developing problem with a piece of equipment), and self-learning (recognizing patterns of use that can improve system performance). These capabilities have led to a profusion of predictions and claims regarding the advent of smart or intelligent buildings. At this time intelligence seems to be in the eye of the beholder, but the potential for amazing control systems is on the horizon. Well-designed and functional controls are essential for a high-performance HVAC system. Controls serve as the mediator between actual needs for HVAC system services and the provision of those services. They need to be accurate, adequate for the task, and appropriate to the owner's needs.

8.8 EMERGING HVAC SYSTEM TRENDS

There are a number of HVAC systems (or subsystems) that are getting current attention by incorporation into high-performance building projects (Grondzik and Kwok, 2015). Several of these systems help to illustrate the principles of high performance addressed above.

Ground-source heat pumps: an air-to-air (conventional) heat pump is more efficient than on-site combustion or electric resistance heat sources because the energy input to the system is used to move heat rather than to produce heat. The closer together the source and destination temperature conditions, the easier it is to "move" heat from point A (say the outdoor air) to point B (say an indoor classroom). The energy consumed in this process is that necessary to produce a thermal "lift" that can overcome the inherent tendency for heat to flow from high to low temperature. The lower the lift, the lower the energy requirement and the higher the COP. A ground-source heat pump takes advantage of this relationship by providing a heat source (the ground) that will at virtually all times during the winter be warmer than the outdoor air – thus requiring less lift to move the heat from source (ground) to receiver (a building space). On a cold winter day (say −12°C [10°F]) the advantage of trying to move heat from an area of soil at around 13°C [55°F] versus the much colder outdoor air is substantial.

There is no serious increase in cooling season efficiency for an air-to-air heat pump versus a vapor compression refrigeration system (because they are the same system). There are serious efficiency advantages to be gained by using a ground-source heat pump for cooling (instead of an air-source heat pump). The efficiency improvement is again brought about by reducing (or actually reversing) the lift required for the system to operate. A number of ground-source heat pump configurations are readily available, including horizontal and vertical loop exterior heat exchangers and closed and open loop water flow configurations.

Dedicated outdoor air systems (DOAS): these systems are exactly what they sound like. The intent behind a DOAS is to decouple the indoor air quality function of an HVAC system from the thermal control function. Thus, a separate (dedicated) system is designed to provide for IAQ mitigation. A DOAS would be paired with a dedicated thermal control system. The combination of these two systems would constitute the HVAC system for the building. The benefit of this approach is the ability to independently optimize the thermal comfort and IAQ functions required in the majority of modern buildings by reducing the need for functional compromises. The benefits of a DOAS system will normally be seen in day-to-day control decisions that affect system operations. The key impediment to greater adoption of the DOAS approach is the real potential for increased first costs. This is also true for the adoption of ground-source heat pump systems. Life-cycle cost analyses must be undertaken to remove the first cost "problem."

Chilled beams: are an interesting alternative approach to heating/cooling delivery. The term chilled beam describes a radiant surface that is used to introduce heating or cooling effect into a space. Rarely are these surfaces actually beams. More commonly the "beam" is an area of ceiling surface or a manufactured device that replaces a portion of the ceiling plane (similar to a light fixture). Manufactured chilled beams can be passive or active – an unfortunate choice of adjectives that indicates whether air is introduced through the device (active) or not (passive) but which also tends to confuse the more critical distinction between active and passive approaches to climate control as a whole. Currently, chilled beams are more popular in Europe than in North America. A radiant floor slab is an alternative configuration for a radiant ceiling panel. The design of any radiant cooling system must careful consider latent cooling and the potential for condensation on the radiant cooling surfaces. The performance benefits of a chilled beam delivery approach reside primarily in the energy efficiency of a water-based distribution system.

Underfloor air distribution (UFAD): is an alternative means of introducing supply air to a space that is promoted as having green attributes. In a UFAD delivery approach supply air is distributed to an underfloor plenum and then delivered to the conditioned space via strategically located floor registers. The benefits ascribed to this approach include improved air quality (pollutants are moved away from occupants by the upward flow of air rather than being pushed back into the occupied zone by air currents) and improved thermal comfort (for roughly the same reason). Both of these benefits would translate into reduced energy use (lower outdoor air ventilation rates for the same net IAQ effect and the potential for more energy-efficient thermostat setpoints). Maintenance of underfloor air plenums to assure cleanliness seems important to the success of this approach.

Ventilation air heat exchanger: the typical building exhausts a fair amount of air in order to make room for an inflow of ventilation air. This exhaust air is warm in the winter and cool in the summer. Energy has been expended to bring this exhaust air into conformance with the comfort conditions maintained within the building. Simply dumping this conditioned air to the outdoors is a waste. With proper planning the incoming ventilation air stream can be brought into adjacency with the outgoing exhaust air stream. Placing a heat exchanger at this point of intersection allows the exhaust air to preheat incoming ventilation air in the winter and precool it in the summer. The heat exchanger may be a sensible-heat only device or a total energy (sensible and latent) device. Efficiencies for such devices can be well over 50%. More critically, all this captured heat is free energy that would otherwise have been thrown away.

8.9   HVAC SYSTEMS INTEGRATION

Systems integration is vitally critical in a high-performance building. There are several aspects to integration that should be considered. The first is integration of the HVAC system with the owner's project requirements. Comparing the characteristics of the many possible options for HVAC systems with the criteria contained in the OPR document will ensure that the system can provide the outcomes anticipated by the project client (ASHRAE, 2011). The capabilities of a potential HVAC system should be compared with the demands and opportunities presented by the project site climate. Among other considerations, this might include looking at solar resources relative to heat source options, soil temperatures relative to ground-source heat pump options, and relative humidity relative to the potential for evaporative cooling and/or the applicability of cooling towers versus air-cooled condensers. A third area for HVAC system integration is with respect to available energy sources. Energy resource availability is site specific and will often preclude the use of otherwise desirable fuels.

Integration of the HVAC system with the many and diverse aspects of the architecture of a building is exceedingly important to development of a high-performance building. One of the responsibilities of an HVAC system is to mitigate the effects (heat flow, infiltration, solar radiation, water vapor) of the exterior environment that are allowed to enter the building through the building envelope. The less successful the envelope design the greater the load on the HVAC system. Architectural design will also affect interior loads such as electric lighting. To a lesser extent, plug loads may also be influenced by architectural design decisions. Conceptually speaking, the loads that are handled (and the energy that is consumed and the carbon that is emitted) by an HVAC system can be categorized as heating related, cooling related, lighting related, and plug load related. The sum of these four broad categories represents the total HVAC system energy (excluding some smaller heat gains due to elevators and such). The relative magnitudes of these four load blocks will vary with climate and building function. Understanding the likely distribution of these loads (is cooling roughly 50% of annual HVAC energy consumption?) can assist in the design of appropriate and efficient HVAC systems.

Another major area of integration of HVAC systems is with other building support systems such as the electrical, structural, and fire protection systems. Electrical systems integration is fairly straightforward and primarily technical. Fire protection systems integration can occur at several levels, the more complex of which may be conceptual coordination of system inter-actions during emergency events. Coordination with the structural system may be simply spatial, but can also involve design philosophies regarding exposure of support system components. The commissioning process can be exceptionally useful in all areas of HVAC system integration.

REFERENCES

ASHRAE: ASHRAE Standard 55: Thermal Environmental Conditions for Human Occupancy. American Society of Heating, Refrigerating and Air-Conditioning Engineers, Atlanta, GA, 2010.

ASHRAE: ASHRAE Handbook – HVAC Applications. American Society of Heating, Refrigerating and Air-Conditioning Engineers, Atlanta, GA, 2011.

ASHRAE: ASHRAE Handbook – HVAC Systems and Equipment. American Society of Heating, Refrigerating and Air-Conditioning Engineers, Atlanta, GA, 2012.

ASHRAE: ASHRAE Standard 62.1: Ventilation for Acceptable Indoor Air Quality. American Society of Heating, Refrigerating and Air-Conditioning Engineers, Atlanta, GA, 2013a.

ASHRAE: ASHRAE Guideline 0: The Commissioning Process. American Society of Heating, Refrigerating and Air-Conditioning Engineers, Atlanta, GA, 2013b.

ASHRAE: ASHRAE Standard 202: Commissioning Process for Buildings and Systems. American Society of Heating, Refrigerating and Air-Conditioning Engineers, Atlanta, GA, 2013c.

ASHRAE: ASHRAE Handbook – Fundamentals. American Society of Heating, Refrigerating and Air-Conditioning Engineers, Atlanta, GA, 2013d.

ASHRAE: ASHRAE Handbook – Refrigeration. American Society of Heating, Refrigerating and Air-Conditioning Engineers, Atlanta, GA, 2014.

Grondzik, W.: *Principles of building commissioning.* John Wiley and Sons, Hoboken, NJ, 2009.

Grondzik, W. & Kwok, A.: *Mechanical and electrical equipment for buildings.* 12th edition, John Wiley and Sons, Hoboken, NJ, 2015.

ISO: ISO 7730: Ergonomics of the Thermal Environment. International Organization for Standardization, Geneva, Switzerland, 2005.

ISO: ISO 16814: Building Environment Design – Indoor Air Quality. International Organization for Standardization, Geneva, Switzerland, 2008.

CHAPTER 9

On-site renewable energy

Robert J. Koester

9.1 INTRODUCTION

Colleges and universities are uniquely positioned to employ on-site renewable energy because they have under singular control all the elements needed for whole-systems design. They have: a hierarchical administrative structure (which *can* serve to bridge the silo-like nature of academic disciplines); operational policies (which can serve to integrate the incentives for operational energy conservation with the budgeting for capital improvements); an inventory of campus buildings of wide-ranging vintage; an extensive energy conversion/distribution infrastructure; and measurable acreage of unoccupied land. In addition, they interact with a near-surround community of citizens who value the economic impact of their local institution but at the same time often feel quite separated from its day-to-day operation.

The only way to engage this opportunity effectively is to look to the long term; to understand the roles and influences of the members of the academic community and the near-surround public community by integrating more fully the institution's capital facilities planning with its operational academic planning – in strategic five-year cycles.

In that context, a reasonable assessment of the potential magnitude for on-site renewable energy production by differing technologies and an understanding of the "order of demand" resulting from the patterns of the day-to-day presence of students, staff, faculty, and administrators must be made. Operational factors of climate, the pricing within energy markets, and the embedded investments in energy infrastructure add complexity.

The following descriptions cover the nature and capacity of varying renewable energy technologies. Most importantly however, is the attendant discussion of how such awareness can be used to inform an integrated approach to the anticipation, facilitation, and celebration of achievement in moving a campus toward becoming a more resource-conserving, environmentally-benign, renewable-energy-using operation.

To reinforce these points, a case study story of Ball State University is provided; emphasizing that all decisions have cumulative impact. If we can cut in half our demand, double the efficiency with which we use energy, and reduce by half the carbon content of the energy used, we can cut our carbon footprint to 1/8th of current levels ($\frac{1}{2} \times \frac{1}{2} \times \frac{1}{2}$).

9.2 ON-SITE ENERGY RENEWABLE ENERGY: THE PRESUMPTIONS

So then, what are the presumptions regarding on-site renewable energy? It is helpful first to distinguish the terms of "stored" sourcing from "real-time" harvesting.

9.2.1 *"Stored" sourcing or "real-time" harvesting*

The acquisition of materiel within an 805 km (500-mile) radius is the metric used by the US Green Building Council (USGBC) to recognize an acceptable embedded environmental and economic cost for resource acquisition and transport. Anything that can be "mined" or "manufactured" within that radius is considered a "local" resource. However, for the sake of this chapter discussion,

Table 9.1.   Strategic distinctions of tactical opportunities.

"Stored" sourcing	"Real-time" harvesting
Open-loop (GSHP) heating and cooling	Closed-loop (GSHP) heating and cooling
Bio-mass-based heating and cooling	Solar thermal-space Heating
Methane-based heating and cooling	Solar thermal-space cooling
Combined heat & power	Solar thermal-DHW heating
Hydro-based electricity	Solar-based electricity
	Wind-based electricity

"local" resources – the stored and the harvestable – are those available within the boundary of property ownership under the administrative authority of the institution; i.e., the physical campus.

"Stored" sourcing for example, would involve the use of groundwater from the underlying aquifer for energy exchange; or the growth of renewable energy feed stocks such as switch grass to power boilers for heat energy distribution; or the use of methane from solid waste – even solid-waste combustion itself – to supply combined heat and power generation. These all involve the extraction of a stored resource.

In contrast, campus operations can exploit the day-to-day "real-time" harvest of nature's widely-available free energy; the radiation from the sun and the solar-driven variability of wind and hydrologic flows for mechanical, thermal, and/or electrical conversion.

As noted in Table 9.1 these two strategic distinctions frame numerous tactical opportunities that face any given college or university as it looks to the energy resources available within its campus boundary.

Each of these technologies is described below, and more detailed descriptions of typical outputs and efficiencies for each system type are summarized in Appendix A.

9.2.1.1   *The "stored" sourcing technologies*
Open loop geothermal ground-source heat pump (GSHP) heating and cooling: the use of ground-source heat pump (GSHP) heating and cooling for individual building or campus-wide district heating and cooling is treated by convention as a "geothermal" technology.[1]

Since the earth and ground water temperatures below the frost line remain stable year round, the capacitance of the earth and its aquifers can be used to "source" and/or "sink" thermal energy – using heat pump exchange. (see: Grumman, 2003; Suozzo, 2000; USDOE, 2005; USEPA, 2014a).

Extractive open-loop withdrawal of water from the aquifer can deplete the stored water source. And thermal pollution effects can result from chilling or heating that water before discarding it to a surface stream.

Conversely, using injection wells to reintroduce the extracted groundwater "back into" the aquifer can be even more problematic; not only is thermal pollution a concern, but this can short circuit the filtration effect of the earth's mantle and can lead to silting of the recharging wells, necessitating additional drilling over time as these sediments accumulate.

In short, open-loop technology can yield unintended consequences. Given the extractive nature of open-loop ground-source heat pump technology, initial consideration requires the bounding of decision space – running calculations of likely flow rates needed to yield daily, seasonal and/or annual heating and cooling energy delivery. Assumptions regarding the conversion rates of heat

---

[1]To clarify, "geothermal" is a term most-appropriately ascribed to "hot rock" energy sourcing; those few locations throughout the world in which the magma of the earth is close enough to the earth's surface to yield steam from the groundwater flow or seepage. However, as it would apply to most campuses, "geothermal" references the thermal capacitance of the earth's soil and groundwater strata.

pump chillers and especially the magnitude of campus load provide the starting point for design conversations about this technical application.

Bio-mass-based heating and cooling: biomass feedstocks can be used as a primary energy source or as an amendment to a fossil-based coal-fired system to reduce the pollution "up the stack" and "in the ash". If an institution owns the very land on which the plant fiber can be grown, the immediacy of production/consumption can benefit the system performance by minimizing transport energy. The renewable nature of such feedstock is tied to the sequestration and release of carbon in the nutrient cycling of the plant fiber (absorbing CO_2e as it grows; releasing it as it is combusted).

If indeed on-campus production of switch grass is intended, one must start with the bounding of the decision space by running calculations of the acreage of production, the likelihood of multiple cycles of annual yield, and the comparison of this magnitude of extraction with the building energy loads, individually and/or collectively; the fact that such biomass is likely to be used as an amendment to fossil-fuel combustion must also be considered. Both technical and financial analyses are framed by the fact that no more than 15% of the feedstock should be mixed with such fuel.

Methane-based heating/cooling: a number of corporations and municipalities are using the methane emissions of landfills to provide a consistent combustible fuel source for heating and cooling. Scaling is the key issue. As noted in Appendix A, the predictable production of methane is directly proportional to the volume of the landfill.

Not only must that productive capacity be determined, but the embedded energy cost of extraction and transport to the campus must be considered.

Combined heat and power: the combustion of a fuel for the generation of electricity is a well-established technology. The residual waste heat from the electricity production becomes a secondary benefit of the system. High-quality electrical power can be distributed on-site and the waste heat from the generating facility can be used to heat water and/or air for distribution to near-by buildings or even the full campus. Often these combined heat and power (CHP) systems are labeled as "total energy" or "co-generation" facilities. Typically, they must be schemed to integrate with the regional electrical distribution grid and in the U.S. come under the permitting authority of a local regulatory commission. Nonetheless, if interchange agreements can be reached with the local electrical utility, such systems can be very beneficial in that they approximately double the extraction of useful work from the fuel source. Of course, these typically are powered by conventional fossil fuels or landfill methane and so still result in GHG emissions.

The scaling of waste heat must be compared to the heat energy needs of individual buildings and/or the campus as a whole, with consideration given to the transport and/or distribution of the waste heat recovery to point of use (CHPA, 2014; USDOE, 2014b; USEPA, 2002; USEPA, 2014a; UTC, 2014).

Hydro-based electricity: it is unlikely that any campus would engage in large-scale damming of surface water to achieve utility-scale electrical conversion. However, micro-hydro technology is worth examining as a means of shaving load on the regional/national electrical grid. A modest pressure head from water flowing to a lower point can yield useful production.

Micro-hydro is most readily applied as a distributed energy generation technology, and yet the pressure head used to drive the turbines is likely to be localized to selected pockets of the campus grounds. Thus, the scaling of productivity against load may dictate use of this mechanical conversion for service of nearby buildings, only.

9.2.1.2 *The "real-time" harvesting technologies*
Closed-loop geothermal (GSHP) heating and cooling: as described in the subject case study of this chapter, Ball State University is in the process of completing the largest closed-loop district-scale geothermal heating and cooling installation in North America. This is a clear-water system whose manifolds are well below frost line and whose closed exchange loops are located in bore-holes

grouted solid to the earth to connect contiguously the capacitance of the closed-loop-circulating water with the capacitance of the earth's material substrate.

This is a fully-benign, totally-inert system whose only impact on the earth (and any groundwater present) is designed to change the temperature no more than $-16°C$ ($2°F$) up or down as a result of "sourcing" and "sinking" energy for campus heating and cooling. To assure performance, the bore-hole fields are located at a depth of 137 m (450 ft) and a density of spacing at 15' on center. The loops within the north-campus and south-campus bore-hole fields are manifolded in pods and brought through headers to the north-campus and south-campus energy stations wherein heat pump chillers are used to heat and cool the separate (clear water) chilled and heated distribution loops that service the 47 buildings of this fully-integrated district-scale application. (More detail on the system is presented later in the chapter.)

Typically thermisivity studies are used to profile the substrate strata of soil in those areas to be devoted to placement of the bore-holes. Knowing the geologic layering also helps in determining installation costs of drilling. But most importantly, the estimate of capacitance of the substrata is the first step in determining the balance of supply and demand. Thorough assessments of individual and total building loads provide the boundary of decision space regarding the number of likely bore-holes required and once the magnitude of demand and supply are determined, the selection of the heat pump chillers can be scaled to that interface.

Solar thermal – space heating: the most immediate experiences with solar thermal space heating (at least in the west) is the shock of opening a car door on a sunny winter day only to find that this small contained space has been "cooking" in the sun; the cabin air at a very high temperature – its seating, steering wheel and dashboard too hot to touch.

The 1970s were filled with numerous published diagrams of the sectional organization of buildings trying to maximize such solar heating.

Such diagrams, however, provided only that single-moment-in-time profiling of the solar interaction; not circumscribed were the complex understanding of the dynamics of hourly solar movement in 15° angular increments nor were the variations in altitude from winter to summer seasonal changes fully accounted. Fortunately, software tools can be used now to simulate that variability; the hourly, monthly, seasonal and annual solar thermal harvest as a product of orientation and aperture size.

Weaving this consideration into campus-scale design is largely a building-by-building, if not space-by-space consideration. However, designers must keep in mind that there are times during the summer season when such thermal load is undesirable and that eastern and western facades present the biggest challenge for controlling low-angle incoming solar thermal radiation – especially so in summer.

Moreover, the challenge in campus-scale design is to consider the range of building types and the associated variations in occupant activities, space sizes, usage patterns, and levels of criticality (labs, museums, classrooms, or assembly halls) when scheming the use of the solar thermal resource. The dynamics of solar thermal load often are best isolated in "buffering" spaces – separated from the day-to-day, hour-by-hour occupied rooms; in movement zones (perimeter-loaded corridors) or in other public spaces that are not continuously occupied (atria, galleria, greenhouses, and entry lobbies).

The primary benefit of direct solar thermal space heating is the opportunity to capitalize on the distributed nature of the solar resource. Every building, after all, is a solar building, i.e., it sits in the sun and is heated. The primary energy flows analysis must occur at the space scale for direct solar gain and at the building scale for indirect solar gain.

Solar thermal – space cooling: the use of heat energy to provide space cooling is a well-established technology. Gas-fired absorption-chilled refrigerators date to the earliest stages of the industrial revolution. Solar thermal energy can be used as a substitute for the heat from combusted fossil fuel.

Solar-driven cooling requires localized capture of high temperature harvests from the sun. This is typically best achieved with evacuated tube collectors that can drive temperatures high enough to perpetuate the absorption-chilling cycle of this technology.

However, designers must examine early on the magnitude of opportunity from such absorption chilling, not only matching the quantities of thermal harvest to load but also matching the pattern of availability with that of demand. Since the cooling load is at its maximum typically during the time of year when the sun is at its highest altitude, hot-water-based solar thermal conversion is ideally located on the roof surfaces to maximize capture during that time of year.

Solar thermal – DHW heating: historically solar thermal domestic hot water heating has been a localized conversion option as well – a small-scale collector with thermo-siphon storage placed on an individual dwelling unit. For a more contemporary building-scale or campus-wide application, the sizing of the collector field and storage is achievable with (straight-line) scaling-up.

Climate, however, plays a significant role given the need for freeze protection, using either a closed-loop glycol system and/or evacuated-tube collector field. A simple analysis of harvesting area can differentiate those buildings on campus that are candidates for such full service from those for which the harvest will only partially meet the occupant need.

As in the case of the solar thermal space heating, converting this "free" energy from the sun has "upstream" implications for reduced electrical load and long-term potential for carbon-credit marketing.

Similarly, domestic hot water heating is best schemed as a localized building-by-building conversion option. This conversion of the distributed harvest of solar energy can be scaled differentially based on the kinds and patterns of load/demand per building type and occupancy activity.

Solar-based electricity: conversion of the full-spectrum radiation from the sun to electrical current is a well-established practice. On-campus harvesting of that energy necessarily presents the opportunity to match the decentralized nature of the solar energy availability with decentralized nature of electrical demand.

A simple acreage assessment of conversion potential, at differing collector tilts, compared against single and double-axis tracking options, overlaid with ranges in conversion efficiency, with attention to the mechanical detailing of structural support, comprise the design-decision space when considering this technology.

Cell types vary considerably in performance efficiency and require different mechanisms for physical attachment. Monocrystalline, amorphous crystalline and thin-film technologies require differing types of substrate and/or racking support. Most modulations use micro-inverters, attached to each panel at the point of harvest, to convert the direct-current to alternating-current supply. These then are aggregated to feed into the campus grid providing an immediacy of supply at point-of-use load.

Attendant to this technology is the leveraging effect of upstream demand reduction. By reducing the draw on the regional electrical grid – particularly in those sectors of the country in which electricity production is primarily from coal-fired generation (brown-power) – the leveraging effect is approximately 3:1. That is, 1 unit of power captured from the sun avoids some 3 units of CO_2 equivalent emissions upstream due to the inefficiencies of the fossil-fired electrical generation and the accounted transport losses from distributing that production over sizeable distances.

Placement of photovoltaic arrays across the campus grounds offers many design options: to provide a shading value integrated with a parking lot – or the top deck of a parking garage; as an entry canopy to shelter from rains; or as an area-wide building rooftop application to maximize summertime production and to buffer against the weathering of the roof membrane.

Radiation striking buildings of course presents the option of direct physical integration of PV panels with the building skin – although there are some technical arguments in favor of stand-off attachment of PV panels so as to maximize their efficiency by cooling from the ambient air. However, if the photovoltaic technology is integrated with a building skin, residual benefits can accrue, such as solar-thermal buffering and/or light filtration. And annual production can be synchronized with the annual electrical load (typically peaking with the peak summer cooling demands of the building).

Exploring the parameters of solar electric conversion must start with the bounding of the decision-space by running calculations of likely monthly and annual outputs.

In the USA, one of the most convenient tools for bounding such potential is PV Watts. Version 2.0 available on the web indexes climate zones and uses real-time weather data to project day-to-day, monthly and annual production.

For more global application, the suite of software offered through METEONORM is recommended (IEA, 2014; USDOE, 2014c; 2014d).

Wind-based electricity: in broad terms, wind conversion falls into two categories of use; low-speed, high torque mechanical leveraging for water pumping (i.e., the classical Stewart wind mill in the USA) or high-speed, low torque conversion to inverted alternating electrical pulses to feed the national grid.

The high speed, low torque wind machine blades literally "fly" in the wind stream using the very principles of the airplane wing to create "rotational lift".

The two design criteria for these devices are the disk area and the wind speed. Since the extractable energy is a function of the cube of the wind speed, that is the most critical factor. The higher one can extend the disk area into the wind stream the faster and more consistent the wind gradient and thus extractable power.

Depending upon scale of application, wind conversion can be substantial. Quick assessment of the prevailing wind patterns and resource conversion can be made using published wind distribution maps. In the United States these profiles are available from the National Oceanic and Atmospheric Administration (NOAA). (AWEA, 2013; USDOE, 2014a).

9.2.2   Accepting principles: the design-decision space

These "stored" *sourcing* and "real-time" harvesting technologies then set the design-decision space for whole-systems, campus-wide intervention. More specifically, the differing scales of applications, the bundling of energy conversion options, and the integration of these with strategic planning for the day-to-day and long-term operations of the institution.

And, to acknowledge that the design of systems is integral to the larger issues of planning, administration and engagement will require that designers accept essential principles for mapping this decision space; as described in the following sections.

9.2.2.1   Bounding options/opportunities

Establishing the institutional boundary and inventorying the demand and resource capacities therein is critical to scaling the interventions in space and time and to identifying those stakeholders who can contribute to planning and implementation. The most immediately useful technique for bounding decision space is by classification of GHG emissions; identifying Scope 1, Scope 2 and Scope 3 impacts.

Scope 1 emissions comprise all on-campus/within-building conversion of combustible fuel. With regard to renewable "free" energy harvest, this essentially is the boundary of the campus setting and the respective foot printing of its many buildings. To the degree that such on-site conversion is achieved, the tributary impact on upstream (Scope 2 emissions) is immediate and highly leveraged.

Scope 2 emissions are those off-site emissions associated with regional or national energy distribution networks. Most typically this applies to the electrical grid, but can include district-scale steam, or hot and/or chilled water production which may well be available to an urban campus. And with the onset of hybrid-electric vehicles, the on-campus recharging of commuter vehicles have measurable "upstream" impact, as well.

Scope 3 emissions relate to tributary impacts of the day-to-day business of higher education; conference attendance or workshop participation, the day-to-day commuting and the seasonal travel to-and-from campus by administrators, faculty, staff and students.

9.2.2.2 *Balance of systems – global and local*

On-site campus-scale energy management must be the foundation for efficient use of renewable energy conversion; this then equates to a balance-of-systems question. Typically the metric for such assessment is scaled to an annual cycle in which zero carbon impact or zero net energy use engenders trade-offs between the heating and cooling seasons. Thus the energy use intensity (EUI) which is the annual operational energy per square foot (equivalent to $= 0.092903$ m^2), (oftentimes weather-normalized per °C (°F)), serves as the benchmark for such discussion.

As will be noted for the case study, we have found a serendipitous benefit of balance-of-system performance occurring in day-to-day energy management. Specifically, during the swing seasons of the annual cycle, it is possible to trade energy among buildings rather than rely on strictly "sourcing" or "sinking" energy from/to the ground as a capacitor.

The balance-of-system question, of course, will be complicated significantly by the variety of building types and occupant activity, and – especially for new system installation – with the sequencing/scheduling of transformation from current on-site fossil-fuel-based energy use to on-site renewable energy use.

At the scale of a single building, the balance-of-system questions comprise the thermal envelope, passive-solar thermal capture, natural ventilation, occupant behavior and use of automated controls. A simple mapping of magnitudes as a starting point for conversation is essential to developing strategies for intervention. As noted at the start of the chapter, three layers of consideration are involved: the carbon content of the source energy, the efficiency with which energy is used, and the magnitude of demand placed on such energy supply; to wit: a 50% reduction of each yields a leveraged reduction of some 88% in cumulative impact.

9.2.2.3 *Building-integrated application*

Integrating on-site conversion technologies with campus buildings will often be a post-facto retro-fit challenge; determining where to "bolt on" the electrical production and/or hot water heating modules. An advantage of such retroactive PV application is that stand-off assembly will tend to maximize photovoltaic panel performance but a conflicting issue is the duplication of the materiel of solar harvest with the materiel of climate boundary. To the extent that PV panels, especially, in new building construction can be integral to spandrel systems, or roof closure, the avoided double-cost of climate envelope and energy technology can improve the overall economic feasibility.

9.2.2.4 *Stand-alone campus placement*

Within the campus boundary, placement of photovoltaic arrays or evacuated-tube hot water heating modules can be made an integral site feature. The shading of parking lots or parking structures, integration with entry porticos or as sunscreens on east, south or west building surfaces and (out of sight) placement on rooftops involve decisions that can be independent of the actual scaling of conversion potential. However, scoping the magnitude of conversion is the first step. Once an understanding of the on-site Scope 1 resource opportunities is determined, then distribution and placement issues can be considered.

9.2.2.5 *Cost-benefit: return on investment*

Conventional back-of-the-envelope assessments of cost-benefit involve a simple calculation of years-payback. Whatever incremental cost one is considering for energy conservation or alternative energy system installation, the annual savings are divided into that numerator to arrive at some number of years at which point the project is considered "break-even". Such analysis, of course, does not include discount factors, rates of inflation and other "financing" values. Nonetheless, it is the most popular "quick" assessment typically used in preliminary conversations about options.

However, this is a pessimistic metric. No other decisions made by an institution are based on a years-payback calculation. Thus it is recommended that all analyses undertaken for any energy conservation or alternative energy intervention use only the return-on-investment calculation.

And as a first-pass analysis, it is recommended that the simple numerator/denominator calculation be used. Without engaging in the more complicated analysis of lost opportunity costs, discount factors, rates of inflation, net present value, etc., this simple calculation provides a useful equitable assessment of the significance of an intervention. For example, a 40-year payback which would be considered far too long at first glance still represents, in simple terms, a 2.5% return-on-investment, which in current economic conditions is actually an attractive value. More importantly, the more typical "payback" of 10, 12 or even 14 years would yield very favorable 10, 8, or 7% returns-on-investment, respectively; any one of which would outstrip the current performance of many a university's endowment holdings.

9.2.2.6   *Cost-benefit: carbon tax avoidance and/or carbon credit sales*
Finally, the benefit of any and all conservation and alternative energy technology enterprise is the measurable reduction in GHG emissions which, in-and-of themselves, have value in the carbon market.

Although in the USA, currently, that market is voluntary and somewhat regionalized, carbon credits can be sold at a value that represents an added income stream for an institution. What is most important, of course, is that this income stream in an annualized benefit for the life of the project. As the carbon markets in the U.S. become more established – possibly even aligning with those of Europe in which carbon is currently selling at about US\$ 14.00/metric ton – the effective income stream from reduction of GHG emissions can be substantial.

Alternatively, one can run a quick spreadsheet on current emissions levels, and project the future-avoided costs for any potential carbon tax, cap-and-trade-rate or even fee-and-rebate obligation; to monetize the true long-term financial value of up-front investment in conservation and alternative energy technology, these all must be considered.

9.3   PLANNING FOR CLIMATE ACTION AT BALL STATE UNIVERSITY

The following section of this chapter summarizes the climate action planning undertaken at Ball State University (BSU) in Muncie, Indiana, USA. The story told derives from: (i) a near thirty-year history of involvement in the green campus movement; (ii) a fortuitous timing of opportunity with the need to replace coal-fired boilers that had lived well beyond their projected useful life; (iii) an administrative commitment to "doing the right thing" by taking a leadership position in the signing of the American College and University Presidents Climate Commitment; and (iv) the

Table 9.2.   Example avoided costs in the case-study project described in Section 9.3.

Possible carbon tax rates [US\$ per metric ton]	Annual carbon tax current [CO_2e] 171,983	At the 1990 level [CO_2e] 129,936	20% Below 1990 [CO_2e] 103,949	80% Below 1990 [CO_2e] 25,987
\$ 3.50	\$ 601,941	\$ 454,776	\$ 363,821	\$ 90,955
\$ 10.00	\$ 1,719,830	\$ 1,299,360	\$ 1,039,488	\$ 259,872
\$ 25.00	\$ 4,299,575	\$ 3,248,400	\$ 2,598,720	\$ 649,680
\$ 50.00	\$ 8,599,150	\$ 6,496,800	\$ 5,197,440	\$1,299,360
\$ 75.00	\$12,898,725	\$ 9,745,200	\$ 7,796,160	\$1,949,040
\$100.00	\$17,198,300	\$12,993,600	\$10,394,880	\$2,598,720

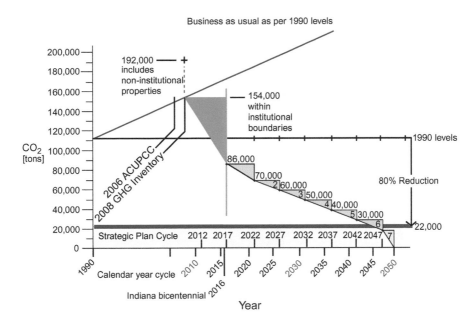

Figure 9.1. Ball State University campus CO_2 emissions.

inventive opportunity with new systems installation of having faculty and facilities management personnel working collaboratively to map the (strategic) steps of the symbiotic integration of the university's academic and operational planning.

In large part the details of the story are still unfolding and so this chapter section presents only a snapshot-in-time; a statement of "where we are", and a projection of "next-steps" as we pursue the goal of becoming a climate-neutral campus.

9.3.1 A five-year time block planning for GHG emissions reduction

The BSU Climate Action Plan was intentionally organized as a living instrument; for implementation by the academic community over a 40-year time horizon:

- It proposes continual staged reductions in GHG emissions.
- It uses five-year blocks for the targeting and benchmarking of achievements.
- It acknowledges the needs for modification to the physical campus, its day-to-day operations and the behavioral practices of the members of the university community.
- It identifies the future purchase of renewable energy credits as a means of completing the totality of GHG mitigation.
- It opens the door to the issuance of carbon credits as a long-term revenue stream.

The action steps in the BSU Climate Action Plan are framed fundamentally by the decision of the Board of Trustees in February of 2009 to sanction installation of a closed-loop geothermal (ground-source heat-pump-chiller) district heating and cooling system. This geothermal system is designed to serve all 47 buildings on campus and is the first major action step in reducing our Scope 1 GHG emissions.

The GHG reduction recommendations, as shown in Figure 9.1, use a time-based diagram for accounting CO_2e impact from energy-conservation practices and green energy sourcing and/or harvesting. Each 5-year block provides a convenient alignment with the five-year increments of

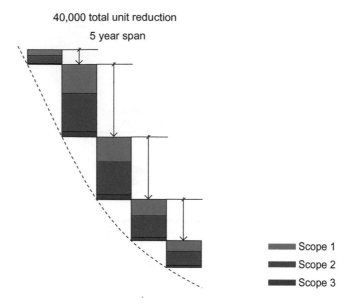

Figure 9.2.   Equally proportioned annual reductions by scope.

well-established cycles of Strategic Planning and allows us to indicate 10, 20 and 30-year time horizons for projecting returns-on-investment for GHG emissions reduction.

9.3.2   *Continually telling the story*

As shown in Figures 9.2, 9.3 and 9.4, each of the five-year blocks can be broken into one-year histograms showing a reduction in scopes of emissions or in the tactical mix of mitigations used to achieve emissions reduction. And although the opportunity for conservation and energy source substitution for each of the 47 buildings on campus is substantial, these histograms are not building-centric.

Moreover, any five-year block and/or any one-year histogram can have internal variability; in each case, reflecting the realities of technology intervention and day-to-day operational practices as well as the impacts of behavioral change resulting from education of the campus members.

As of this writing, the university has completed installation of the first phase of its geothermal project and has integrated the increments of GHG emissions reduction with the setting of goals and objectives for its newest Strategic Plan. This switch to geothermal means that all building loads in turn will shift to Scope 2 GHG emissions; that is to say, the heating and cooling, the lighting and plug loads for any given building, will combine as a load on the electrical grid. And that will be the target for the next wave of reductions.

9.4   THE CONTENT OF THE CLIMATE ACTION PLAN

The reduction of GHG emissions at BSU does not fall solely on the shoulders of facilities planning and management staff. Rather it is a task for the full membership of the university community – and even extends to the near-surround community.

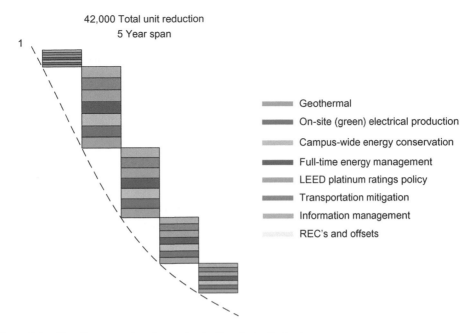

Figure 9.3. Equally-proportioned annual reductions by <u>tactical mix.</u>

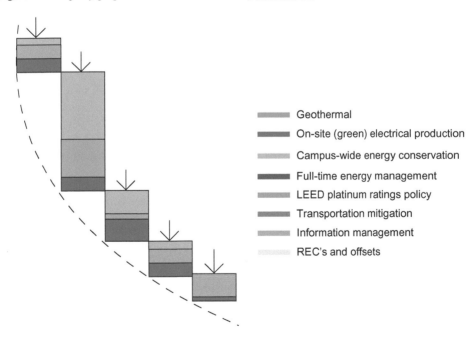

Figure 9.4. Differentiates from year-to-year varying amounts of impact.

9.4.1 *Involve the full breadth of the university community*

Certainly administrators have the opportunity to serve as champions of the cause by working collaboratively to adopt policies or to employ best practice guidelines for energy conservation and emissions reduction.

Faculty, also, can be a champion of the cause in their education endeavors; facilitating open dialogue about the complexities of social, environmental and economic interactions.

Staff members, of course, in their day-to-day activities, have perhaps the best sense of the "pulse" of the institution. Routine campus operations are a significant avenue for energy conservation and GHG emissions reductions. These opportunities are integral to the goals outlined in the university's 102 unit-level sustainability plans, adopted in 2009.

Students, during their years on campus, have the opportunities for immersion in these issues and contribution to a reduction in energy demand and resulting GHG emissions.

Finally, the near-surround community can contribute by encouraging use of its hybrid-electric municipal transportation through free ridership.

9.4.2   Nine tactical areas

We have targeted reductions within nine tactical areas as listed below and have scripted for each, a pilot project to demonstrate and leverage success. Those descriptions will follow:

Scope 1: Energy conservation

- Geothermal district H&C.
- Buildings.
- Information technology.
- Operational: policies.

Scope 2: Electrical production

- On-site solar-photovoltaic (PV) production.
- Off-site wind-energy production.
- Off-site PV production.

Scope 3: Information transparency

- Real-time monitoring and reporting.
- Transportation.

9.4.2.1   *Scope 1*
9.4.2.1.1   Geothermal district heating and cooling – pilot project: mount a geothermal
            education, research and outreach program

BSU is currently completing the installation of a closed-loop geothermal (ground-source heat-pump chiller) district heating and cooling system serviced by 3600 geothermal boreholes and two energy stations (Fig. 9.5).

We are on schedule to completely eliminate coal-fired combustion on or before March 2014 and currently have all 47 buildings on campus connected to the district-scale network for cooling and some 22 buildings in the northern half of campus connected for heating. In the next year and a half the remaining buildings will be brought online as the south energy station is completed and connected to the south bore-hole field already in place. Once the system is fully networked, the university will undergo commissioning activities to further balance and fine-tune the campus-wide operation.

As a complement to installation of this district-scale system, we are integrating the education and research activities of our students and faculty, and scripting ways to make information available to the public, not only through web links, and hand-held information technologies but also with on-site campus displays.

More specifically, students and faculty in the geological sciences have been monitoring the performance of the north bore-hole fields using instrument wells that were placed at the perimeter of the bore-hole areas at the time of their installation. Research data are being used as a platform for undergraduate course activities, graduate-level study, and research publications by faculty.

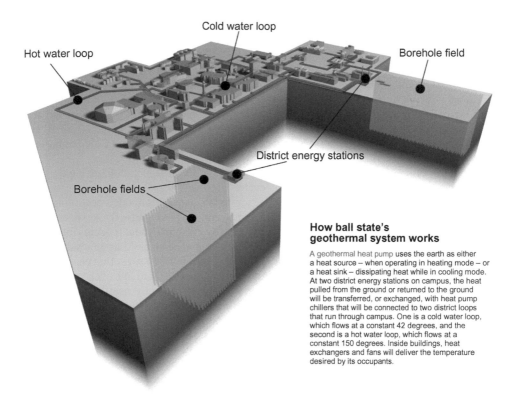

Hot water loop

Cold water loop

Borehole field

District energy stations

Borehole fields

How ball state's geothermal system works

A geothermal heat pump uses the earth as either a heat source – when operating in heating mode – or a heat sink – dissipating heat while in cooling mode. At two district energy stations on campus, the heat pulled from the ground or returned to the ground will be transferred, or exchanged, with heat pump chillers that will be connected to two district loops that run through campus. One is a cold water loop, which flows at a constant 42 degrees, and the second is a hot water loop, which flows at a constant 150 degrees. Inside buildings, heat exchangers and fans will deliver the temperature desired by its occupants.

Figure 9.5.   The BSU "geothermal" GSHP district-scale heating and cooling system.

Figure 9.6.   The LEED Silver David Letterman building.

Instrument wells also have been placed at the perimeter of the south bore-hole fields and to the west of that field a specialized test pattern of bore-holes and equipment types are slated for installation with the specific goal of serving as a research tool for use by faculty and students in the sciences.

9.4.2.1.2   Buildings – pilot project: require LEED-Gold Building Certification

Energy consumption on the BSU campus is driven largely by the needs of some 20,000 administrators, faculty, staff and students occupying some 47 buildings. Those buildings undergo continual repair and rehabilitation wherein facilities personnel have been able to upgrade insulation levels of roofs, replace and/or enhance performance characteristics of windows, modify electrical-motor drives on air handling and pumping, and install more energy efficient lamps and luminaires.

In the end, however, it is the behavior of the occupants that drives a significant piece of the electrical energy demand of these buildings, especially the plug-loads and lighting. To inform the occupants meaningfully, it is necessary to index the university's cumulative GHG emissions in units that are understandable.

Therefore, as previously noted, we provide a diagram of GHG reduction blocking tied to the five-year increments of the strategic planning cycle.

As a publicly-funded state institution, BSU is required to manage its budgets according to state law. One of the confounding aspects of this is the split incentives that attend to the separate line item budgeting process wherein the appropriation of day-to-day operating budgets occurs independent of the bonding authority for capitalization of new buildings and/or building renovations. As a result, the capitalization methods tend to bias design decisions toward lowest first cost rather than life-cycle assessment of operational expense. And certainly not factored into such split incentive is the less-easily-identified evaluation of building occupant productivity, health and happiness.

One means of working around this inherent conflict is to consider the use of an internally-constructed green-revolving-loan fund using endowment monies. The cost-benefit then can be treated as earnings on the capital investment. More specifically, if the cost differential of a more effective green building can be "covered" by supplemental funding from the endowment, then that annual savings and operating expense can (in theory) be returned to the endowment as a return-on-investment. And certainly within the lifetime of a typical building, the annual savings to accrue over that life will typically yield higher percentages of ROI than currently available in most investment markets.

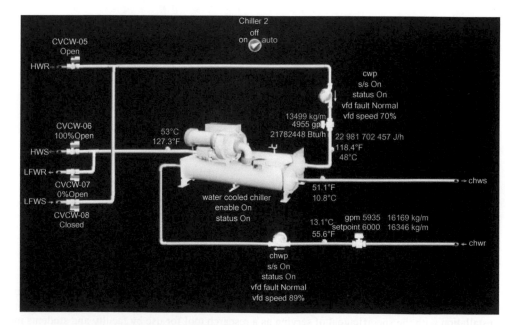

Figure 9.7.   Tabular monitoring of system.

9.4.2.1.3   Information technology – pilot project: adopt web-based dashboards for information
         management (Fig. 9.7)

In this tactic we are focusing on energy use reduction through modified equipment acquisition usage practices and behavioral change.

Electrical load reduction can be achieved through information technology conservation; not only by modification to computing equipment but also the ever-important use of network servers for web-based display of real-time building energy use/campus energy performance.[2]

As of this time, we have built a screen display interface which tracks the real-time operational performance of the heat pump chillers in the north energy station and the connected loads to the district-scale distribution loops as well as the north bore-hole field thermal energy exchange. An advantage of displaying the schematics of the operational system with the attendant instrument readouts is not only educational but also this provides a starting point for data feeds to a web-based dashboard that would be publicly available. This particular display at the moment is only available on a screen in the north energy station. As we gain more experience with the data-tracking and information-sharing, we will be able to shape the dashboard display in a more informed manner. It is likely that we would provide data-sets over time to track the accumulated performance of the system. This would align with our need to report to the ACUPCC the annual reductions in GHG emissions associated with our improved energy efficiency resulting from the use of the geothermal district-scale heating and cooling system and related building weatherization to reduce on-campus Scope I emissions.

9.4.2.1.4   Campus energy conservation policy – pilot project: capture split-incentives for
         energy conservation

Develop, through a participative process, a policy for how energy is used on campus including: temperature set points, hours of operation, laboratory vent hood usage, computer laboratory practices, powering down of lights and equipment during non-use times, eliminating space-heater use, and adopting purchasing standards for energy-efficient technology upgrades.

These needed policies fall into three broad categories of focus: prescriptive standards, operational practices, and behavioral change. Expanding and reinforcing the use of LEED certification for all new building construction and renovation can be coupled with the indexing of energy use intensities for building types throughout the campus. Setting the climate neutrality goal of net zero energy use by a future date, especially when tied to the five-year strategic planning cycles, will determine the benchmarks of performance that must be achieved. In the USA, using the CBECS database as a touchstone and integrating the indexing produced by the Clean Air-Cool Planet GHG Calculator reinforces the importance of benchmarking and standard setting.

Operational practices of course must include careful attention to ongoing maintenance and system upgrades, continuing focus on opportunities for modest investment with immediate short-term impact and the careful management of energy flows during the spring and fall "swing" seasons.

Behavioral change is the most challenging task – given the large number of constituents, each of whom has a differing daily/weekly schedule and sense of commitment to the institution. But in every case, the only means of affecting behavioral change is through effective communication. Not only must the campus community be apprised of the importance of the President's commitment to climate neutrality on or before 2050, but must become informed about the progress to date and the availability of next steps for their respective participation. Indexing the energy-per-square-foot profile for reduction over five-year increments is a stepping stone in such communication (Fig. 9.8). More importantly, care must be taken to assure that the campus community adopts conserving practices ranging from the use of daylight to minimize electrical

---

[2]This of course relates to Tactic 1.1 wherein we have built a web interface for the display of STARS-related information presentation and management.

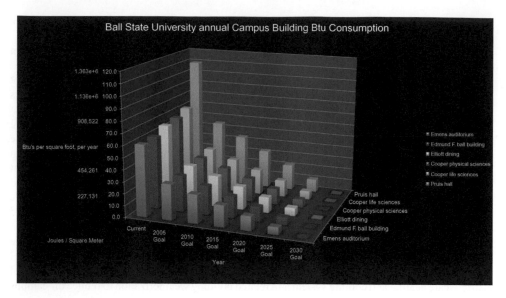

Figure 9.8.   Profiling energy use reduction by 2030.

consumption, to turning off lights when not needed, to discontinuance of desk-by-desk unit space heaters.

Finally, requests for energy-star-rated equipment replacement must be part of the reduction of electrical demand from plug loads. Of course each building is unique and so setting targets involves thorough analysis of usage patterns, equipment and plug loads, the effectiveness of the overall building shell, and the efficiency with which air handling and hydronic distribution are achieved. Moreover, such planned management must recognize the cycles of renovation for scheduled repair and rehabilitation and the use of such funding to achieve long-term operational effectiveness.

One of the challenges is to avoid the standard practice of energy service companies (ESCOs) which is to seek out the low-hanging fruit; the switch-outs of lamps and luminaires and other easily accessible equipment for a quick-order reduction in energy use is problematic. While of value, this focus on the most immediately doable short-term reductions builds in a pessimistic evaluation of longer-term more substantial capital investment. In short, it is best to bundle strategies in which high-return-on-investment and low-return-on-investment are packaged together so as to average out the cost-benefit of intervention. Such analysis of bundling is essential to achieving hybrid performance. As a further complement to this effort, the university has been conducting for the last several years a spring and fall campus energy reduction competition among the residence hall complexes. The BSU Energy Action Team (BEAT) is the lead agent in this effort and comprises student interns working under the direction of facilities management staff to invoke social media technology to encourage participation by the 6400+ residents of the campus residence hall complexes. The residence hall complex which achieves the greatest percentage reduction in a one-month time frame of competition receives temporary custody of the Energy Challenge Trophy.

This past spring BEAT expanded its efforts to include academic buildings. In this competition, the constituents in the "winning" building received additional funding for faculty travel; a very meaningful financial incentive to spur action on the part of students, faculty and staff to the benefit of their respective departments. The information display used in these competitions comprises a weekly web-based histogram-reporting as constituents move toward ever-greater energy

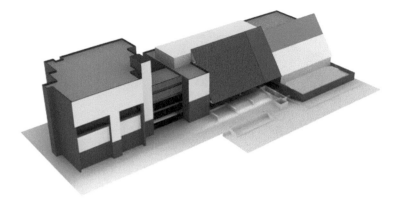

Figure 9.9.   Patterning studies of PV "filtration" on the architecture building.

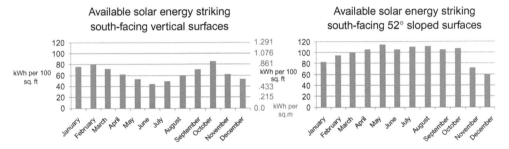

Figure 9.10.   The monthly availability of convertible solar energy during a typical meteorological year (TMY) is shown in the two histograms; the respective areas for potential placement of photovoltaic panels are shown on the building massing.

reduction. Over time this reporting will likely be integrated with the other dashboard displays to be developed.

9.4.2.2   *Scope 2*

9.4.2.2.1   On-site photovoltaic (PV) electrical production – pilot project: install building integrated PV electrical production on architecture building (see Figs. 9.9, 9.10, and 9.11)

Electrical production from photovoltaic panels has the benefit of offsetting approximately 3 times the GHG impact of any given energy conservation technique primarily because the source of electrical power for our region of the Midwest comes from coal-fired generating facilities. Some 2/3 of the input energy from that combustion is lost through inefficiencies of conversion and distribution. By contrast, on-site electrical conversion of solar radiation provides power at point-of-use, with no attendant distribution losses.

The placement of photovoltaic technology on campus involves choices between stand-alone or stand-off armatures on which PV panels are "racked" and oriented to idealized exposure for the given latitude or made integral to the skins of existing and new buildings. In these latter cases, while the orientation may be readily dictated, the angling of panels for optimal production becomes more complicated. Certainly placement on the roof affords exposure to the full altitude of the sky vault as well as the varying solar path angles from hour-to-hour during the differing seasons.

Figure 9.11.   Axonometric of the architecture building with a PV system.

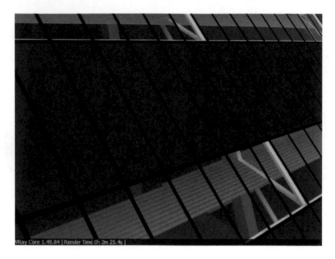

Figure 9.12.   First-solar panels.

We have a fortuitous opportunity on our campus of scheduling one of the next retrofits of an existing building to be that of the facilities for College of Architecture & Planning and Center for Energy Research/Education/Service. The east wing of this facility especially is well-suited to photovoltaic placement largely because of the expanse and pitch of the south-facing roof-slope. Given the existing building design which uses extensive greenhouse glazing to promote air circulation through a solar chimney and permits extensive views to the outdoors, the retrofit of photovoltaic panels raises the question of modifying such transparency for the sake of electrical production. As shown in Figure 9.12, we have scoped a range of covering percentages to maintain some continuance of indoor/outdoor views but with the thought of maximizing potential electrical production from this idealized sloping surface.

However, we have gone beyond that simple abstract assessment and profiled varying production rates of differing solar panel types ranging from amorphous silicon to free-floating "solar ivy" (Fig. 9.13).

And operating on available utility rate and installed cost assumptions, we have profiled a range of returns on investment. An advantage to this integral upgrade of an existing facility is that the photovoltaic panels themselves can replace some of the otherwise necessary materiel for the "reroofing" of the sloped surface.

We also recognize this as another opportunity for integrating the education and research efforts of the faculty and students. One of our goals would be to provide systematic monitoring of

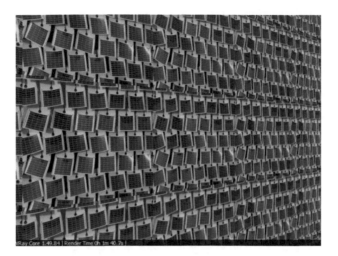

Figure 9.13.   Solar ivy cells.

performance and use a range of cell types to cross-compare those performances over time. The extensive patio-lab area of the Energy Center can be used, as well, for standalone placement of PV panels as new technologies come onto the market. Thus the building could provide long-term data-tracking on productivity and more immediate market response research and assessment once the new technologies arise.

9.4.2.2.2   Off-site wind-energy production – pilot project: acquire dedicated electrical
                   production from a local wind turbine
"Green" power distributed through the electrical grid provides an additional option for avoiding GHG Emissions. The purchase of the emissions value of that production (through RECs) is one option. Alternatively, dedicated production from a stand-alone megawatt wind machine located near campus in the east central Indiana region would provide a source of green power under university "ownership", and yield marketable "carbon credits" income (Fig. 9.14).

Our campus location in east central Indiana places the university near prevailing wind corridors which are beginning to be exploited by wind farm installation. In fact, in the last couple of years Indiana has had one of the faster-growing machine placement records in the USA.

Numerous options exist ranging from negotiating the purchase of electricity produced by a near-campus wind machine owned by a third party, to the potential for fund raising to support the installation of a wind machine owned by the university, to even the opportunity for the donation of such unit to the institution. In this latter case, the federal tax laws would permit the a manufacturer to claim the retail value of the installed system as a credit against earnings when in fact the real cost to the manufacturer would be wholesale. These necessarily longer term opportunities that will require considerable ground work to align interested parties into the scheduling of implementation. Nonetheless, it is important to begin thinking at this time of the options and to scale the magnitude of such wind energy conversion to the demands of the campus. Ideally as campus energy conservation yields ever greater demand-side reduction, the potential "fit" of stand-alone wind power conversion to that demand should come into optimal alignment.

9.4.2.2.3   Off-site PV production – pilot project: acquire dedicated electrical production from a
                   solar PV array
Off-site electrical production from a photovoltaic solar farm has been used extensively in the desert (southwest region of the USA) to provide supply to the national grid. We are in conversation

Table 9.3. Spreadsheet of performance and cost.

Brand	Type	Efficiency [%]	Units	Area [m²]	Area [sq.ft]	Annual production [kWh]	Cost Savings at 0.10/kWh [US$]	Installation cost at US$ 5.50/W [US$]	R.O.I [%]	CO$_2$e annually [Mt]	Carbon tax savings** [US$]
SunPower	Monocrystalline	19.5	511	831.5	8950	1 835 349	183 535	492 250	37.2	38 775	378 750
Sharp	Monocrystalline	14.7	612	761.8	8200	1 161 403	16 140	451 000	25.7	24 537	245 367
SunTech	Monocrystalline	15.7	638	603.9	6500	779 407	77 941	357 500	21.8	16 466	164 663
Sanyo	Monocrystalline	17.7	794	696.8	7500	1 169 859	116 986	412 500	28.3	24 715	247 153
SunTech	Polycrystalline	15.0	667	743.2	8000	1 128 000	112 800	440 000	25.6	23 831	238 310
Sharp	Polycrystalline	13.7	648	787.8	8480	1 160 957	116 096	466 400	24.8	24 527	245 273
BP Solar	Polycrystalline	13.8	640	799.0	8600	1 199 261	119 926	473 000	25.3	25 337	253 365
Kyocera	Polycrystalline	–	704	817.5	8800	–	–	484 000	–	–	–
First Solar	Thin-film	11.1	–	492.4	5300	366 364	36 636	291 500	12.5	7740	77 401
Abound Solar	Thin-film	12.5	662	492.4	5300	412 572	41 257	291 500	14.1	8716	87 163
Super Sky	BIPV	–	204	492.4	5300	–	–	600 270	–	–	–
Super Sky	BIPV	–	408	507.0	5457	–	–	300 135	–	–	–
SMIT	Monocrystalline	–	–	1013.9	10914	–	–	–*	–	–	–
SunPower	Monocrystalline	19.5	419	682.8	7350	1 237 791	123 779	404 250	30.6	26 151	261 505

*SMIT products currently priced at US$ 10 per leaf. Number of leaves in dependent on project/client.

**Annual savings, assuming a US$ 10 per Mt carbon tax.

Estimates of electrical production were developed using the Natural Resource Energy Labe (NREL) software PV Watts which uses state-of-the-art modeling algorithms for determining climate optimized performance. (http://rredc.nrel.gov/solar/calculators/PVWATTS/version1/)

Figure 9.14.   A wind farm in Indiana.

Figure 9.15.   The IND solar farm at the Indianapolis International Airport.

with the near-surround community leadership about installing a demonstration solar farm on a brownfield property in south Muncie. We are currently working with the developers of the IND Solar Farm located at the Indianapolis International Airport. This 30 ha (75-acre) 30-MW production facility is a stand-alone free-field placement of PV panels on non-tracking armatures (Fig. 9.15).

- 30 ha (75 acre) solar farm
- Consist of more than 41,000 panels
- 12 MW solar system
- Generates approximately 17,000,000 kWh/year, enough to power 1800 homes
- Airport is owned and operated by Indianapolis Airport Authority (IAA)
- GES, Telamon, and Johnson Melloh developed and will finance the project. Cenergy Power will engineer, procure and construct the system
- Project is estimated to cost US$ 35–45 million, create up to 140 temporary jobs and 12 permanent positions

- Solar energy produced will be sold to Indianapolis Power & Light through a 15 year power purchase program
- Environmental equivalents of the clean power produced by the solar system annually
- Carbon sequestered by 10,000 acres (4947 ha) of forest
- CO_2 emissions from 27,000 barrels (4293 m^3) of oil

Our current arrangements with the developers are to engage faculty and students in the documenting, illustrating and monitoring the system design, installation and performance with outlying research opportunities to be engaged over the coming years. This experience could be replicable in the years to come on some of the remote land holdings of the university. We have several farmsteads and this would only add to their use as home-bases for education, empirical research and community outreach.

9.4.2.3   Scope 3

9.4.2.3.1   Real-time communication, monitoring and reporting – pilot project: implement "get on the map" using attributes of the STARS framework

As a first step in providing campus-wide transparency of operations and energy performance, we have provided a web-based tool for constituent reporting of sustainability-related activities, events and achievements (Fig. 9.16). This follows the structure of the Sustainability Tracking Assessment and Rating System (STARS) as developed by the Association for the Advancement of Sustainability in Higher Education (AASHE) and most importantly provides the linkage of that information structure to the actual physiography of the campus. For every given building, one can inquire about activities within the seventeen categories of the STARS rating system and/or one can look into the specific unit-level sustainability plans attributed to the administrative entities within a given building.

Separate from this multi-scaled web-user interface for "mapping sustainability" at BSU (www.bsu.edu/sustainability/), we also have surveyed the range of information-sharing technologies that can inform the university community about our progress. Included are the conventional dashboard systems as developed by commercial companies such as Lucid Technologies or the more academically-rooted reporting as developed by faculty staff and students at Arizona State University (http://www.asu.edu/dashboard/).

In addition, we have explored the idea of building-specific Kiosks, and through the Department of Technology have developed geo-caching apps for smart phones which enable campus community members to "dial-up" information about the university's district-scale geothermal system, its energy practices, and building-specific performance.

Moreover, we provide for each building a histogram profiling of energy use reduction targets which must be met for the university to become climate-neutral on or before 2050. These histograms correlate to the five-year strategic planning as noted earlier in the chapter and provide a quick indexing of that information (Fig. 9.17).

The key in all such information sharing is to continue to index the magnitude of performance against the five-year incremented reduction histograms as noted earlier. By "blocking" reduction benchmarks in five-year timeframes, the 20,000 some members of the university community can understand their individual roles. For example, if in a given five-year time frame the goal of

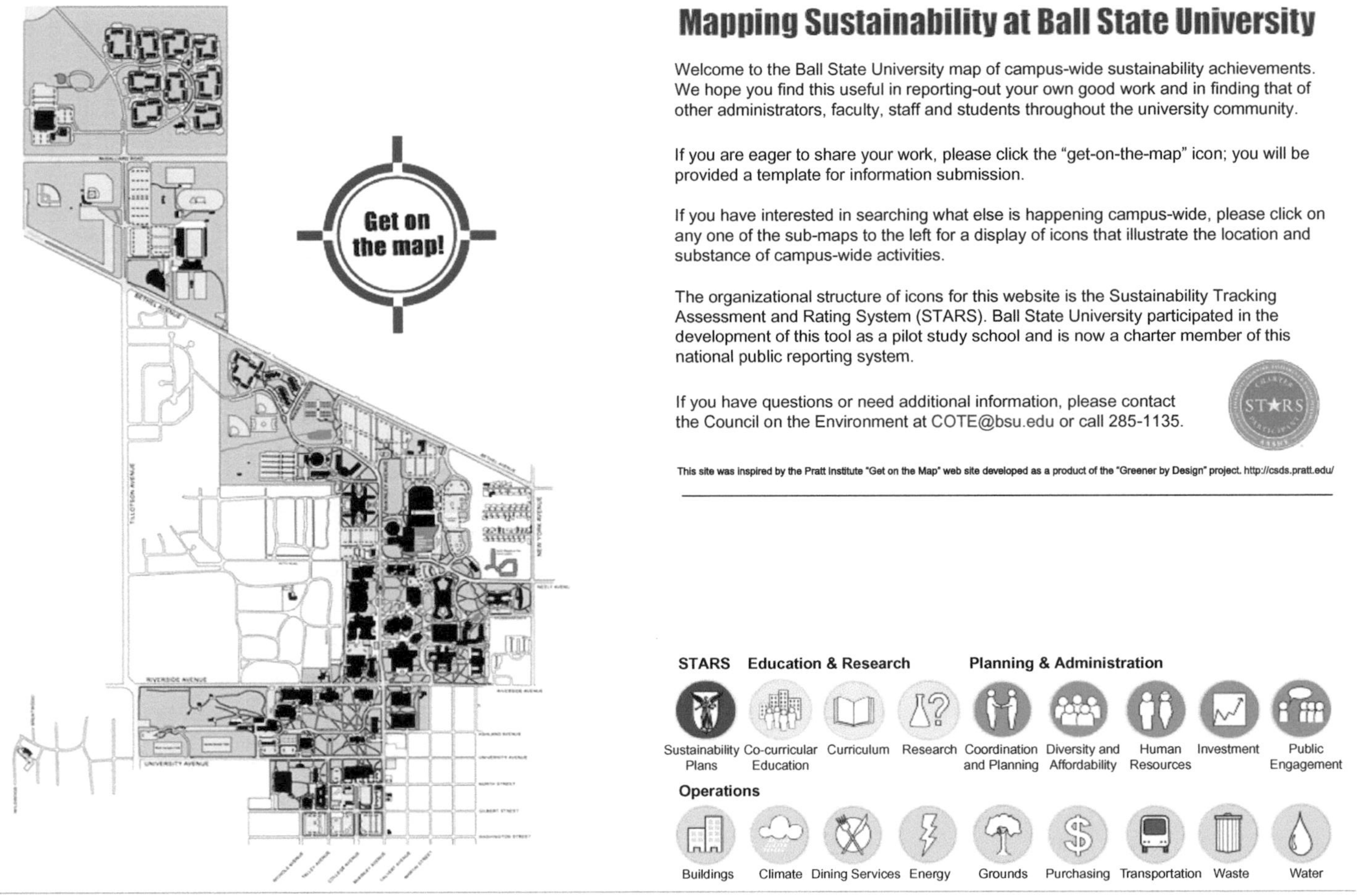

Mapping Sustainability at Ball State University

Welcome to the Ball State University map of campus-wide sustainability achievements. We hope you find this useful in reporting-out your own good work and in finding that of other administrators, faculty, staff and students throughout the university community.

If you are eager to share your work, please click the "get-on-the-map" icon; you will be provided a template for information submission.

If you have interested in searching what else is happening campus-wide, please click on any one of the sub-maps to the left for a display of icons that illustrate the location and substance of campus-wide activities.

The organizational structure of icons for this website is the Sustainability Tracking Assessment and Rating System (STARS). Ball State University participated in the development of this tool as a pilot study school and is now a charter member of this national public reporting system.

If you have questions or need additional information, please contact the Council on the Environment at COTE@bsu.edu or call 285-1135.

This site was inspired by the Pratt Institute "Get on the Map" web site developed as a product of the "Greener by Design" project. http://cads.pratt.edu/

STARS   **Education & Research**   **Planning & Administration**

Sustainability Plans   Co-curricular Education   Curriculum   Research   Coordination and Planning   Diversity and Affordability   Human Resources   Investment   Public Engagement

Operations

Buildings   Climate   Dining Services   Energy   Grounds   Purchasing   Transportation   Waste   Water

Get on the map!

Figure 9.16.   "Get on the Map" dashboard using the attributes of the STARS Framework.

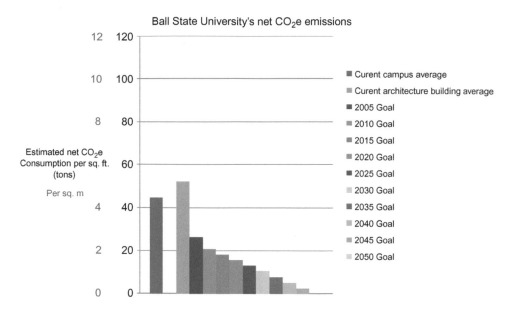

Figure 9.17.   Ball State University's net CO$_2$e emissions.

reduction was 20,000 metric tons of reduction, every member of the community would have a role to play in reducing some 1000 metric tons on a per capita basis.

Similarly, the aggregate scaling of emissions reductions can be assigned to departments, colleges or other administrative units – or even the quadrants of the campus itself; the more newly constructed buildings in the north half of campus or the more substantial vintage of buildings in the southern half of campus.

As shown in Figure 9.16, these various display techniques can be accessible in real-time and can be used to provide annual summative reports on the reduction of energy use and the attendant GHG emissions (http://cote.iweb.bsu.edu/stars/).

9.4.2.3.2   Transportation – pilot project: create a web-based gateway for professional travel
Perhaps the most difficult metric to account in compiling GHG inventories for colleges and universities involves the capture of Scope 3 emissions. Aggregating the transportation and tributary impacts resulting from the professional activity of administrators, faculty, staff and students, and their day-to-day commuting.

Campus commuting, professional and sports-related air and ground travel, campus security, and ground travel in support of physical operations accounts for about 7% of total GHG emissions. The bulk of these Scope 3 emissions originate from the combustion of gasoline, diesel, bio-fuel, and jet fuel. A variety of technical, informational and social strategies can quickly and cost-effectively reduce carbon emissions from these transportation systems. These include promoting incentives for the adoption of carpooling and alternative vehicles, endorsement of the use of electric vehicles, biking and walking, and purchase of carbon offsets for travel and the use of public transportation.

For purposes of gathering these data, a centralized web-based gateway for professional travel would be the most effective approach (Fig. 9.18). As a public university, BSU must remain open to competitive pricing for travel services but, any more, most faculty, staff, administrators and students book flights and hotels directly through hand-held devices. Nonetheless, if the university were to provide a singular gateway for the booking of travel arrangements, the institution would be able to aggregate data as part of the day-to-day activities of its constituents. Most importantly this information could be reflected back to the academic community in quarterly reports (if not

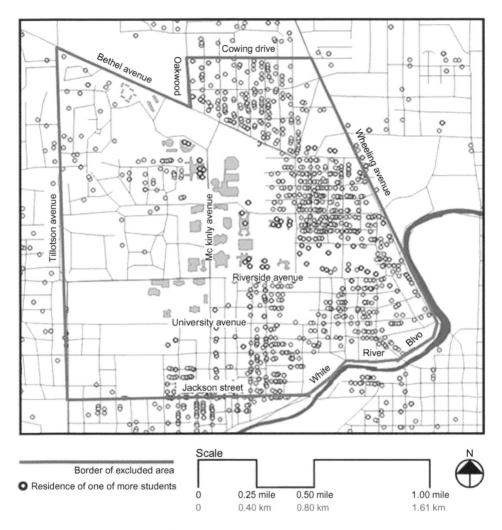

Figure 9.18.   Mapping commuter factors.

in real-time) so as to help the administrators, faculty, staff and students understand their tributary environmental loads. To the extent that this could modify subscription and foster asynchronous and web-based participation, the institution would benefit in reducing its Scope 3 GHG emissions. Much of this, however, is beyond the capacity of the institution to control; rather it requires influence and encouragement. As regards day-to-day and to/from campus commuting, however, the institution can institute policies that would benefit more effective use of transportation fuels to dial-down GHG emissions. For example, preferential parking spaces made available to those academic community members driving hybrid-electric vehicles, or to those participating in certified carpooling commutes or incentive financial discounts for those relying primarily on bicycles or walking for campus commutes.

Additional on-site energy management can be achieved with the switch of fossil-fuel-based transportation to bio-fuel and/or hybrid-electric vehicles. At BSU, we have secured free use of municipal buses for all students, faculty, staff and administrators; these buses are hybrid-electric, bio-fuel vehicles. In addition, we have switched a large percentage of the university fleet to hybrid-electric, and we use hybrid-electric, and bio-diesel-based shuttle buses for public on-campus and near-campus transportation.

Finally, we acquired an all-electric vehicle for use in mail-drop delivery and have been monitoring its cost-benefit. Of course any use of hybrid-electric or completely-electric vehicles requires sourcing electrical power from the grid and if such power is fossil-fueled, there is a 3:1 upstream impact; for every unit of energy consumed on campus, some 3 units of energy would be tapped to generate that end-use supply. Nonetheless, leveraging the switch from fossil-based conversion to green-power within the regional grid is an added opportunity for affecting change.

9.5   THE IMPLEMENTATION OF THE CLIMATE ACTION PLAN

9.5.1   *Initiate and secure funding for kick-starting the nine pilot projects*

Of the nine pilot projects proposed, all but three involve little or no cost using existing campus personnel and financial resources. As noted, the first-listed pilot project which focuses on information transparency is already operational. This "get on the map" opportunity is available to all members of the university community. Everyone is able to self-report success within the categories of the STARS framework – education and research, operations, and planning and administration.

9.5.2   *Establish immersive education and research activities for each pilot project*

Under the university branding of Immersive Learning, students are to be actively recruited and involved in the implementation of each of the pilot projects.

A qualifying factor is the semester-scale modulation of student availability so careful attention must be given to the start and stop times for these projects. Nonetheless, the potential is there for connection of this work with classes already on the books and the opportunity for service clubs, student government and residence hall associations to contribute to the university's success.

9.5.3   *Monitor and report pilot project performance (outreach)*

Real-time display of performance will affect human behavior. To the extent that we can maintain a transparency of information about the workings of the institution, its energy consumption and its GHG emissions, we are confident that all members of the university community will actively contribute to BSU becoming a climate neutral campus.

Although we have proposed that achievement over a 40-year time horizon, we can only get there with routine annual reporting of our progress and the integration of the planning and execution of our five-year strategic plans. We issue annual reports through ACUPCC, STARS, GRI and now the ISCN. In addition we've mounted the residence hall and academic building energy competitions whose results are posted on the web.

Finally, we anticipate opportunities to link some of this work with the growth of the green economy in the state of Indiana and especially in the east central Indiana. Discussions are underway. BSU is demonstrating climate action planning as a form of a systemic ecology – a system of systems. We have balanced the structure and content of our approach; identifying targets of reduction and adopting institutional structure to foster, implement and sustain continuing reduction.

9.6   CONCLUSION

On-site renewable energy is an area of unique opportunity for institutions of higher education. As noted in the opening paragraphs, colleges and universities are ideally suited for taking a whole systems view of the design and application of this technology. Implementation can reach beyond the mere placement of equipment, involving aspects of research and education as well as community outreach. The impact of implementation can have impact on the local economy and certainly by emulation can lead to a broader deployment in the neighborhoods, business districts and civic facilities of the near-surround community.

However, to borrow descriptive phrasing from architect William McDonough, the focus for assessment of approach must be on effectiveness and not simply efficiency; and to paraphrase a correlating observation from Amory Lovins, to optimize only a single system component will tend to pessimize the performance of the system as a whole.

Thus colleges and universities must seek whole system performance using a highly-integrated approach to serve not only the mission of energy conservation and on-site renewable energy production but also the larger mission of education and research. Our choices must recognize interactive impact, we must acknowledge that all decisions are cumulative over time, and we must avoid the short-term view. Administrators must recognize that post-poning decision-making merely exports to the future the many choices which will be no less difficult – if not even harder – to make; saddling future administrations with the burden of that "non-choice". Thus it is in everyone's interest to take action as soon as possible, to understand the campus physiography and citizenry as elements of a systemic, operational "ecology".

Universities must find ways to integrate the budgeting of capital and operational planning, must focus on demand-side management; must stay tuned to the opportunities for research and education embedded in campus transformation; must provide the visualization of options so as to transcend the forced-boundaries of disciplinary language; all to set the stage for an effectiveness of intervention. Finally, returns-on-investment must be seen as more than monetary; comprising the very institutional reputation and branding.

ACKNOWLEDGEMENTS

GHG inventory participants: Robert A. Fisher (Professor of Architecture, Resident Fellow CERES), Debra Atkinson (Institutional Research Analyst), Debbra Bear (Accounting Representative), Gregory Graham (Facilities Planning), Brenda Kearns (Transportation, Kevin Kenyon (Facilities), Larry King (MITS), John Lewis (University Dining), Tammy Sue Neal (Student Affairs), Michael Planton (Facilities), Frank Sabatine (Associate Provost Economic Development), David Schoen (Department of Urban Planning), Randy Sollars (Director of University Budgets), Matt Stephenson (Director of Purchasing), John Taylor, (Field Station and Environmental Education Center), Rhonda Thomas (Human Resources), Michael Twigg (University Libraries), Amy Wagner (University Dining), Sue Weller (Facilities), and Nancy Wray (Parking).

BSU Cabinet: Jo Ann M. Gora (President), Terry King (Provost), Kay Bales (Vice President for Student Affairs and Dean of Students), Hudson Akin (Vice President for University Advancement), Randy Howard (Vice President for Business Affairs and Treasurer), Phillip Repp (Vice President for Information Technology), Tom Taylor (Vice President for Enrollment, Marketing and Communication), Bill Scholl (Director of Intercollegiate Athletics), Gretchen Gutman (Associate Vice President of Governmental Affairs), and Sali Falling (Vice President and General Counsel).

BSU Board of Trustees: Hollis E. Hughes, Jr., Frank Hancock, Rick Hall, Barbara Phillips, Thomas C. Bracken, R. Wayne Estopinal, Marianne Glick, Matt Momper, and Michael Miller.

REFERENCES

AWEA (American Wind Energy Association): The wind energy fact sheet. Last updated 2013, www.awea.org (accessed 21/10/2013).
CHPA, US Combined Heat and Power Association, last updated 2014: www.uschpa.admgt.com/ (accessed 21/10/2013).

FSEC, Florida Solar Energy Center, Photovoltaics, last updated 2014, www.fsec.ucf.edu/pvt/ (accessed 21/10/2013).

Grumman, D.L. (ed): *ASHRAE GreenGuide*. American Society of Heating, Refrigerating and Air-Conditioning Engineers, Atlanta, GA, 2003.

IEA (International Energy Agency): IEA Photovoltaic Power Systems Programme, last updated 2014, www.iea-pvps.org (accessed 21/10/2013).

Suozzo, M.: *Guide to energy-efficient commercial equipment*. 2nd edition, American Council for an Energy Efficient Economy, Washington, DC, 2000.

USDOE: Buildings energy databook. US Department of Energy, Office of Energy Efficiency and Renewable Energy, Washington, DC, 2005.

USDOE: A consumer's guide to energy efficiency and renewable energy: sizing small wind turbines. US Department of Energy, Energy Efficiency and Renewable Energy, last updated 2014a, www.eere.energy.gov/consumer/your_home/electricity/index.cfm/mytopic=11010 (accessed 21/10/2013).

USDOE: Hydrogen Program. US Department of Energy, last updated 2014b, www.hydrogen.energy.gov/fuel_cells.html (accessed 21/10/2013).

USDOE: Micro hydropower systems. US Department of Energy, last updated 2014c, www.eere.energy.gov/consumer/your-home/electricity/index.cfm/mytopic=11050 (accessed 21/10/2013).

USDOE: National Center for Photovoltaics; US Department of Energy, last updated 2014d, www.nrel.gov/ncpv/ (accessed 21/10/2013).

USEPA: Catalogue of CHP technologies. United States Environmental Protection Agency, Combined Heat and Power Partnership, Washington, DC, 2002, www.epa.gov/chp/project_resources/catalogue.htm (accessed 21/10/2013).

USEPA: Combined Heat and Power Partnership. United States Environmental Protection Agency, last updated 2014a, www.epa.gov/chp (accessed 21/10/2013).

USEPA: Energy Star Program. United States Environmental Protection Agency, last updated 2014b, www.energystar.gov/ (accessed 21/10/2013).

UTC Power, last updated 2014: www.utcfuelcells.com (accessed 21/10/2013).

APPENDIX A

Table 9.4. Typical oputputs and/or efficiencies.

Bio-mass-based heating and cooling
- Typical heating value is 5.6 kWh/kg (8500 Btu/pound), dry weight.
- When co-fired with coal, only a small amount of biomass is typically added (no more than 15% of the total amount of fuel going into the boiler) to maintain the boiler's efficiency.

Methane-based heating and cooling
- ~8.5 m^3/min (300 std.ft^3/min) of landfill gas (LFG) is available for utilization for every million tons of WIP (waste in place).
- Methane content of LFG is 50%.
- Methane heat content is ~10.36 kWh/m^3 (1012 Btu h/std.ft^3) methane.
- Weighted average heat rate for LFG-fired engines, turbines, and boiler/steam turbines is ~3.428 kW/h (11,700 Btu/kWh).
- 1 MW capacity = 7.446 million kWh/year at net capacity factor of 0.85 – which accounts for availability, operating load, and parasitic losses of generating unit(s).
- An LFG collection well will recover approximately ~105–315 W/m^3 (10–30 Btu h/std. ft^3) of LFG.
- Global warming potential (GWP) for methane is 21.

Combined heat and power (CHP) [... or "co-generation"; ... or "total-energy"]
- Heat & electricity are produced "in synch".
- Can serve both base loads and peaking loads.
- Can be used to "shave" peak loads and related demand charges.
- Useful for large-scale continuous demand.

    5 TYPES:

Gas Turbine:	500 kW–250 MW (... or, can use landfill methane)
Micro turbine:	30 kW–350 kW
Reciprocating engine:	4 MW–65 MW
Steam turbine:	50 kW–250 MW
Fuel cell:	200 kW–250 kW

    NOTE for fuel cells:

    High power density (proton exchange membrane (PEM) technology)
    Requires *pure* hydrogen
    Operating core temperature at 79°C (175°F)
    Minimum time before maintenance 5000 h
    Production range 3–5 kW

    Low power density (phosphoric acid fuel cell (PAFC) technology)
    Can tolerate *impure* hydrogen
    Production range: 200 kW–1 MW
    Recovering waste heat "doubles" the overall operating efficiency to 80%
    NOTE for pre-packaged (modulated) micro-turbines:
    increase production by "adding" modules

Closed-loop GSHP
- Highly dependent on soil characteristics and climate-driven demand. For our campus system, we have installed 3600 bore-holes to service some 668901.8 m^2 (7.2 million square foot) of offices, classrooms, auditoria, labs, libraries, gymnasia, cafeterias, and residence halls.

Hydro-based electricity
- Energy is a product of pressure (head) and volume (flow rate).
- ~1 m (3 ft) of pressure head is minimum.
- Production range: 10 kW–15 MW.

(*continued*)

Table 9.4.   Continued.

Solar-based electricity
- Not useful for heating and cooling.
- Stand-alone batteries needed; or often available.
- Grid-based net metering (preferred).
- Peak demand matching @ mid-day.
- 8–16 hour day-time production.
- 35–50% increased insolation with tracking.
- Some potential for cell-plate heat recovery.
- Cost-benefit influenced by time-of-use demand charges.
- Need for controllers, inverters, meters.
- Production ranges – 1 kW–20 MW.
- Efficiencies: 10–15%; with lab-bench studies of "layered cell types" approaching 40%.

Wind-based electricity
- Micro-systems 20 W–500 W.
- Small systems 20 W–100 kW.
- 16 km/h (10 mi/h) cut-in wind speed typically needed.
- 24 hour (intermittent) production.
- Modest noise factor.
- Harvestable energy a function of the cube of the wind speed (doubling the speed, yields eight times the energy).

Solar thermal space heating
- The solar constant is 41 Wh/m^2 (442 Btu/ft^2); when factoring atmospheric effects of moisture and contamination and diurnal and seasonal angles of incidence, that value can range widely. A useful starting point for calculation is ~10 Wh/m^2 (~100 Btu/ft^2) as this can be readily converted later in the design stages. Alternatively, the use of *typical-meteorological-year* (TMY) weather data provides the most complete profiling.

Solar thermal space cooling
- Requires high temperature concentration to achieve thermal levels needed to drive absorption chillers.

Solar thermal DHW heating
- Requires modulation of "cascaded" storage in large installations to assure temperature regimes that match harvestability of the diurnal resource.

CHAPTER 10

Shifting agendas

Raymond J. Cole & Amy Oliver

10.1 INTRODUCTION

Building design involves the participation, interaction and coordination of a wide range of professions and trades, within a multitude of regulatory agencies and jurisdictions, and typically within demanding time and cost constraints (Cole, 2005a). Building design priorities as such do not exist in a vacuum, but are shaped by the prevailing paradigm and value system of the societal and cultural context within which they emerge (Du Plessis, 2011; King, 2004; Rapoport, 1969). Similarly, the technologies deployed by society reflect its culture, its worldview, and how it understands and engages natural systems (Mang, 2012).

The "worldview" held by a society operates silently to "channel attention, filter information, categorize experience, anchor interpretation, orient learning, establish moods, secrete norms, and legitimates narratives, ideologies, and power structures." (Gladwin *et al.*, 1997, p. 245). Worldviews have, historically, embodied different notions of the relationship of humans to the natural world and vary markedly across cultures. They have also taken centuries to mature and become manifest in the shaping of human settlement and building practices. Western societies remain entrapped in the dominant Cartesian Newtonian mechanistic worldview that emerged in the mid-seventeenth century wherein nature was still largely understood and managed by reducing it to its parts. This expansionist worldview places human enterprise dominant over and essentially independent of nature and considers that through technology and science, biophysical limits can be exceeded (Bazerman *et al.*, 1997; Capra, 1996; du Plessis, 2012; Rees, 1999). Within this overarching value frame, the current ways and extent that environmental issues are emphasized in building design is further influenced by immediate societal concerns, such as in the aftermath of significant events like social insecurity or economic instability. Indeed, a widely held position is that until natural disasters resulting from environmental instability set in, ecological issues will be compromised in the political realm by economic, social and military priorities (Ingersoll, 1991). More directly, environmental priorities in design are influenced by the types of tools and methods deployed in practice.

Through time, norms and conventions develop within the building industry that all participants adhere to within their respective realms of responsibility, and, when a new agenda emerges, these norms and conventions are invariably challenged. Here, building environmental assessment methods have been enormously successful in forging a "common language" for various stakeholders and providing a common measure of successful green performance. However, although the urgency to address climate change and environmental degradation require fundamental changes to current building design, practice, new ideas and requirements are most often assimilated within the existing approaches (and can be referred to as creating 'incremental change'). Fundamental shifts or 'quantum changes', cannot be readily or quickly assimilated within the design and production of buildings[1]. Bernard Tschumi suggests that "architectural and philosophical concepts do not disappear overnight… ruptures always occur within an old fabric which is constantly

---

[1]This distinction between 'incremental change' and 'quantum change' is one made by Montreal-based architect, Daniel Pearl of L'OEUF s.e.n.c.

dismantled and dislocated in such a way that its ruptures lead to new concepts or structures" (Tschumi, 1988, p. 35). This implies that environmental theories and strategies will only be assimilated partially and selectively within an existing design context, and only seldom result in a complete paradigm shift.

Several of the earlier chapters in this book have explored the introduction of environmental strategies and technologies, each of which challenge many aspects of contemporary practice. This final chapter of the book, by contrast, is concerned with shifts in overarching notions and capabilities that directly and indirectly shape design. In particular, it examines the shift from green building design to regenerative approaches that, while deploying many current green strategies and technologies, accept a much broader, holistic framing of design responsibilities. It also investigates the ways in which human and automated intelligence, and the role of information and communication technologies (ICT) may reinforce or challenge the regenerative agenda. This is particularly the case in reinforcing the importance of engaging and experiencing place.

Other significant advances are occurring in parallel with our emerging understanding of environmental and sustainability issues and the possible directions that society could take to address them. If and how we transition toward a more sustainable future and the time it will take are uncertain; however, it is reasonably safe to argue that the growing role of a knowledge-based economy will indeed be of great significance. Rueda (2007), for example, argues that in order for cities to achieve complexity, stability, and sustainability, they will need to redefine competitiveness based on exchange of information, rather than consumption of resources. In particular and in addition to significantly transforming product design and manufacture – including buildings – the widespread adoption of information and communications technologies will likely transform human settlement through "demobilization" and "dematerialization." (Mitchell, 1999) Since the "intelligence embodied within an infosphere" is potentially less harmful to natural systems than previous industrial practices (Brand, 1999), the direct and indirect implications are critical from an environmental standpoint. Although the social and behavioral implications of widespread adoption of information and communications technologies are currently uncertain, they will profoundly alter our perceptions of obligations to one another (Gumpert and Drucker, 1996, p. 363), to the built environment, and, one can speculate, on the perceived limits of human possibility.

INTELCITY – a one-year (2002–3) roadmap project funded by the European Union Information Societies Technology Programmed – was directed at the role that Information and ICTs would play in future sustainable urban design scenarios. (Curwell *et al.*, 2005) In an *e-Democracy City*, for example, ICT was considered "an enabling mechanism, changing peoples' opinions and behavior patterns through information provision and empowerment" that could provide "new ways of decision-making and negotiation through inclusiveness and accessible participation in the decision-making processes that affect the community." Similarly, the *Virtual City* scenario represents a knowledge society of networks and flows, where citizens are able to work and live anywhere in the city, supported by intelligent environments that are 'lean, green and SMART' (economically efficient and ecologically sound). (Lombardi, 2002) This chapter similarly examines the increasing capabilities and widespread use of information and communication technologies with the specific objective of exploring its potentially complementary and reinforcing role with regenerative approaches to building design. In doing so, however, the authors are aware that a host of other interconnected developments and societal concerns – globalization, food security, public health, aging, and the economy – have equal, if not greater, significance and consequence for any discussion related to buildings and sustainability.

10.2   SHIFTING FROM GREEN TO REGENERATIVE

10.2.1   *Green design*

The term 'green design' has been used fairly consistently over the past decade or so to emphasize the environmental performance of buildings and "green building" to describe those that

Table 10.1.   Key green building attributes.

1.	Reduces damage to natural or sensitive sites
2.	Reduces need for new infrastructure
3.	Reduces impacts to natural features and site ecology during construction
4.	Reduces potential environmental damage from emissions and outflows
5.	Reduces contributions to global environmental damage
6.	Reduces resource use – energy, water, materials
7.	Minimizes discomfort of building occupants
8.	Minimizes harmful substances and irritants within building interiors

have a higher environmental performance compared to that of typical buildings. Green building practices have become increasingly commonplace over the past two decades shaped, in part, by the emergence and widespread use of building environmental assessment methods such as the UK's Building Research Establishment Environmental Assessment Method (BREEAM) and the US Green Building Council's Leadership in Environmental and Energy Design (LEED®). These have profoundly influenced the range of considerations deemed important in design and are now embedded within the parlance of building procurement, design and construction, and operation. Moreover, performance targets such as LEED-Gold are increasingly incorporated into institutional requirements for new buildings. However, despite their considerable success, such approaches alone will be insufficient to support a transition to a sustainable future.

The key attributes of green building performance typically included in most assessment tools are shown in Table 10.1 and cover resource use, emissions/waste and health and comfort issues. The language of these generic performance attributes reflects the primary characteristic of green design as one of reducing resource use and adverse environmental impacts and improving the health and comfort conditions for building occupants. As has been argued in many other publications (McDonough and Braungart, 2002; Reed, 2007), green design is primarily directed at "doing less harm" or, more generally, reducing the degenerative consequences of human activity on the health and integrity of ecological systems. Implicit within green building performance, therefore, is the goal of net zero impact. Indeed, the notions of net zero energy and carbon neutrality are increasingly referenced performance aspirations for buildings rather than those evidenced in the assessment methods, e.g., BREEAM Outstanding or LEED Platinum.

10.2.2   *Regenerative design*

Regenerative design involves approaches that support the co-evolutionary, partnered relationship between human and natural systems. The building is not "regenerated" in the same sense as the self-healing and self-organizing attributes of a living system. Rather, the act of building and inhabiting a *system* consisting of the building, its inhabitants and the biophysical and socio-cultural context is regenerative by being a catalyst for positive change within the unique "place" it is situated (Mang and Reed, 2012). Regeneration carries the positive message of considering the act of building as one that can give back more than it receives and thereby over time builds social and natural capital. Within regenerative approaches, built projects, stakeholder processes and inhabitation are collectively focused on enhancing life in all its manifestations – human, other species, ecological systems – through an enduring responsibility of stewardship (Mang and Reed, 2012).

Whether projects are realized or not, current green practice is premised on measurable performance targets and on perceived certainty in the outcome. By contrast, given the systems approach, accepting uncertainty and longer timeframes of engagement, the benefits of regenerative design and development cannot be fully understood at the design stage. As such, the measure of success in regenerative design can perhaps only be represented in terms of the capacity invested in a building at the outset and stakeholder input that endow it with an ability to support a future co-evolution

Table 10.2.   Key regenerative design attributes.

1.	Restores and enhances local ecosystem function capacity
2.	Creates positive synergistic connections between resource cycles & local ecological systems
3.	Improves effectiveness of life-cycle resource use
4.	Builds resiliency to undesirable natural and human stresses
5.	Connects inhabitants to ecological systems and processes
6.	Enhances health, comfort and wellbeing of building inhabitants
7.	Improves health and wellbeing of local community inhabitants
8.	Generates opportunities for social engagement and education
9.	Generates opportunities for cultural development
10.	Generates economic wealth within local community
11.	Acts as a catalyst to generate change beyond site boundary

of human and natural systems. However, determining if and to what extent a capability has been invested in a project will be based on the collective experience of the design team, continued stakeholder engagement and, what Reed (2007) emphasizes, "conscious processes of learning and participation through action, reflection and dialogue" (p. 678), rather than evaluating the achievement of specific, easily quantifiable features or measures. Table 10.2 highlights some of the key capabilities of regenerative design, which, in comparison to green design, emphasis the potentially positive contributions that buildings can offer. (Cole, *et al.*, 2012c)

In contrast to green design and assessment, regenerative design shifts the emphasis toward systems thinking – the process of understanding how things influence one another within a whole (Meadows, 2008) and thereby logically draws on the systems theory (Skyttner, 2006; von Bartalanffy, 1969) and living systems theory (Capra, 1996). A systems-based approach is premised on the belief that the component parts of a system can best be understood in the context of relationships with each other and with other systems, rather than in isolation. Gasparatos, *et al.* (2007), for example, reinforce the notion that "completely understanding the constituent parts of a complex adaptive system does not allow a complete description of it because the interrelations between its parts also have a significant effect on its overall behavior…" (p. 54).

In contrast to fragmented, technology-biased green-building practices, Reed (2007) character-izes regenerative design being a whole, living systems approach. Regenerative design requires a fundamental re-conceptualization of the role and impact of an individual building; primarily in terms of imagining, formulating and enabling its role within a larger context. While Cole (2012) emphasizes that the performance requirements of both green and regenerative design are both necessary, the overall positive framing of the latter will likely prove more attractive to clients, designers and other stakeholders. There is, of course, a need to maintain focus and engagement on current pressing environmental issues such as climate change and loss of biodiversity, while consciously laying the foundation for the future benefits emphasized through regenerative design and development.

Planning processes, and to a lesser extent, building design have increasing recognized the importance of engaging stakeholder input. Mang and Reed (2012) and Hoxie *et al.* (2012) empha-size how regenerative design maintains and solidifies the need to create 'common ground' with diverse stakeholders and the potential that the regenerative development process holds in this regard. But most significant is the garnering of stakeholder engagement over the long-term. Here, Mang and Reed make a critically important distinction between regenerative "design" and regenerative "development." While regenerative design builds the regenerative, self-renewing capacities of designed and natural systems (the designed interventions), regenerative development creates the conditions necessary for its sustained, positive evolution. Regenerative development and design, they suggest, "does not end with the delivery of the final drawings and approvals, or even with construction of a project" but design responsibilities include "putting in place, dur-ing the design and development process, what's required to ensure that the ongoing regenerative

Table 10.3.   Key differences between green and regenerative design.

	Green design	Regenerative design
1.	Buildings designed to do less harm to natural systems	Buildings considered integral to natural systems and processes
2.	Reduces social and natural capital	Builds social and natural capital
3.	Implicit aspiration of being net zero	Explicit aspiration of being net positive
4.	Focuses on performance of individual buildings as isolated entities	Focuses on relationship of building and community setting
5.	Anthropocentric view with an emphasis on managing the environment	Fosters a co-creative, coevolutionary process between human and natural systems
6.	Relies on incremental change	Promotes fundamental values shift and reassessment
7.	Fragmented, reductive approach	Whole/living systems approach
8.	Efficient technologies & strategies	System of technologies & strategies based on understanding of ecosystems
9.	Based on belief in knowing certainty of future performance and outcomes	Operate within the uncertainty of complex, dynamic systems

capacity of the project, and the people who inhabit and manage it, is sustained through time." (p. 34). This form of active and reflective stewardship builds the capacities of people to design, create, operate and evolve regenerative socio-ecological systems in their place.

A regenerative project underscores the need for inhabitants to understand their building, both from the standpoint of "learning to engage appropriately with building systems and controls" and in terms of "psychological benefits of knowing how the building works and comfort is provided, and feeling a sense of responsibility around its performance, which can in turn shape comfort and comfort related behavior." (Cole *et al.*, 2010, p. 345). Similarly, regenerative projects have a need for incorporating feedback loops within their designs. Cole, *et al.* (2008), for example, convey that communication and interaction are bi-directional, where the experience of comfort and the building systems performance are both dependent on a form of ongoing dialogue. Such notions assume that inhabitants have a degree of familiarity with the building and its systems. Finally, regenerative projects emphasize the idea of "inhabitant," rather than occupant. The term 'inhabitant' captures more accurately the active participation and potential agency of building users with building systems than the term 'occupant' used in conventional comfort research to portray a passive recipient of universalized comfort standards (Cole *et al.*, 2008).

10.2.3   *Green building assessment methods*

The contrast between traditional 'green' approaches and regenerative approaches may be evidenced in building assessment tools. Building environmental assessment methods evaluate performance across a range of resource use, ecological loadings and indoor environmental quality criteria. They generally have recognizable 'frameworks' that organize or classify environmental performance criteria in a structured manner with assigned points or weightings. More importantly, assessment methods are managed by and operate within known organizational contexts – for example, BREEAM is operated by the UK Building Research Establishment and the LEED®rating system by the US Green Building Council. Although parts of an assessment method may be used selectively by design professionals at their discretion, full engagement of a method involves some form of registration or certification. This characteristic represents a critical distinction between assessment tools and assessment methods since the third-party verification and scrutiny invariably brings additional layers of constraints, bureaucracy and costs to the process (Cole, 2005b).

All green building assessment systems have a scope and structure that represents their developer's understanding and priorities of environmental performance issues. The characteristics of green building assessment tools relevant to this paper have been articulated by Birkeland (2007), Cole (2012) and others, and include:

- Individual performances are evaluated relative to a benchmark, either implicitly or explicitly, rather than in their absolute consequence on human and natural systems.
- Criteria are technically framed and based on metrics that are quantifiable, measurable and comparable and which, in aggregate, are assumed to offer an accurate measure and understanding of overall green building performance.
- The need for clear, unambiguous assessment and avoiding double-counting has required the assessment criteria to be kept discrete.
- Overall success is measured through the simple addition of the weighted (either implicitly or explicitly) scores attained for the individual performance issues.
- Process related criteria (e.g., commissioning) are incorporated alongside performance issues (e.g., percentage reduction in operating energy).
- While assessing the performance of green buildings has received considerable and increasing scrutiny over the past five years and remains controversial, there is still the fundamental belief that long-term performance can be predicted with some certainty.
- Although maintaining or enhancing the health of natural systems is implicit in many of the performance criteria, the conceptual underpinnings and structure of the assessment tools do not typically emphasize or communicate it.
- They do not go beyond a linear approach toward conserving resources and fail to preserve resources through a conscious cyclical process of regeneration (Fisk, 2009).
- They are primarily aimed at forging a broad market transformation and, as such, are premised on creating gradual, incremental change.
- By being generic in their formulation, they have struggled to adequately accommodate regional priorities.

The way that building environmental assessment methods identify discrete performance requirements often translates into design as a series of isolated gestures rather than encouraging creative synergies, closing loops and responding appropriately to the local ecological and social contexts. For these reasons, Reed (2007) criticizes current green design as being reductive and fragmented. However, attaining the higher levels of environmental performance (e.g., LEED "Platinum", BREEAM "Outstanding", etc.) inevitably requires the design team to begin to consider the building as an integrated system and account for the opportunities afforded by its location. While building environmental assessment tools were initially conceived to provide a distinct role, their scope and application are increasingly being expanded. A priority in the early generation environmental assessment tools' was to engage industry and ensure their widespread adoption. Today these same assessment tools must accommodate the larger scale issues of communal impact and climate change. This expansion and the attendant redefining of boundaries bring a host of new opportunities that should legitimately shape the next generation of tools.

While the performance requirements of both green and regenerative design are both necessary, the overall positive framing of the latter will likely prove more attractive to clients, designers and other stakeholders. There is, of course, a need to maintain focus and engagement on current pressing environmental issues such as climate change and loss of biodiversity, while consciously laying the foundation for the future benefits emphasized through regenerative design and development.

10.2.4 *Regenerative design support tools*

Regenerative design requires a fundamental re-conceptualization of the act of building design primarily in terms of imagining, formulating and enabling its role within a larger context. It would therefore seem appropriate that the representation of regenerative design should reflect this interplay. While the aspirations and key principles of regenerative design can be readily understood, its operation and practice are less clear than green building design. With the new emphasis to design, new tools emerge to begin to represent the key characteristics and assist practitioners and stakeholders. Fisk (2009), Svec *et al.* (2012), Plaut *et al.* (2012) and Cole *et al.* (2012a) provide examples of some of the emerging regenerative design support tools and frameworks. These are qualitatively different from those for green building assessment tools which were conceived to provide a measure of performance and, in practice, are also used to guide design by communicating what are deemed priority environmental issues.

Plaut *et al.* (2012) argue that current green building tools 'offer little guidance in the way of guiding people through the creation, implementation, and operation of projects' and by focusing on 'measuring the performance of an end result or product' can be described as 'product-based'. By contrast, their LENSES and the other regenerative frameworks are best described as what they call 'process-based' and are primarily directed at guiding design. Moreover, whereas the product-based tools keep individual environmental performance requirements discrete, the graphic organization of the emerging regenerative design tools expand the issues to include social, cultural, economic and ecological systems and processes but also emphasize the relationship between them. In short, they accept the built environment as a complex socio-ecological system and attempt to offer guidance to designers and other stakeholders in situating projects within it. However, these tools have yet to incorporate different building programs (residential, commercial, institutional, etc.), different living and work styles, and the prevalence of ICT.

10.3   ICT CAPABILITY AND AUTOMATION

Technological capabilities clearly play a significant role in affecting building environmental performance. A host of green technologies – photovoltaic systems, rainwater harvesting, onsite wastewater treatment systems, etc. – directly shape buildings. Other technologies, by contrast, influence human capabilities and how, what and where they perform various activities and, in doing so, influence human expectations. Of particular interest to this chapter are those related ICT and building automation.

10.3.1   *Information & communications technology*

Coinciding with the increasing realization to understand and respect biophysical limits, we live at a time where we expect greater choices and freedoms permitted by unprecedented information and communications technology (ICT) capability. ICTs integrate telecommunications and computers to facilitate the accessing, storing, transmitting, and manipulating of information. This capability of near instantaneous widespread dissemination of, and access to, information through personal handheld devices has created a wealth of social networks that are radically changing how people communicate with each other.

Almost every human enterprise employs and, indeed, now depends on this ubiquitous technological capability. While this is evident in almost every human endeavor, this chapter is primarily directed at building design and performance. The workplace is one area that clearly illustrates how ICTs are altering relationships with the built environment and carry a host of implications for comfort provisioning and performance. As such, this section focuses on the radical transformation in how, what and where knowledge-based work is conducted.

Whereas past changes to the nature of the workplace have been largely about flexibility and reconfiguration of interior systems in response to changes in organizational requirements, the current nature of the office "workplace" is qualitatively different. First, developments in information and communications technologies permit new forms of working, as well as new ways of engaging with public and private spheres. Work can now be undertaken in transit between different venues, e.g., planes, trains, and boats, which are increasingly being outfitted with personal access plugs and access to internet to support travelling knowledge-workers. Mobile technologies have not only allowed for increased mobility and flexibility in the workplace, but have permitted knowledge-workers to combine work and leisure applications in a variety of locations, from the office, to the home, to the public realm in the city. Moreover, "[w]hile information technology has enabled greater distribution, it has also created a situation where the mechanisms employed for communication, coordination and awareness differ from those employed in a face-to-face setting…" (Doherty *et al.*, 2012, p. 2).

There now exists a variety of forms of remote working that illustrate the emerging trends of greater workplace mobility and the blurring of work and leisure. The practices of teleworking and hot-desking are increasingly commonplace. In contrast to the traditional fixed office, non-territorial approaches such as "hot-desking" and "hoteling" are increasingly adopted strategies and offer numerous benefits for both organizations and their employees. The benefits for organizations typically include: reduced office space usage that can translate into smaller accommodation requirements and infrastructure costs; reduced employee transport costs; increased productivity of workers, increased organizational commitment and attraction of skilled personnel who prefer working in a less-constrained work environment (Kelliher and Anderson, 2010) The benefits for employees are less clear but typically include: effective correspondence between mobile employees and fellow coworkers and clients; greater satisfaction by choosing ones preferred work location; more effective balancing of work and personal life; increasing ability to collaborate; potentially reduced travel time and ability to access any number of facilities. The disadvantage is that 'workers are likely to experience more work, at an intense pace, under greater time pressure with more stress and heavier use of the mobile phone, as a single package' (Bittman *et al.*, 2009, p. 687).

Secondly, the "workplace" now embraces a wide range of possibilities that extend beyond the domain of the "office" to the home and to a host of "hot-spots" in public venues available within the city. The once fairly static workplace is now fluid and mobile and, with this shift, emerges the need to redefine a host of spatial and environmental control requirements and strategies (Harrison, 2001). Whereas in the fixed office, where notions of comfort and worker satisfaction are traditionally associated with issues pertaining to air quality and thermal comfort, the emphasis in the mobile workplace shifts to access to information and communications technologies, as well as the nature of social interactions, and a collective sense of agency and identity (Cole *et al.*, 2012b). Such developments in ICT have a number of implications for buildings design and environmental performance:

- *Personalisation*: as the workplace shifts to being non-territorial and outside the traditional office, these ideas take on a qualitatively different nature. Given increased mobility within the workplace, knowledge-based workers can locate to a variety of different situations, but presumably do not attach the same level or type of loyalty as they would to dedicated workplaces. Whereas the literature clearly reinforces the human desire to personalize the workplace (Brunia and Hartjes-Gosselink, 2009; Burke 2011; van der Voordt and van Meel, 2002), the notion of personalization takes on a different form and emphasis between the home, the office, and public locations in the city (Cole *et al.*, 2012a).
- *Comfort expectations*: comfort provisioning in buildings requires the design of systems and accompanying controls to match requirements and expectations of building inhabitants. It is widely understood that comfort provisioning and experienced comfort are context-dependent (Ackerman, 2002; Cooper, 1998; Crowley, 2001). Comfort provisioning derives from a host of design requirements, priorities, and assumptions about building occupants and the types and costs of available environmental control technologies. Meanwhile, experienced comfort depends on the intersection of technical comfort provisions and the psychological and social realms of experience, movement (mobility), interaction and behavioral factors. (Brager and de Dear, 2001) This evolution of the workplace as not only flexible but now the mobile workplace becomes the antithesis of the fixed office, and all the subset human/environmental relationships and issues of comfort, satisfaction and comfort (Cole *et al.*, 2012a).
- *Knowledge of building systems*: the importance of inhabitants' understanding of their building, both from the standpoint of "learning to engage appropriately with building systems and controls" and in terms of "psychological benefits of knowing how the building works and comfort is provided, and feeling a sense of responsibility around its performance, which

can in turn shape comfort and comfort related behavior" (Cole *et al.*, 2010, p. 345) is an increasingly important issue. Cole *et al.* (2008) explain that communication and interaction are bi-directional, where the experience of comfort and the building systems performance are both dependent on a form of ongoing dialogue. Such notions assume workers have a degree of familiarity with the building and its systems. Because workers are becoming more mobile, they are becoming less attached to the office workplace and are less likely to become actively involved in the environmental systems and controls of the building. Further research is required, however, regarding the role of wireless sensor devices and mobile environmental applications that can save the mobile worker's preferences as he/she moves from office to coffee shop to city.

- *Blurring of work and leisure*: ubiquitous information and communications technologies are transforming spatial boundaries and blurring traditional notions of work and leisure, e.g., the workplace has become increasingly domesticized and the home increasingly 'officized.' "Work" is often defined in terms of 'obligated time, whether paid or unpaid' (Lewis, 2003, p. 344) and "leisure" is often constructed as the antithesis of work, meaning non-obligated time activities that are chosen freely, and intrinsically motivated (Iso-Ahola, 1997; Lewis, 2003). This blurring is creating different expectations of the office workplace and its culture with both positive and negative consequences for organizations and their employees. Increasing freedoms and choices in the workplace (in terms of time and space) contribute to a situation in which work is becoming indistinguishable from leisure (Lewis, 2003, p. 344). The blurring between 'work' and 'leisure' is most apparent for knowledge-based workers, who 'have increasingly more permeable boundaries between their work and the rest of life, although they are also likely to have more personal control over these boundaries than other workers' (Lewis, 2003, p. 346). These knowledge-workers, though permitted much more work-leisure flexibility tend to use this flexibility to work more intensely and for longer hours (*ibid*, p. 347). What is more, Bittman *et al.* (2009) argue that the increase in work and intensity of work is especially prominent amongst knowledge-based workers that rely heavily on 'work extending technologies'. With the use of mobile devices, knowledge-workers can now Skype or check Facebook at the office on the one hand, and check work emails in the home during what used to be classified as 'leisure hours.'

Access to environmental controls and the opportunity to adjust thermal comfort, natural ventilation, daylight and glare, as well as deal with excess noise, are central in this debate. A critical distinction is the ability to control the ambient conditions in the workplace is representative of personal adjustments at the workstation. Again, opportunities for workers to engage environmental controls – either ambient or personal – varies considerably across the workplace contexts, typically being greatest in the home and least within public venues within the city. It is within this domain that the notions of automated and human intelligence in the provision of ambient or personalized comfort requirements are most evident. However, there remain several unknowns with regards to personalization across various public workplace contexts.

Table 10.4 highlights the key distinctions between the conventional and mobile workplace. These distinctions are a result of the pervasive use of ICT.

Thirdly, knowledge-based remote work depends on the quality and capability of information and communication technologies, support resources (e.g., printers), web connection and access to power/ electricity. A related issue within this domain is the type of technical support available to knowledge-based workers, with a clear distinction between physical (IT personnel visiting) and remote (telephone or direct access) support. What is increasingly likely is that at indoor public venues within the city, access to power outlets and internet often outweighs comfort criteria (Cole *et al.*, 2012a). For semi-outdoor or outdoor public spaces, however, the quality of the microclimatic conditions will typically always be the most important driver of a preferred location. The workplace is but one example of how the pervasive use of ICT is altering relationships between people and the built environment. This relationship is also evidenced in the home, in civic buildings, in public spaces, and in modes of transit; however, the workplace most clearly exemplifies how technologies alter spatial configurations and human behavior patterns.

Table 10.4. Distinctions between conventional and mobile workplace.

	Conventional workplace	Mobile workplace
1.	Fixed, dedicated workstation	Non-territorial approaches – "hot-desking" and "hoteling" permitting a variety of workplace locations – home, office, city, trains
2.	Opportunity to personalize workplace	Diminished personalization and control associated with non-territorial workplace
3.	Potentially greater familiarity and understanding of building systems and their operation	Less aware and engaged in building systems and operation
4.	Potentially greater interactivity with co-workers, workplace social interactions and dynamics	Potentially greater autonomy and opportunity for informality
5.	Adaptive opportunities through clothing dependent on office workplace culture and dress codes	Optimal adaptive opportunity: free movement of mobile worker permits seeking places that offer the best overall workplace experience.
6.	Inhabitants: Active participation and potential agency of building users with building systems	Occupants: passive recipient of universalized comfort standards

10.3.2 *Building automation*

A central issue in the operational efficiency of buildings and their effectiveness to provide functionality and occupant comfort is where 'intelligence' is assumed – either implicitly or explicitly – particularly as it relates to automated technologies or the decisions and actions made by building occupants (Cole and Brown, 2009). In practice, this involves knowing which aspects of building control to implement automatically and those to be made accessible to the individual occupant (Cohen *et al.*, 1999). Currently, this debate is polarized between advocates of these two distinct positions – automated intelligence and human intelligence – although almost every building deploys some combination of both. The decision regarding the type and extent of 'automated' and 'human' intelligence in buildings can have consequences for performance issues ranging from energy and operational efficiency to inhabitant satisfaction, productivity, security and privacy – issues that cover a broad range of domains of knowledge and disciplines (Clements-Croome, 2004; Himanen 2004). Moreover, and of relevance to this chapter, the notions of human and automated intelligence have consequences for regenerative design as discussed later in Section 10.4.

Automated controls can be defined here as those mechanisms that provide a regulatory service for building inhabitants without their direct input (Cole *et al.*, 2012a). The performance is derived from either sensing a condition in real time – temperature, lighting levels, etc. – and making appropriate adjustments, or based on the pre-programming of the system to deliver an anticipated set of required conditions. Advocates of intelligent buildings consider greater automation as being equally capable and necessary to provide high environmental performance. In "Bright Green" buildings, "fully networked systems transcend the simple integration of independent systems to achieve interaction across all systems, allowing them to work collectively, optimizing a building's performance, and constantly creating an environment that is conducive to the occupants' goals" (CABA, 2008, p. 5). Furthermore, the "fully interoperable systems in these buildings tend to perform better, cost less to maintain, and leave a smaller environmental imprint than individual utilities and communication systems" (CABA, 2008, p. 6).

Human intelligence, by contrast, places responsibility in the building inhabitants allowing them to make appropriate adjustments to suit their specific needs. Human intelligence is manifest through base-level technologies that can be human-controlled to immediately provide a certain condition or perform a task, e.g., a light switch that humans can operate in order to immediately access more light; blinds which can be manually opened to admit more natural light/control glare, or a thermostat that is manually operable to increase or decrease ambient space temperatures.

While the key differences between human and automated intelligence and the polarity between their advocates are clear, what remains less clear is how they are employed in practice. In practice

Table 10.5. Key differences/implications of human and automated intelligence.

Human intelligence	Automated intelligence
Direct effects/consequences	
Assumes that inhabitants are most equipped to evaluate and make adjustments to control environmental conditions in the workplace	Assumes that all users cannot be relied upon, or not sufficiently willing and able to manage systems in all manner of situations
Inhabitants take a more active role in shaping indoor conditions through improved access, and engagement with, personal control	Building inhabitants free to pursue more productive and useful tasks
Personal controls can permit significantly diverse thermal responses for building inhabitants	Integrated, intelligent management systems respond in real time to inhabitant functional needs and preferred comfort requirements
Greater tolerance for a wider range of comfort conditions	Less tolerance if comfort conditions are not met
Human body integrates environmental responses in establishing satisfaction with ambient environmental conditions	Simple sensing technologies inadequate to fully represent or provide experienced comfort
Direct engagement with environmental controls to create immediate changes extends range of user comfort	Complex systems create greater instances of performance variability and variety of unforeseeable outcomes
Indirect effects/consequences	
Greater controls: personal control enhances a sense of empowerment	Invisibility and intractability of complex technology may result in inhabitant estrangement.
Enhances inhabitant's sense of ownership of building and responsibility for building performance	While override capability is provided, users remain largely uninvolved with comfort provisioning and awareness of systems
Instill a sense of environmental responsibility by directly influencing natural lighting and ventilation	Few opportunities to reinforce building users connection and engagement with external conditions

the situation is much more complex and blurred. Almost every building employs some degree of human and automated controls and, in more specific ways, single control strategies can rely on both (e.g., lights turned on by occupants when required and automatically switched by occupancy sensors). Approaches to intelligent buildings vary considerably, and clearly there is no single correct approach to achieving improved environmental performance. Moreover, the ways and extent that individual occupants can or are able to access controls is different according to varying organizational and managerial contexts (e.g., authority of pupils in schools relative to teachers; patients in hospitals, etc.).

Where intelligence is invested or assumed in buildings – human or automated – has a number of implications for building design and environmental performance:

- It will be necessary to engage and understand the relationship between individual and collective human intelligence in relationship to automated systems. There is an increasing significance in framing the issue of human and automated intelligence to contrast the direct and indirect consequences of shared responsibility and engagement with controls.
- The range of different workplace settings now permitted with advances in ICT – the home, office or venues within the city – will again have profound consequence for individual and group engagement with control systems and/or adaptation opportunities.

Table 10.5 highlights the key differences/implications of human and automated intelligence.

Callaghan *et al.* (2009) provide a 'socio-technical' framework for intelligent building research that depicts key factors associated with human interaction with pervasive computer technology within intelligent buildings and environments. The framework uses two key dimensions: the 'functionality' of the system and the way it is derived, for example, programs, rules derived from user behavior, etc., and the 'topology' of the system, its components and the way they are networked together. The extremes of these dimensions represent conditions where, at one end, a user explicitly determines both the functionality and the topology, and at the other end, where a system whose functionality and topology are totally controlled by intelligent 'agents'. A third dimension is used to characterize occupant behavior and response to these approaches, ranging from 'technophilia' to 'technophobia'. Embedded within this framework lies its authors' interest in how intelligent systems might 'stimulate or constrain human creativity and the consequences that might flow from this' and their view that the 'less understanding of, and control over, their environment that people have the more resistant or fearful they will be of it and vice versa' (Callaghan *et al.*, 2009, p. 71).

10.4   IMPORTANCE OF PLACE

Sections 10.2 and 10.3 above have explored the emerging notion of regenerative design and the increasing capabilities and deployment of Information and Communication Technologies respectively. Of the manifold ways that they directly and indirectly complement each other in nurturing/supporting environmental change in buildings, their reinforcement of the importance of place is paramount.

Regenerative design accepts and promotes "place" as the primary starting point for design and "... connecting people back to the spirit of place in a way that they are vitalized by it and become intrinsically motivated to care for it" (Mang, 2009, p. 5). The notion of place has been variously part of architectural discourse since Vitruvius over two thousand years ago. The modernist movement broke with this understanding and replaced the significance of place with the more anonymous and abstract notion of space. Leatherbarrow (2009) presents that, in modernist theory, "space was presented as the all-embracing framework of every particular circumstance, the unlimited container of all possible contents" and "...possessing a selfsameness congenial to intellectual mastery because of the conceptual character of its attributes." By contrast, the "topography in which buildings perform" is [p]olytropic, heterogeneous, and concrete, it regions contrast, conflict, and sometimes converse with one another" (p. 63).

During the 1960s and 1970s bioclimatic design and bioregionalism emerged as powerful notions to reestablish connection to place. Bioregionalism, for example, was committed to developing communities integrated with their surrounding ecosystems. Rather than legally defined regions, bioregionalism considered geographical province with a marked ecological and often cultural unity, often demarked by the watersheds of major river systems. What distinguished bioregionalism from other movements and theories was its firm base in the right of a group to self-determination and decision-making. But, the greatest challenge of bioregionalism was "the crucial and perhaps only and all-encompassing task of understanding place, the immediate specific place where we live" (Sale, 1985, p. 42).

Breaking with the modern movement tradition in an attempt to reconnect with the emphasis on the specificity of place, 'regionalism', for example, was posited within mainstream architectural discourse as a potential remedy to the "homogeneity and mediocrity of the current built environment" (Buchanan, 1983; 1984). In *Towards a Critical Regionalism: six Points for an Architecture of Resistance* (1983), Frampton laid down criteria deemed relevant to a regionalist architecture and attempted to focus the architectural debate in the notion of 'place'. 'Critical regionalism' was offered as a "strategy to mediate the impact of universal civilization with elements derived indirectly from the peculiarities of a particular place" (p. 21).

FICTION

It's 8:30 am when Erik's alarm snooze goes off. He didn't mean to sleep in so late, but he was up late working on a deadline. He sits up in bed, puts his glasses on, and reaches out for his iPad. After a couple of minutes of browsing through the cover stories on Flipboard, he selects a play list and hops in the shower. Next, he texts his boss and says he will spend the morning working from a coffee shop and come into the office after lunch. Erik works for an up-andcoming software development company – a growing firm where the average employee is in his or her mid twenties, earns a decent salary, and can wear jeans to work. Erik grabs his iPad and laptop and jogs over to the coffee shop down the street, 'The Zone'. The coffee shop he frequents is filled with mobile workers and freelancers, many of whom he has grown to know on a first name basis. Erik spots someone leaving from a workstation in the corner. He grabs the seat and from the coffee shop's app, he restores the workstation to his own comfort preferences. The lamp and fan turn on, and the dividing screen between him and his neighbour flips up. Next, with the same app, he orders his standard cappuccino, fruit salad, and cranberry oat muffin. No need to pay at the cashier. It's time to work! From here, he can access his company's Cloud and all the necessary files. Erik puts on his headphones, and restores his internet pages: the news, Facebook, Linkedin, and his email. After staying up until 2 am for yesterday's deadline, he doesn't feel guilty easing slowly into today's work. Besides, he did very well in last week's performance reviews, and his bosses emphasize results over hours spent working.

At noon, Erik's app advises him that his three hours at 'The Zone' are up. If he wants to stay longer, he must pay a service fee for use of the workstation, or buy another meal. Erik packs his bags and walks over to the light rail station. He has exactly thirty minutes until he arrives at his office, just enough to catch up on some outstanding emails and to have a quick Skype call with his team leader, who works from home half the time. When Erik reaches the office, he sees a few of his colleagues playing ping pong and grabbing some snacks from the office vending machines. They invite him to join, and he agrees, but "only for a few minutes." When Erik's boss sees that he's there, he calls a team meeting. The team members expediently move their desks and chairs (both on wheels) and create an informal seminar table. Some teammates prefer to sit on bean bag chairs or fitness balls. Erik's boss explains that the meeting will be quick. He congratulates the team for making yesterday's deadline and explains that they will be launching a new project in a few weeks. In the meantime, everyone is to find himself or herself useful and assist other teams with the projects that the office has on the go. When the meeting is over, Erik wheels his desk back next to the window. It's sunny out and quite warm. He wonders if it is nice enough out to work outside, and consults his Climate App. The app stores his comfort preferences, as well as his favourite and most recent locations. It also connects to the local meteorological centre and is able to access data on cloud cover, temperature, humidy, and dew points. He clicks on the office's communal rooftop and the app tells him, "Erik, on the rooftop it is currently 22 degrees with low humidity. On the ground, it is 26 degrees. The rooftop is sunny and will be unshaded until 4:13 pm. You should wear a T-Shirt". The building superintendant has encouraged all her tenants to use this app. There are at least one week in the summer and two weeks in the winter when the building's passive heating and cooling systems cannot create adequate interior temperatures. During these weeks, the tenants are notified and encouraged to work from home or at a coffee shop. Erik grabs his iPad, sunglasses and a sketchbook and decides to do some concept development work at the same time as catching some sun.

When the rooftop becomes shaded from the building next door, Erik heads back down to the office. It's early, but he decides to head home and cook dinner for his girlfriend. His Climate App notifies him of the interior office conditions and notifies him of automated night flushing from 9 pm – 4 am. Even though Erik is only at his desk half the time, he likes being able to control his environment – both in terms of thermal comfort and access to information, resources, and his teammates. After a quick video game to help unwind, he heads out the door to pick up some groceries for dinner. There are a few things he didn't get done today at the office, but he will have to do them tonight on his laptop while he and his girlfriend watch a movie.

A key distinction exists between green design and regenerative design with respect to place. Most green building assessment tools have wrestled with accommodating regional distinctions and cultural differences, as they are increasingly deployed outside their countries of origin. Being largely technocratic and conceived as a generic, top-down approach, they typically lack the specificity and social-ecological engagement central to a regenerative approach. The need for discrete performance criteria in green assessment methods also carries the potential consequence of fragmentation. Regenerative design and development, by contrast, seeks understanding of whole systems. Mang and Reed (2012) illustrate how the concept of place is used in regenerative design and development as a "coalescing context in that it serves as the basis for illuminating what has shared meaning for all human and natural stakeholders, bigger than any one issue or cause, and thereby for discovering how a project can become truly meaningful." (p. 28) They also emphasize the potency of using the "story of place", together with "pattern literacy" as a mean of providing "a coherent organization of information, and the relationships and connections between discrete pieces of information and different types of information" and wherein an "underlying narrative structure enables relating this information and these relationships and connections in a way that reveals a holistic, understandable picture" (p. 29).

The destructive and homogenizing nature of globalization on regionally specific building practices are widely acknowledged. Buchanan, however, suggests that "perhaps (and hopefully) what we are witnessing is the traumatic period prior to the birth of a viable global civilization in which networks of communication and trade will no longer be homogenizing and destructive agents but will have such abundant capacity as to allow regional peculiarities to survive and be savored". He continues, "a truly universal civilization will not only envelop the world and give its citizens the breadth of experience so ordered. It will also encourage that depth of experience that comes from being rooted in, and caring for, local issues and which is another dimension to the idea of being universal.". One could anticipate that architectural outcomes within a regenerative approach emerge from a thoughtful response to the unique social, cultural and ecological opportunities and constraints of place, drawing equally on the appropriate use of broader contemporary technological capabilities. That a fully matured global information system and culture may instigate and permit the creation of regionally and place-based practices is central to the notion of GLOCAL. Derived from global and local, GLOCAL recognizes the need for balance between "the invisible global forces" and the "actual sense of place and culture" (Nagashima, 1999).

The reemergence of the notion of place is clearly not confined to architecture and may also be a reaction and manifestation of people wanting to reclaim more control over their lives. Localism, for example, supports local production and consumption of goods, local control of government, and promotion of local history, local culture and local identity. Which aspects, and their extent, can be reestablished and maintained at a local level and which remain within the domain of national and global production, trade and exchange, will clearly evolve according to the constraints and opportunities afforded by place.

While proponents of regenerative design emphasize place as a starting point and major focus of design, exploring and understanding the ecological and other systems for the broader region in which the project is situated clearly also offers potential value for design: what, for example, are the ecological and other physical assets? What are the ecological and physical constraints? What are the ecological services offered by the place – to adjacent regions and the biosphere, to other species, to humanity? While current design may have some familiarity and engagement with some aspects of place and may readily extend their commitment to place-specific approaches, it typically has little, if any, understanding of the larger ecological and social systems that are the longest lasting features of the physical environment and set the context for future long-term possibilities.

Pedersen-Zari (2012) argues that regenerative developments "cannot exist in isolation from their larger surrounding contexts" with the implication that "there is a need to understand ecosystem services at a larger scale (city, region, or ecosystem boundary) when devising goals and targets for individual buildings or small developments." (p. 62). As with the need for both green and regenerative approaches, both scales of understanding are valuable and necessary.

The notion of place is also being fundamentally transformed through information and communication technologies. The discussion of human versus automated controls in comfort provisioning presented in this chapter is clearly set within the larger discourse of the pervasiveness, consequence and design responsibilities of computers and other digital technology in building, design and operation, all of which are diminishing the attachment of occupants with a specific, fixed place. Callaghan *et al.* (2009), for example, present a possible future where intelligent buildings and environments are based on the use of "numerous 'invisible', omnipresent, always-on, communicating computers embedded in everyday artifacts and environments" (p. 56) and which "promises to connect not just every citizen of this planet, but ultimately every artifact in the world" (p. 58). McCullough (2004) characterizes how digital systems have already become so embedded into almost every human endeavor that they now form a background – "present but unobtrusive." For McCullough, the "placelessness" resulting from having digital access from anywhere to anywhere requires an even greater need to situate ourselves in identifiable and meaningful physical places designed for social interaction. For Kupfer, this "placelessness" deprives people of aesthetic experiences particular to a place, it causes people to lose touch with themselves and with others, and it causes people to depreciate place (Kupfer, 2007, p. 39). It would seem the case, therefore, that the ways in which humans interact together as a result of pervasive computing technology is equally, if not more important, as the ways in which humans interact with technology. Thus, the discussion takes on a decidedly public dimension, transcending preoccupations with a private realm that technology currently permeates and influences.

10.5 CONCLUSIONS

This chapter has explored the potential complementary relationship between the emerging notion of regenerative design and development and considerable capabilities of information and communication technologies. A key common feature within this exploration was the multiple ways that each reinforces the importance of the engagement with and experience of place. While place has a long history in architectural discourse, in an increasingly homogenized and abstract world, it offers a necessary and tangible counterpoint. Other parallels between regenerative design and ICTs can also be made regarding the embracing of complexity and uncertainty and improving comfort and well-being.

Regenerative projects engage with complex systems. Salvador Rueda, for instance argues that the more complex an urban system, the less energy is at the heart of its organization, and the more information becomes primordial. While resource necessities are only accumulating, information is multiplying and is participating in the complexity and in the stability of the city. Interestingly, "With the strategy of increasing the complexity in urban ecosystems, it must be taken into account that adding a similar amount of information to two different systems is more enriching for systems that already have more information. Information is not added, but multiplied". (Rueda, 2007, p. 17) Regenerative projects also tend to stress the importance of adaptive learning. Within such a period of transformation, adaptive learning will be central to 'capacity building' necessary to deal with 'complexity, uncertainty and surprise.' (Glaser *et al.*, 2008, p. 79) Here, the governance of complex adaptive systems 'requires flexibility and a capacity to respond to environmental feedback' (Glaser *et al.*, 2008, p. 79) and a need for continuously testing, learning, and developing knowledge and understanding in order to cope with their change and associated uncertainty (Carpenter and Gunderson, 2001). Folke (2004) further emphasizes that 'knowledge acquisition of complex systems is an ongoing, dynamic learning process, and such knowledge often emerges with people's institutions and organizations' and, moreover, that this 'seems to require institutional frameworks and social networks nested across scales to be effective' and to create the necessary responsible stewardship. Hence, these issues suggest the need to question the performance requirements of buildings within an emerging period of unprecedented, rapid and uncertain transformation.

Intelligent buildings place considerable emphasis on the design and integration of automated control systems and technologies with the intent of freeing occupants of control tasks. As such, the emphasis of the design of automated approaches has been in resolving complex technical systems with significantly less consideration deemed necessary on the human, social and behavioral issues and consequences. The more complex a system, the greater instances of performance variability and greater the variety of unforeseeable outcomes. The inherently complex systems of intelligent buildings have potential implications for occupant well-being (improved or damaged psychological state of mind) and improved or hindered group dynamic interactions, redefinitions of social hierarchies, and economic and environmental consequences. An automated system, for example, that replaces manual controls over space temperature requires a potentially large set of operations and operators, ultimately governed by a building systems manager. If the temperature does not meet an individual occupant's personal preferences, considerable effort may be required to restore that occupant's comfort level. Unintended indirect consequences of automated HVAC systems can therefore be added frustration, hostility towards building management, decrease in occupant productivity, and severance of environmental-psychology relationship between human and place. Buildings that employ automation are not only complex from a technical standpoint, but also display a need for adaptive learning and for incorporating feedback mechanisms within the design.

The regenerative agenda and the automation/information communications technologies agenda also tend to mutually reinforce each other when it comes to improving thermal and social comfort within the context of an individual building. The emerging shifts toward more mobile and transient workplaces requires rethinking traditional interpretations of inhabitant ownership, engagement and agency considered necessary to support higher degrees of satisfaction. Whereas in the fixed office, where notions of comfort and worker satisfaction are traditionally associated with issues pertaining to air quality and thermal comfort, the emphasis in the mobile workplace shifts to access to information and communications technologies, the nature of social interactions, and a collective sense of agency and identity. Regenerative projects aim to achieve greater social comfort and well-being by creating a collective understanding of experience comfort through engaging stakeholders early on in the design process, through incorporating educational opportunities both throughout the design process and life of a building, through co-developing agency for achieving comfort, and by supporting the active participation of inhabitants in the ongoing maintenance and negotiation of comfort-related issues of a building. In both cases, the role of social comfort becomes increasingly important, while at the same time, there is a move from passive occupants to active inhabitants and stewards.

ACKNOWLEDGEMENTS

This chapter draws heavily on previously published work by the authors in *Building Research and Information* and *Intelligent Buildings International*.

REFERENCES

Ackerman, M.E.: *Cool comfort: America's romance with air conditioning.* Smithsonian Institution Press, Washington, DC, 2002.
Bazerman, M.H., Messick., D.M., Tenbrunsel, A.E. & Wade-Benzoni, K.A. (eds): *Environment, ethics, and behavior: the psychology of environmental valuation and degradation.* New Lexington Press Management Series, New Lexington Press, San Francisco, CA, 1997.
Birkeland, J.: Positive development: designing for net positive impacts. BEDP Environment Design Guide, August, Gen 4, Royal Australian Institute of Architects, Melbourne, Australia, 2007.

Bittman, M., Brown, J.E. & Wajcman, J.: The mobile phone, perceptual contact and time pressure. Work. *Emp. Soc.* 23:4 (2009), pp. 673–691.

Brager, G.S. & de Dear., R.: Climate, comfort & natural ventilation: a new adaptive comfort standard for ASHRAE Standard 55. *Proceedings: Moving Thermal Comfort Standards into the 21st Century*, Oxford Brookes University, Windsor, UK, April 2001.

Brand, S.: *The clock of the long now, time and responsibility*. Weidenfeld & Nicolson, London, UK, 1999.

Brunia, S. & Hartjes-Gosselink, A.: Personalization in non-territorial offices: a study of a human need. *Journal of Corp. Real Estate* 11:3 (2009), pp. 169–182.

Buchanan, P.: With due respect: regionalism. *Archit. Rev.* 1035 (1983), pp. 15–16.

Buchanan, P.: Only connect. *Archit. Rev.* 1052 (1984), pp. 23–25.

CABA: Bright green buildings convergence of green and intelligent buildings. Research report, Continental Automated Buildings Association, 2008.

Callaghan, V., Clarke, G. & Chin, J.: Some socio-technical aspects of intelligent buildings and pervasive computing research. *Intell. Build. Int.* 1:1 (2009), pp. 56–74.

Capra, F.: *The web of life: a new scientific understanding of living systems*. Anchor Books, New York, 1996.

Carpenter, S.R. & Gunderson, L.H.: Coping with collapse: ecological and social dynamics in ecosystem management. *BioScience* 51 (2001), pp. 451–457.

Clements-Croome, D.: Building environment, architecture and people. In: D. Clements-Croome (ed): *Intelligent buildings: design management and operation*. Thomas Telford, London, UK, 2004, pp. 53–100.

Cohen, R., Ruyssevelt, P., Standeven, M., Bordass, W. & Leaman, A.: *Building intelligence in use: lessons from the Probe project*. Usable Buildings, UK, 1999.

Cole, R.J.: Energy and urban buildings. In: N. Munier (ed): *Fundamentals and principles of urban sustainability*. Springer Press, The Netherlands, 2005a, pp. 389–438.

Cole, R.J.: Building environmental assessment methods: redefining intentions and roles. *Build. Res. Inf.* 33:1 (2005b), pp. 455–467.

Cole, R.J.: Transitioning from green to regenerative design. *Build. Res. Inf.* 40:1 (2012), pp. 39–53

Cole, R.J. & Brown, Z.: Reconciling human and automated intelligence in the provision of occupant comfort. *Intell. Build. Int.* 1 (2009), pp. 39–55.

Cole, R.J., Robinson, J., Brown, Z. & O'Shea, M.: Re-contextualizing the notion of comfort. *Build. Res. Inf.* 36:4 (2008), pp. 323–336.

Cole, R.J., Brown, Z. & McKay, S.: Building human agency: a timely manifesto. *Build. Res. Inf.* 38:3 (2010), pp. 339–350.

Cole, R.J., Bild, A. & Matheus, E.: Direct & indirect consequences of automated & human activated controls. *Intell. Build. Int.* 4:1 (2012a), pp. 4–14.

Cole, R.J., Bild, A. & Oliver, A.: The chancing nature of knowledge-based work: consequences of comfort, satisfaction and productivity. *Intell. Build. Int.* 4:3 (2012b), pp. 182–196.

Cole, R.J., Busby, P., Guenther, R., Briney, L., Blaviesciunaite, A. & Alencar, T.: A Regenerative design framework: setting new aspirations & initiating new discussions. *Build. Res. Inf.* 40:1 (2012c), pp. 95–111.

Cooper, G.: *Air conditioning America: engineers and the controlled environment, 1900–1960*. Johns Hopkins University Press, Baltimore, MD, 1998.

Crowley, J.: *The invention of comfort: sensibilities and design in early modern Britain and early America*. Johns Hopkins University Press, Baltimore, MD, 2001.

Curwell, S., Deakin, M., Cooper, I., Paskaleva-Shapira, K., Ravetz, J. & Dominica Babicki, D.: Citizens'expectations of information cities: implications for urban planning and design. *Build. Res. Inf.* 35:5 (2005), pp. 55–66.

Doherty, G., Karamanis, N. & Luz, S.: Collaboration in translation: the impact of increased reach on cross-organizational work. *Lect. Notes Comput. Sc.* 21:6 (2012), pp. 525–554.

du Plessis, C.: Towards a regenerative paradigm for the built environment. *Build. Res. Inf.* 40:1 (2012), pp. 7–22.

du Plessis, C. & Cole, R.J.: Motivating change: changing the paradigm. *Build. Res. Inf.* 39:5 (2011), pp. 436–449.

Fisk, P.: The eco-balance approach to transect-based planning: efforts taken at Verana, a new community and university in San Antonio, Texas. Center for Maximum Building Potential Building Systems, Austin, TX, 2009.

Folke, C.: Traditional knowledge in social-ecological systems. *Ecol. Soc.* 9:3–7 (2004), http://www.ecology andsociety.org/vol9/iss3/art7/.

Frampton, K.: Towards a critical regionalism: six points for an architecture of resistance. In: H. Foster & H. Port Townsen (eds): *The anti-aesthetic: essays on postmodern culture*. Bay Press, 1983.

Gasparatos, A., El-Haram, M. & Horner, M.: The argument against a reductionist approach for assessing sustainability. In M. Horner, C. Hardcastle, A. Price & J. Bebbington (eds): *Proceedings: SUE-MOT Conference 2007, International Conference on Whole Life, Urban Sustainability and its Assessment,* 27–29 June 2007, Glasgow, UK, 2007, pp. 52–65.

Gladwin, T.N., Newberry, W.E. & Reiskin, E.D.: Why is the northern elite mind biased against community, the environment, and a sustainable future. In: H. Bazerman, D.M. Messick, A.E. Tenbrunsel & Wade-Benzoni (eds): *Environment, ethics, and behaviour*. The New Lexington Press, San Francisco, CA, 1997, pp. 234–274.

Glaser, M., Krause, G., Ratter, B. & Welp, M.: Human-nature-interaction in the anthropocene. Potential of Social-Ecological Systems Analysis, 2008, www.dg-humanoekologie.de/pdf/DGHMitteilungen/GAIA200801_77_80.pdf (assessed 11 June 2012).

Gumpert, G. & Drucker, S.J.: Media technology as a determinant of urban form. In M. Gray (ed): *Proceedings of Evolving Environmental Ideals, 14th Conference of the International Association for People-Environment Studies,* Stockholm, Sweden, July 30–August 3rd, 1996, pp. 362–368.

Harrison, A.: The city is the office. *ECIFFO* 38 March 31, 2001, www.eciffo.jp/en/issue/38/fromeciffo38-2_e.html.

Himanen, M.: The intelligence of intelligent buildings. In: D. Clements-Croome (ed): *Intelligent buildings: design, management and operation*. Thomas Telford, London, UK, 2004, pp. 25–52.

Hoxie, C., Berkebile, R. & Todd, J.A.: Stimulating regenerative development through community dialogue. *Build. Res. Inf.* 40:1 (2012), pp. 65–80.

Ingersoll, R.: Unpacking the green man's burden. *Design Book Review*, Spring 1991, pp. 19–26.

Iso-Ahola, S.: A psychological analysis of leisure and health. In: J.T. Haworth (ed): *Work, leisure and well-being*. Routledge, London, UJ, 1997.

Kelliher, C. & Anderson, D.: Doing more with less? Flexible working practices and the intensification of work. *Hum. Relat.* 63:1 (2010), pp. 83–106.

King, A.: *Spaces of global cultures: architecture urbanism identity*. Routledge, London, UK, 2004.

Kupfer, J.A.: Mobility, portability, and placelessness. *J. Aesthet. Educ.* 41:1 (2007), pp. 38–50.

Leatherbarrow, D.: *Architecture oriented otherwise*. Princeton Architectural Press, New York. 2009.

Lewis, S.: The integration of paid work and the rest of life. Is post-industrial work the new leisure? *Leisure Stud.* 22 (2003), pp. 343–355.

Lombardi, P.: INTELCITY Thematic Network of South-Mediterranean Regions. Report, 2002, http://areeweb.polito.it/ricerca/ict-sud/results/index.html.

Mang, N.S.: *The rediscovery of place and our human role within it*. PhD Thesis, Saybrook Graduate School and Research Center, San Francisco, CA, 2007.

Mang, N.S.: Toward a regenerative psychology of urban planning. Saybrook Graduate School and Research Center, 2009.

Mang, P.: Regenerative development and design. In: R.A. Meyers (ed.): *Encyclopedia of sustainability science and technology*. Springer, 2012.

Mang, P. & Reed, B.: Designing from place: a regenerative framework and methodology. *Build. Res. Inf.* 40:1 (2012), pp. 23–38.

McCullough, M.: *Digital ground: architecture, pervasive computing, and environmental knowing*. MIT Press, Cambridge, MA, 2004.

McDonough, W. & Braungart, M.: *Cradle to cradle: remaking the way we make*. North Point Press, New York, NY, 2002.

Meadows, D.H.: *Thinking in systems – a primer*. Chelsea Green Publishing, White River Jct., VT, 2008.

Mitchell, W. J.: *e-topia*. MIT Press, Cambridge, MA, 1999.

Nagashima, K.: Global approach toward architecture of the future. Union Internationale des Architectes and the Japan Institute of Architects, Zushi, Japan, 1999.

Pedersen Zari, M.: Ecosystem services analysis for the design of regenerative built environments. *Build. Res. Inf.* 40:1 (2012), pp. 54–64.

Plaut, J.M., Dunbar, B., Wackerman, A. & Hodgin, S.: Regenerative design: the LENSES framework for buildings and communities. *Build. Res. Inf.* 40:1 (2012), pp. 112–122.

Rapoport, A.: *House form and culture*. Prentice Hall, Englewood Cliffs, NJ, 1969.

Reed, B.: Shifting from 'sustainability' to regeneration. *Build. Res. Inf.* 35:6 (2007), pp. 674–680.

Rees, W.E.: Achieving sustainability: reform or transformation? In: D. Satterthwaite (ed): *The Earthscan reader in sustainable cities*. Earthscan, London, UK, 1999, pp. 22–52.

Rueda, S.: Barcelona, a compact and complex Mediterranean city: a more sustainable vision for the future. Barcelona City Council, Barcelona, Spain, 2007.

Sale, K.: *Dwellers in the land: the bioregional vision.* Sierra Club Books, San Francisco, CA, 1985.

Svec, P., Berkebile, R. & Todd, J.A.: Inspiring generous behaviour: the evolution of the US GBC's REGEN tool concept. *Build. Res. Inf.* 40:1 (2012), pp. 81–94.

Skyttner, L.: *General systems theory: problems, perspectives, practice.* World Scientific Publishing Company, 2006.

Tschumi, B.: Parc de la Villette, Paris. *Archit. Design* 58:3/4 (1988), pp. 33–39.

von Bartalanffy, L., von Bertalanffy, L. & Bertalanffy, L.: *General system theory.* George Braziller, New York, NY, 1969.

van der Voordt, D.J.M. & van Meel, J.J.: Psychologische aspecten van kantoorinnovatie. Faculty of Architecture and the Built Environment, Delft University of Technology, Delft, The Netherlands, 2002.

Subject index

Sustainable Energy Developments

Series Editor: Jochen Bundschuh

ISSN: 2164-0645

Publisher: CRC Press/Balkema, Taylor & Francis Group

9. Advanced Oxidation Technologies – Sustainable Solutions for Environmental Treatments
   Editors: Marta I. Litter, Roberto J. Candal & J. Martín Meichtry
   2014
   ISBN: 978-1-138-00127-5 (Hbk)

10. Computational Models for CO_2 Geo-sequestration & Compressed Air Energy Storage
    Editors: Rafid Al-Khoury & Jochen Bundschuh
    2014
    ISBN: 978-1-138-01520-3 (Hbk)

11. Micro & Nano-Engineering of Fuel Cells
    Editors: Dennis Y.C. Leung & Jin Xuan
    2015
    ISBN: 978-0-415-64439-6 (Hbk)

12. Low Energy Low Carbon Architecture: Recent Advances & Future Direction
    Editor: Khaled A. Al-Sallal
    2016
    ISBN: 978-1-138-02748-0 (Hbk)

T - #0181 - 111024 - C310 - 246/174/14 - PB - 9780367574840 - Gloss Lamination